Improving the Productivity and Sustainability of Rice–Wheat Systems: Issues and Impacts

Improving the Productivity and Sustainability of Rice–Wheat Systems: Issues and Impacts

Proceedings of an international symposium cosponsored by Division A-6 (International Agronomy), the International Rice Research Institute (IRRI), and Centro Internacional de Mejoramiento de Maiz y Trigo (CIMMYT), held at the 2001 Annual Meetings of the American Society of Agronomy, Crop Science Society of America, and Soil Science Society of America, Charlotte, NC, 22 Oct. 2001.

Editors
Jagdish K. Ladha, James E. Hill, John M. Duxbury,
Raj K. Gupta, and Roland J. Buresh

Organizing Committee
Jagdish K. Ladha and Larry W. Harrington

Editor-in-Chief ASA
Kenneth A. Barbarick

Editor-in-Chief CSSA
Jeffrey J. Volenec

Editor-in-Chief SSSA
Warren A. Dick

Managing Editor
Lisa Al-Amoodi

ASA Special Publication Number 65

American Society of Agronomy
Crop Science Society of America
Soil Science Society of America
Madison, Wisconsin, USA

2003

Cover: Farmers, national agricultural research systems, and international agricultural research institutes are working together in a participatory mode to sustain the rice–wheat systems and livelihoods of its stakeholders in South Asia. Line drawing is by Mr. A.V. Prasanth, a creative illustrator, cartoonist, and painter, and the photograph is by Dr. S.V.R.K. Prabhakar, a CIMMYT Research Affiliate, both of New Delhi, India.

American Society of Agronomy, Inc.
Crop Science Society of America, Inc.
Soil Science Society of America, Inc.
677 South Segoe Road, Madison, WI 53711-1086 USA

Library of Congress Control Number: 2003102422

Printed in the United States of America

CONTENTS

FOREWORD

Global food demand is projected to increase by 50% in the next thirty years. This increase comes about in large part due to population growth, but at a time when cereal production on a per capita basis has been decreasing. At the beginning of the 21st century, cereal utilization was slightly higher than production. This does not mean that population growth is outpacing food production, but rather that other factors such as low world cereal grain prices along with a changing food consumption pattern have resulted in a shift of cropland from cereal production to other foodstuffs. Wheat and rice, the subject of this publication, account for more than 50% of the world cereal production, and combined they produce grains for about 20% of the world population.

Rice–wheat growing areas have shrunk in per capita terms. The intensifying competition for water may well enhance this decrease in land area available for flooded paddy rice culture, the system responsible for 75% of the total production. Loss of production of flooded paddy rice is crucial not only from the standpoint of loss of irrigation water, but this method of production has been an effective weed control technique. Without the ability to flood, producers will be forced to use other weed control techniques, including the more expensive systems that involve herbicides.

This publication identifies the problems that agronomists will need to solve to ensure that rice and wheat production continue to increase at the rate necessary to sustain the world food supply, while at the same time preserving natural resources for future generations. This publication will serve as an excellent reference for future policy makers as well as for those that conduct research aimed at meeting the world cereal needs.

The American Society of Agronomy is grateful to the Editors, the organizing committee, and the authors for their untiring efforts at creating this excellent publication. We are also appreciative of the contributions made by the International Rice Research Institute (IRRI) and Centro Internacional de Mejoramiento de Maiz y Trigo (CIMMYT).

ROBERT G. HOEFT
President
American Society of Agronomy

PREFACE

Rice–wheat rotation is one of the world's largest agricultural production systems, occupying 24 million ha of cultivated land on the Indo-Gangetic Plains and in China. It accounts for about one-third of the area of both rice and wheat grown in South Asia, and it produces staple grains for more than one billion people, or about 20% of the world's population. The adoption of Green Revolution technologies led to impressive increases in rice and wheat production from 1965 to 1985, but concerns are arising that future production increases may not keep pace with demand as populations continue to expand. Rice–wheat growing areas have shrunk in per capita terms from 1200 m^2 in 1961 to less than 700 m^2 now. In the many areas that are already double- or even triple-cropped, intensification is not an option. Yield increase is the only remaining strategy for coping with growing demand, but the very high rates of yield growth achieved in some areas during the Green Revolution era are probably not repeatable. Production costs are steadily increasing, and water-induced land degradation, the slow loss of soil fertility, and increasing incidence of pests are other limitations to production. These factors all fuel the fears of agricultural scientists and economists that food production may fail to increase by 2.5% each year, the rate required to feed growing populations, raise incomes, and reduce malnutrition. To meet this challenge, technologies that will enhance productivity, ensure environmental safety, and conserve natural resources are emerging.

This ASA Special Publication is devoted to (i) discussing the challenges facing the sustainability of rice–wheat systems, (ii) reviewing emerging technologies in rice and wheat and their integration into the systems, and (iii) identifying knowledge gaps and priorities for future research. The 10 chapters are co-authored by a diverse group of 36 scientists from 8 countries. The first four chapters (Gupta et al., Grace et al., Ladha et al., and Byerlee et al.) present a comprehensive picture of sustainability dimensions in rice–wheat systems. The next chapter (Sharma et al.) provides a critical analysis of the effects of soil puddling in rice on the performance of the subsequent wheat crop. There follow four chapters (Balasubramanian et al., Hobbs and Gupta; Samra et al., and Connors et al.) describing technological options for rice and wheat in the systems. The last chapter (Gaunt et al.) explores the implications of adopting a research approach that focuses on people and livelihoods.

This publication is the product of a great deal of effort at both IRRI and CIMMYT, and at the ASA–CSSA–SSSA. We thank Ronald P. Cantrell and Timothy G. Reeves, the directors general of IRRI and CIMMYT, respectively, for their overall support for the symposium and for making this publication possible. The editors also wish to thank Erick Fernandez, A-6 division program chair, for organizing the symposium, Timothy Reeves for chairing the sessions, the participants for lively discussions, Bill Hardy for editing the manuscripts, several peer reviewers (V. Balasubramanian, Gelia Castillo, David Dawe, Ken Giller, Peter Grace, Peter Hobbs, Chris van Kessel, Noel Magor, Shaobing Peng, Anil Singh, and Bijay Singh) for their critical input, and Anthony Pulpulaan and Flor Junsay of IRRI, and Lisa Al-Amoodi of ASA Headquarters staff for editorial assistance.

JAGDISH K. LADHA, JIM E. HILL, AND ROLAND J. BURESH
IRRI

JOHN DUXBURY
Cornell University

RAJ K. GUPTA
The Rice–Wheat Consortium, Delhi, India

CONTRIBUTORS

Mubarik Ali — Agricultural Economist, and Head of the Socioeconomic Unit, The Asian Vegetable Research and Development Center (AVRDC), P.O. Box 42, Shanhua, Tainan, Taiwan. Email: mubarik@netra.avrdc.org.tw

Vethaiya Balasubramanian — Agronomist/Soil Scientist, International Rice Research Institute (IRRI), DAPO Box 7777, Metro Manila, Philippines. Email: v.balasubramanian@cgiar.org

John R. Best — Lecturer, The University of Reading, International and Rural Development Department (IRDD), Earley Gate, Whiteknights Road, Reading, RG6 6AR, UK. Email: j.r.best@reading.ac.uk

Lav Bhushan — Research Fellow, IRRI-CIMMYT-SVBPUA&T Rice–Wheat Collaborative Project, Sardar Vallabh Bhai Patel University of Agriculture and Technology (SVBPUA&T), Modipuram Meerut (U.P.) 250 110 India. Email: lavbhushan@yahoo.com

Derek Byerlee — Lead Economist, Rural Development Department, World Bank, 1818 H St. NW, Washington, DC 20433 USA. Email: dbyerlee@worldbank.org

David J. Connor — Professor of Agronomy, The University of Melbourne, School of Agriculture and Food Systems, Victoria 3010, Australia. Email: djconnor@unimelb.edu.au

David Dawe — Economist, Social Sciences Division, International Rice Research Institute (IRRI), DAPO Box 7777, Metro Manila, Philippines. Email: d.dawe@cgiar.org

John L. Gaunt — Senior Research Scientist, Agriculture and Environment Division, Rothamsted Research, West Common, Harpenden, Hertforshire, AL5 2JQ, UK. Email: john.gaunt@bbsrc.ac.uk

Peter R. Grace — Principal Scientist, Natural Research Management Systems International, 1 Chiba St., Ashgrove, Queensland 4060, Australia. Email: prgrace@optusnet.com.au

Raj K. Gupta — Regional Facilitator, Rice–Wheat Consortium for the Indo-Gangetic Plains, CIMMYT-RWC, CG Block, NASC Complex, DPS Marg, Pusa Campus, New Delhi 110 012, India. Email: R.Gupta@cgiar.org

Larry Harrington — Director of Natural Resources, Centro Internacional de Mejoramiento de Maiz y Trigo (CIMMYT), Apartado # 370, Postal 6-641, Mexico 0600, D.F. Mexico. Email: L.Harrington@cgiar.org

Peter R. Hobbs Agronomist, CIMMYT, Cornell University, 611 Bradfield Hall, Ithaca, NY 14853, USA. Email: P.Hobbs@cgiar.org

M. Zahid Hossain Senior Programme Coordinator, Ecological Agriculture Programme, Proshika, A Centre for Human Development, i/1-GA, Section-2, Mirpur-2, Dhaka-1216, Bangladesh. Email: Zhossain@proshika.bdonline.com (and: ayo7l9us@yahoo.com)

Elizabeth Humphreys Senior Research Scientist, Irrigated Agriculture, Commonwealth Scientific and Industrial Research Organization (CSIRO) Land and Water, PMB 3, Griffith 2680, Australia. Email: liz.humphreys@grf.clw.csiro.au

Mahesh C. Jain Professor, Division of Environmental Sciences, Indian Agricultural Research Institute (IARI), A 121, Inderpuri, New Delhi 110 012, India. Email: maheshcjain@hotmail.com

Zheng Jiaguo Professor, Crop Research Institute, Sichuan Academy of Agricultural Sciences (SAAS), Shizishan, Chengdu, Sichuan 610066, China. Email: zhjguo@mail.sc.cninfo.net

Kuldip Kumar Research Associate, University of Minnesota, Department of Soil, Water, and Climate, 1991 Upper Buford Circle-Borlaug Hall, University of Minnesota, St. Paul, MN 55108, USA. Email: K.Kumar@soils.umn.edu

Jagdish K. Ladha Soil Scientist, Crop, Soil and Water Sciences Division, International Rice Research Institute (IRRI), DAPO Box 7777, Metro Manila, Philippines. Email: j.k.ladha@cgiar.org

Randhir Singh Mehla Additional Director Agriculture, Department of Agriculture, Govt. of Haryana, Krishi Bhawan, Sector 21, Panchkula, Haryana 134442, India. Email: rsmehla@rediffmail.com

Ram K. Naresh Officer-In-Charge of Krishi Vigyan Kendra, Sardar Vallabh Bhai Patel University of Agriculture and Technology (SVBPUA&T), Modipuram Meerut (U.P.) 250 110, India. Email: rknaresh@yahoo.com

Patricia Norrish Independent Consultant, Development Communication, 6 White House Close, Chalfont St. Peter, Bucks, UK SL9 ODA. Email: c/o j.norrish@ioe.ac.uk

Richard J. Palmer-Jones Senior Lecturer, School of Development Studies, University of East Anglia, Norwich, NR4 755, UK. Email: r.palmer-jones@uea.ac.uk

Himanshu Pathak Scientist, Division of Environmental Sciences, Nuclear Research Laboratory Building, Indian Agricultural Research Center (IARI), New Delhi 110 012, India. Email: him_ensc@iari.ernet.in

G. Philip Robertson Professor of Crop and Soil Sciences, Michigan State University (MS), Kellogg Biological Station, 3700 East Gull Lake Drive, Hickory Corners, MI 49060, USA. Email: robertson @kbs.msu.edu

Elizabeth Z.J. Robinson Fellow and Lecturer in Economics, St Hugh's College and Department of Economics, University of Oxford. St Hugh's College, St Margaret's Road, Oxford OX2 6LE, UK. Email: elizabeth.robinson@economics.ox.ac.uk

Jagir Singh Samra Deputy Director General, Division of Natural Resource Management, Indian Council of Agricultural Research (ICAR), Krishi Bhawan, New Delhi 110001, India. Email: j.s.samra@ icar.delhi.nic.in

Pradeep K. Sharma Senior Soil Scientist, Department of Soil Science, CSK Himachal Pradesh Agricultural University (CSKHPAU), Palampur 167-062, Himachal Pradesh (HP), India. Email: pradeepksharma2001@yahoo.co.in

Akmal Siddiq Project Economist, Agriculture, Environment, and Natural Resources Division, Mekong Department, Asian Development Bank (ADB), 6 ADB Avenue, Mandaluyong City, 0401 Metro Manila, Philippines. Email: asiddiq@adb.org

Bijay-Singh Senior Soil Chemist, Department of Soils, Punjab Agricultural University (PAU), Ludhiana 141 004, India. Email: bjs_20@sify.com

Yadvinder-Singh Senior Soil Chemist, Department of Soils, Punjab Agricultural University (PAU), Ludhiana 141 004, India. Email: yadvinder16@rediffmail.com

Alistair J. Sutherland Principal Scientist in the Livelihoods and Institutions Group, National Resources Institute (NRI), Central Avenue, Chatham Maritime, Kent, ME4 4TB, UK. Email: a.j.sutherland@ gre.ac.uk

Jagadish Timsina Senior Research Scientist, Commonwealth Scientific and Industrial Research Organization (CSIRO) Land and Water, PMB 3, Griffith 2680, Australia. Email: jagadish.timsina@ csiro.au

Agnes Tirol-Padre Assistant Scientist, Crop, Soil & Water Sciences Division, International Rice Research Institute (IRRI), DAPO Box 7777, Metro Manila, Philippines. Email: A.Padre@cgiar.org

Stephanie K. White Post-doctoral Scientist, Agriculture and Environment Division, Rothamsted Research, Harpenden, Hertfordshire AL5 2JQ, UK. Email: Stephanie.white@bbsrc.ac.uk

Conversion Factors for SI and non-SI Units

Conversion Factors for SI and non-SI Units

To convert Column 1 into Column 2, multiply by	Column 1 SI Unit	Column 2 non-SI Units	To convert Column 2 into Column 1, multiply by
		Length	
0.621	kilometer, km (10^3 m)	mile, mi	1.609
1.094	meter, m	yard, yd	0.914
3.28	meter, m	foot, ft	0.304
1.0	micrometer, µm (10^{-6} m)	micron, µ	1.0
3.94×10^{-2}	millimeter, mm (10^{-3} m)	inch, in	25.4
10	nanometer, nm (10^{-9} m)	Angstrom, Å	0.1
		Area	
2.47	hectare, ha	acre	0.405
247	square kilometer, km^2 (10^3 m)^2	acre	4.05×10^{-3}
0.386	square kilometer, km^2 (10^3 m)^2	square mile, mi^2	2.590
2.47×10^{-4}	square meter, m^2	acre	4.05×10^{3}
10.76	square meter, m^2	square foot, ft^2	9.29×10^{-2}
1.55×10^{-3}	square millimeter, mm^2 (10^{-3} m)^2	square inch, in^2	645
		Volume	
9.73×10^{-3}	cubic meter, m^3	acre-inch	102.8
35.3	cubic meter, m^3	cubic foot, ft^3	2.83×10^{-2}
6.10×10^{4}	cubic meter, m^3	cubic inch, in^3	1.64×10^{-5}
2.84×10^{-2}	liter, L (10^{-3} m^3)	bushel, bu	35.24
1.057	liter, L (10^{-3} m^3)	quart (liquid), qt	0.946
3.53×10^{-2}	liter, L (10^{-3} m^3)	cubic foot, ft^3	28.3
0.265	liter, L (10^{-3} m^3)	gallon	3.78
33.78	liter, L (10^{-3} m^3)	ounce (fluid), oz	2.96×10^{-2}
2.11	liter, L (10^{-3} m^3)	pint (fluid), pt	0.473

		Mass	
2.20×10^{-3}	gram, g (10^{-3} kg)	pound, lb	454
3.52×10^{-2}	gram, g (10^{-3} kg)	ounce (avdp), oz	28.4
2.205	kilogram, kg	pound, lb	0.454
0.01	kilogram, kg	quintal (metric), q	100
1.10×10^{-3}	kilogram, kg	ton (2000 lb), ton	907
1.102	megagram, Mg (tonne)	ton (U.S.), ton	0.907
1.102	tonne, t	ton (U.S.), ton	0.907
		Yield and Rate	
0.893	kilogram per hectare, kg ha^{-1}	pound per acre, lb $acre^{-1}$	1.12
7.77×10^{-2}	kilogram per cubic meter, kg m^{-3}	pound per bushel, lb bu^{-1}	12.87
1.49×10^{-2}	kilogram per hectare, kg ha^{-1}	bushel per acre, 60 lb	67.19
1.59×10^{-2}	kilogram per hectare, kg ha^{-1}	bushel per acre, 56 lb	62.71
1.86×10^{-2}	kilogram per hectare, kg ha^{-1}	bushel per acre, 48 lb	53.75
0.107	liter per hectare, L ha^{-1}	gallon per acre	9.35
893	tonne per hectare, t ha^{-1}	pound per acre, lb $acre^{-1}$	1.12×10^{-3}
893	megagram per hectare, Mg ha^{-1}	pound per acre, lb $acre^{-1}$	1.12×10^{-3}
0.446	megagram per hectare, Mg ha^{-1}	ton (2000 lb) per acre, ton $acre^{-1}$	2.24
2.24	meter per second, m s^{-1}	mile per hour	0.447
		Specific Surface	
10	square meter per kilogram, m^2 kg^{-1}	square centimeter per gram, cm^2 g^{-1}	0.1
1000	square meter per kilogram, m^2 kg^{-1}	square millimeter per gram, mm^2 g^{-1}	0.001
		Density	
1.00	megagram per cubic meter, Mg m^{-3}	gram per cubic centimeter, g cm^{-3}	1.00
		Pressure	
9.90	megapascal, MPa (10^6 Pa)	atmosphere	0.101
10	megapascal, MPa (10^6 Pa)	bar	0.1
2.09×10^{-2}	pascal, Pa	pound per square foot, lb ft^{-2}	47.9
1.45×10^{-4}	pascal, Pa	pound per square inch, lb in^{-2}	6.90×10^3

(continued on next page)

Conversion Factors for SI and non-SI Units

To convert Column 1 into Column 2, multiply by	Column 1 SI Unit	Column 2 non-SI Units	To convert Column 2 into Column 1, multiply by
		Temperature	
1.00 (K – 273)	kelvin, K	Celsius, °C	1.00 (°C + 273)
(9/5 °C) + 32	Celsius, °C	Fahrenheit, °F	5/9 (°F – 32)
		Energy, Work, Quantity of Heat	
9.52×10^{-4}	joule, J	British thermal unit, Btu	1.05×10^{3}
0.239	joule, J	calorie, cal	4.19
10^{7}	joule, J	erg	10^{-7}
0.735	joule, J	foot-pound	1.36
2.387×10^{-5}	joule per square meter, J m^{-2}	calorie per square centimeter (langley)	4.19×10^{4}
10^{5}	newton, N	dyne	10^{-5}
1.43×10^{-3}	watt per square meter, W m^{-2}	calorie per square centimeter minute (irradiance), cal cm^{-2} min^{-1}	698
		Transpiration and Photosynthesis	
3.60×10^{-2}	milligram per square meter second, mg m^{-2} s^{-1}	gram per square decimeter hour, g dm^{-2} h^{-1}	27.8
5.56×10^{-3}	milligram (H_2O) per square meter second, mg m^{-2} s^{-1}	micromole (H_2O) per square centimeter second, μmol cm^{-2} s^{-1}	180
10^{-4}	milligram per square meter second, mg m^{-2} s^{-1}	milligram per square centimeter second, mg cm^{-2} s^{-1}	10^{4}
35.97	milligram per square meter second, mg m^{-2} s^{-1}	milligram per square decimeter hour, mg dm^{-2} h^{-1}	2.78×10^{-2}
		Plane Angle	
57.3	radian, rad	degrees (angle), °	1.75×10^{-2}

	Electrical Conductivity, Electricity, and Magnetism		
10	siemen per meter, S m^{-1}	millimho per centimeter, mmho cm^{-1}	0.1
10^4	tesla, T	gauss, G	10^{-4}
	Water Measurement		
9.73×10^{-3}	cubic meter, m^3	acre-inch, acre-in	102.8
9.81×10^{-3}	cubic meter per hour, m^3 h^{-1}	cubic foot per second, ft^3 s^{-1}	101.9
4.40	cubic meter per hour, m^3 h−1	U.S. gallon per minute, gal min^{-1}	0.227
8.11	hectare meter, ha m	acre-foot, acre-ft	0.123
97.28	hectare meter, ha m	acre-inch, acre-in	1.03×10^{-2}
8.1×10^{-2}	hectare centimeter, ha cm	acre-foot, acre-ft	12.33
	Concentrations		
1	centimole per kilogram, cmol kg^{-1}	milliequivalent per 100 grams, meq 100 g^{-1}	1
0.1	gram per kilogram, g kg^{-1}	percent, %	10
1	milligram per kilogram, mg kg^{-1}	parts per million, ppm	1
	Radioactivity		
2.7×10^{-11}	becquerel, Bq	curie, Ci	3.7×10^{10}
2.7×10^{-2}	becquerel per kilogram, Bq kg^{-1}	picocurie per gram, pCi g^{-1}	37
100	gray, Gy (absorbed dose)	rad, rd	0.01
100	sievert, Sv (equivalent dose)	rem (roentgen equivalent man)	0.01
	Plant Nutrient Conversion		
	Elemental	***Oxide***	
2.29	P	P_2O_5	0.437
1.20	K	K_2O	0.830
1.39	Ca	CaO	0.715
1.66	Mg	MgO	0.602

1 Sustainability of Post-Green Revolution Agriculture: The Rice–Wheat Cropping Systems of the Indo-Gangetic Plains and China

Raj K. Gupta

Rice–Wheat Consortium for Indo-Gangetic Plains
New Delhi, India

Ram K. Naresh

Sardar Vallab Bhai Patel University of Agriculture & Technology
Meerut, India

Peter R. Hobbs

International Maize and Wheat Improvement Center
Kathmandu, Nepal

Zheng Jiaguo

Sichuan Academy of Agricultural Sciences
Chengdu, China

Jagdish K. Ladha

International Rice Research Institute
Manila, Philippines

ABSTRACT

Agricultural policy in the 1960s and 1970s focused on food security through increased coverage of high-yielding varieties. This enabled rice (*Oryza sativa* L.)–wheat (*Triticum aestivum* L. and *T. durum* Desf.) to emerge as a major cropping system in the Indo-Gangetic Plains (IGP), ushering in the Green Revolution (GR). The GR technologies have remained the cornerstone of the South Asian strategy for food security, rural development, conservation of natural resources, and poverty alleviation. Evidence is now appearing that indicates the rice–wheat system has fatigued the natural resource base. Agriculture is now in need of an infusion of technologies to tap new sources of productivity growth. On the basis of driv-

 Improving the Productivity and Sustainability of Rice–Wheat Systems: Issues and Impacts. ASA Special Publication 65.

ing variables for agricultural development, the IGP has been delineated into five relatively homogeneous transects to address location-specific rice–wheat system ecology problems. The integration of crop improvement; water management; nutrient management; weed, disease, and pest management; and policy and social science research in relation to crop establishment within these transects allows sustained production, diversification, and enhanced system productivity. This chapter describes various options available to farmers for crop establishment and diversification in rice–wheat system ecology and the relative potential of improved technologies across the IGP. We highlight new strategies and draw upon some Chinese agrotechniques relevant to the rice–wheat system of the IGP.

In South Asia, Rice–Wheat Consortium (RWC) countries—Bangladesh, India, Nepal, and Pakistan—have devoted nearly half of their total land area of 401.72 million ha to agriculture to feed and provide livelihoods for 1.8 billion people (FAO, 1999). Rice and wheat, the staple food crops for these people, are of great significance, as these crops contribute more than 80% of the total cereal production in these countries (Timsina & Connor, 2001). The rice–wheat system occupies nearly one-fifth of the total area under these crops (Hobbs & Morris, 1996; Ladha et al., 2000). Thus, the intensively cultivated irrigated rice–wheat system is fundamental to employment, income, and livelihoods for hundreds of millions of rural and urban poor of South Asia (Evans, 1993; Paroda et al., 1994). Suitable thermal regimes for rice and wheat crops during the annual cycle, the development of short-duration cultivars, irrigation, and the ever-increasing demand for food remained the driving force for expansion in area of the rice–wheat system during the GR. In the last few decades, high growth rates for food grain production (wheat 3.0%, rice 2.3%) in Consortium countries have kept pace with population growth (Asia Pacific Association of Agricultural Research Institutions, 2000). Evidence is now appearing that the rice–wheat system has fatigued the natural resource base (Abrol et al., 2000). Thus, food security of the region is under continuous threat from new emerging challenges of post-GR agriculture.

The growing realization that agriculture of the post-GR will be guided by the need to produce more high-quality food from more marginal-quality land and water resources, besides sustaining environmental quality, only adds to the challenge. To address the concerns of the rice–wheat system in the IGP, the four national agricultural research and extension systems of Bangladesh, India, Nepal, and Pakistan together with international agricultural research centers (CIMMYT, CIP, ICRISAT, IRRI, and IWMI) commissioned an ecoregional program known as the Rice–Wheat Consortium for the Indo-Gangetic Plains. The major challenge for the Consortium countries is to develop the rice–wheat system to produce more at less cost and improve profitability and sustainability. This suggests that agriculture in Consortium countries will be in need of an infusion of new technologies that are able to tap into new sources of productivity growth and are more sustainable. In recent years, several reviews have focused on technical knowledge and agronomic and crop management of the rice–wheat cropping system (Pingali, 1999; Ladha et al., 2000; Timsina & Connor, 2001). Since tillage systems for crop establishment serve as guideposts for the choice of agronomic and crop management practices for water, nutrients, and pest complexes, this chapter complements earlier work (Paroda et al., 1994; Ahmed & Meisner, 1996; Ladha et al., 2000; Timsina & Connor,

2001; Hobbs & Gupta, 2003, Chapter 7 of this publication). We discuss various tillage and crop establishment options for growing rice and wheat in sequence, describe the limitations and relative potential of these options in a systems perspective, assess prospects for intensification and diversification of the rice–wheat system ecology, enhance its sustainability, and provide insights for upscaling promising technologies among different regions of the IGP.

DISTRIBUTION OF THE RICE–WHEAT SYSTEM

The magnitude and intensity of crops grown in a region depend on soil and bioclimatic conditions, socioeconomic situations of the farmers, and opportunities for marketing the surplus. Although rice and wheat crops have been grown for more than a thousand years in South Asia, the system has generally developed through the introduction of rice in traditional wheat-growing areas and vice versa (Paroda et al., 1994; Tran & Marathée, 1994). Now, most of the rice–wheat system is grown widely in subtropical and warm-temperate regions of South and East Asia characterized by warm, wet summers and cool, dry winters. It extends across the IGP, spanning from the Swat Valley in Pakistan through the states of Punjab, Haryana, Uttar Pradesh, Bihar, and West Bengal in India and into Nepal and Bangladesh. It has been reported that nearly 85% of the rice–wheat system of South Asia is located in the IGP (Fig. 1–1; Timsina & Connor, 2001; Huke et al. 1993a,b;

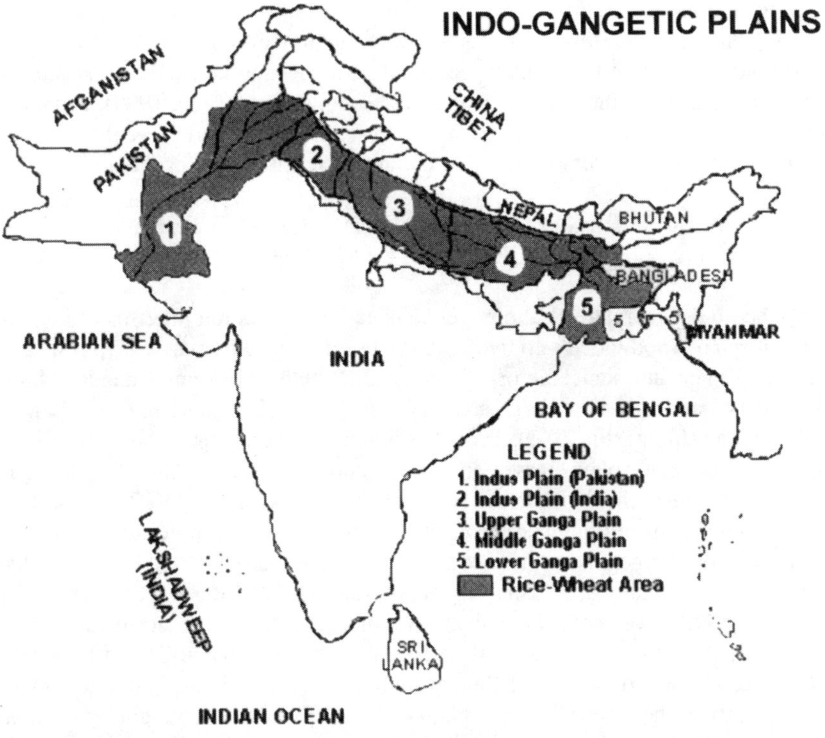

Fig. 1–1. Generalized map of the Indo-Gangetic Plains showing different transects.

Woodhead et al., 1993, 1994). The remaining rice–wheat system is practiced in several other parts of the Indian subcontinent outside the IGP in Madhya Pradesh, Himachal Pradesh, the Brahmaputra floodplains of Assam, and the southwestern parts of India and Bangladesh. In China, the rice–wheat system is widely practiced in Jiangsu, Zhejiang, Hubei, Guizhou, Yunan, Sichuan, and Anhui provinces along the Yangtze River plains (Huke et al. 1993c; Lianzheng & Yixian, 1994; Zheng, 2000). Because the national agricultural crop statistics reported in the region are published for individual crops and not for cropping systems, and there are spatial and temporal variations among crops, most estimates of rice–wheat area are subjective (Paroda et al., 1994; Hobbs & Morris, 1996; Yadav et al., 1998; Timsina & Connor, 2001), although attempts were made to estimate area more accurately and develop RW atlases in the early stages of the RWC (Huke et al., 1993a,b; Woodhead et al., 1993, 1994). The situation is further complicated by the fact that, whereas some farmers grow rice and wheat on the same plot in the same year, others grow break crops in between rice and wheat in different year, and also in different plots. Thus, estimates of rice–wheat area vary among different studies. However, most researchers seem to agree that rice and wheat together contribute more than 70% to total cereal production and the relative importance of the rice–wheat system varies among countries (Table 1–1). On the basis of recent studies, the estimated area of the rice–wheat system in India, Pakistan, Bangladesh, and Nepal is about 10.0, 2.2, 0.8, and 0.5 million ha, respectively, totaling 13.5 million ha (Ladha et al., 2000). The rice–wheat area in China is reported to range from 9.5 to 13.5 million ha (Pan Jiarong, personal communication, 2002; Zheng, 2000; Timsina & Connor, 2001; Huke et al., 1993c). Keeping in view population pressure and good lands being diverted to other sectors of national economies, the prospects for further expansion of the rice and wheat area seem remote (FAO, 1999), and additional sources of productivity growth in rice–wheat would have to come through newer technological interventions that enhance overall system productivity.

CHARACTERISTICS OF THE IGP AND THE RICE–WHEAT SYSTEM

The IGP is a relatively homogeneous ecological region in terms of vegetation but can be subdivided into five broad transects, based primarily on physiography, bioclimate, and social factors (Gupta et al., 2002b). The trans-Gangetic plains (Transects 1 and 2) occupy large areas of east Punjab (Pakistan) and west Punjab and Haryana (India) (Fig. 1–1). Transects 3 and 4 comprise the areas of the upper and middle Gangetic plains in western-central and eastern Uttar Pradesh, Bihar, and the Terai of Uttaranchal in India and Nepal. The lower parts of the IGP in West Bengal, India, and parts of Bangladesh constitute Transect 5. The transect approach helps delineate more homogeneous rice–wheat zones within the IGP and facilitates discussions on enhancing and sustaining the productivity of natural resources.

The IGP have a continental monsoonal climate. In the northwest trans-Gangetic plains, the average annual precipitation ranges from 400 to 750 mm yr^{-1} and increases toward the Bay of Bengal. In the warm and humid Transect 5, comprising parts of the lower Gangetic plains of West Bengal and Bangladesh annual rainfall is as high as 1800 mm yr^{-1}. Nearly 85% of the total precipitation is received

Table 1–1. Rice and wheat area, and its contribution to total cereal and rice–wheat production in various countries.

Country	Rice–wheat area		Area contribution of the system crop to the total area of individual crop: Rice		Wheat		Contribution of rice and wheat to total cereal cereal production‡	Contribution of the rice–wheat system to total rice and wheat production†
	Estimate 1†	Estimate 2‡	Estimate 1	Estimate 2	Estimate 1	Estimate 2		
	million ha		%					
China	--	13.5	--	31	--	35	72	--
India	9.1	10.3	22.3	23	38.7	40	85	25.7
Pakistan	1.6	2.2	80.3	72	22.4	19	92	35.8
Bangladesh	0.5	0.5	4.9	5	84.7	85	100	7.9
Nepal	0.6	0.6	34.3	35	81.7	84	71	77.6

† Hobbs & Morris (1996).
‡ Timsina & Connor (2001).

during the monsoon season from June to September. In winter months, only a few showers are received from December to February. The weather is cool and dry during the wheat growing season (November–March). Rice is grown during the warm, humid or subhumid monsoon season (June–October). Traditionally wheat has been cultivated for a long time in the northwest and rice in the eastern IGP. Because of the availability of suitable varieties released in the 1970s, significant areas are now successfully grown to the rice–wheat cropping pattern in the northwest and eastern parts of the IGP. Whereas the Gangetic plains gradually slope from the northwest toward the Bay of Bengal, the Indus plain drains into the Arabian Sea. There are wide variations in soil types, generally coarser in the trans- and upper Gangetic plains and becoming finer with the run of the river systems. Soils are primarily calcareous and micaceous alluviums with sandy loam to loam in the upper reaches, and becoming finer textured in the distal plains close to the mouth of the river systems. Irrigated (and partially irrigated) and rainfed agriculture coexist in most districts. This has led to a mosaic pattern of agricultural development in the IGP. Most soils in the IGP are deficient in nitrogen. Deficiency of phosphorus is next in order of importance. Because of the micaceous nature of soils, the use of potassic fertilizers has been avoided for a long period, but deficiency of potash is starting to emerge in several areas (Ladha et al., 2003, Chapter 3 of this publication). Iron and zinc deficiencies are common to the IGP. In the subhumid eastern sector of the IGP, crops respond to the application of boron as well (Velayutham et al., 1999).

A significant observation not reflected in the above description is the reliance of farmers on tube-well irrigation in all parts of the IGP. Excessive groundwater development in freshwater aquifer zones has led to a recession of the water table in some areas and to a rise in the water table in some areas underlain with low-quality aquifers. Low-quality aquifers having problems of residual alkalinity [(1 M HCO_3^{1-} + 2 M CO_3^{2-}) – (2 M Ca^{2+} + 2 M Mg^{2+}) mol L^{-1}], high Na adsorption ratio {Na/[(Ca + Mg)/2]$^{-1/2}$ mol L^{-1}}, and/or excess salts are found in many areas of northwest IGP and are being used successfully in crop production by farmers (Gupta et al., 1994). Whereas in the 1970s more than 2.5 million ha of alkali soils occupied larger areas in the northwest, most of these soils are now at various stages of reclamation. The rice–wheat system is a preferred choice during the initial reclamation stage of alkali soils (Gupta & Abrol, 1990).

In the northwest, the rice–wheat system is highly mechanized, input-intensive, and dependent on the conjunctive use of surface and groundwater. In contrast, the rice–wheat system of the eastern IGP is labor-intensive and less mechanized, and farmers use low inputs because of serious problems of drainage congestion and rainwater management (Velayutham et al., 1999). Farm holdings are larger in the northwest than those of resource-poor farmers of the eastern region.

Some farmers in the IGP grow greengram (mungbean [*Phaseolus aureus* Roxb.) and cowpea (*Vigna sinensis* Salvi) during the transition phase between wheat and rice transplanting. To overcome herbicide resistance to isoproturon [3-(4-isopropylphenyl)-1,1-dimethylurea) in little-seed Canary grass (*Phalaris minor* Retz.), farmers rotate wheat with Indian-mustard [*Brassica juncea* (L.) Czern. var. *juncea*], sugarcane (*Saccharum officinarum* L.), or berseem (*Trifolium alexandrinum* L.). In India and Pakistan, farmers intercrop or mix crop mustard with wheat (Hobbs et al., 1985). Many Nepalese farmers grow soybean [*Glycine max* (L.) Merr.]

on the peripheral bunds of rice fields. In the eastern parts of the IGP in Uttar Pradesh, Bihar, West Bengal, and Bangladesh, many farmers grow three rice crops in a year: the *aus* (April–July), *aman* (July–November), and *boro* rice crop (November–April). Boro rice is generally grown in poorly drained lower positions of the landscapes with provision for irrigation. In the low-lying areas of eastern Uttar Pradesh and Bihar (India), parts of Bangladesh, and Jiangsu Province of China, where water recedes late or soils remain wet for a long time, farmers sow lentil (*Lens culinaris* Medik. subsp. *culinaris*), pea (*Pisum sativum* L. subsp. *sativum*), and wheat on drying paddies a couple of days before the harvest of rice (Razzaque et al., 1995; Timsina & Connor, 2001; Gupta et al., 2002b). In comparison with northwestern IGP farmers, eastern IGP farmers diversify the rice–wheat system more to cover risks of drought- and flood-prone agriculture. Many farmers replace wheat with oilseeds (*Brassica juncea* Coss. or *napus*), pulses (pea), grass pea (*Lathyrus* spp.), chickpea (*Cicer arietinum* L.), lentil, potato (*Solanum tuberosum* L.), or sugarcane, and occasionally rice by pigeonpea [*Cajanus cajan* (L.) Millsp.], maize (*Zea mays* L.), sunflower (*Helianthus annuus* L.), soybean, and sorghum [*Sorghum bicolor* (L.) Moench]. Winter maize is also popular in eastern parts of the IGP in Bihar and Uttar Pradesh and Bangladesh. Compared with those of the northwest, temperatures rise early in the east and mean temperatures are higher for longer. This reduces the length of the growing season of wheat in the east. To improve wheat productivity, it is crucial that rice vacate the fields early in the winter season in the eastern IGP.

In China, the rice–wheat system is located primarily in the Yangtze River valley around 30 ± 4° N latitude (Zheng, 2000; Huke et al., 1993c). Annual precipitation varies from 650 to 1400 mm, increasing from the south to southeast. Most precipitation is received during April and October. The annual total sunshine hours are more in the river plains (≈2000 h) than in the plateau region in the south or the low mountainous regions in southeastern China (1100–1400 h). In the plateau region, red earths are low in soil fertility. In the valley and low mountainous regions, purple earths are fertile soils. The growing season for wheat is shorter in the south (early November to mid May) than in the north (early October to mid June). Consequently, longer-duration wheat cultivars are combined with shorter-duration rice cultivars of *japonica* type during the cool dry winter in northern China. In southern and southeastern China, farmers grow a second rice crop in the rice–wheat system and sweet potato [*Ipomoea batatas* (L.) Lam. var. *batatas*] replaces the second rice crop in southeastern and southwestern China (Timsina & Connor, 2001). Whereas the average productivity of the rice crop in China ranges from 6.0 to 8.2 Mg ha^{-1}, wheat yields are 2.1 to 3.2 Mg ha^{-1} (Pan Jiarong, personal communication, 2002).

The above description suggests that gradual transitions in climate and physiography have a significant effect on cropping systems and the agronomic and crop management practices of the rice–wheat system across transects in the IGP and in China.

SUSTAINABILITY DIMENSIONS OF THE RICE–WHEAT SYSTEM

In the rice–wheat cropping system, the two crops have contrasting edaphic requirements. Whereas rice is commonly transplanted into puddled soils and is given continued submergence, wheat is grown in upland well-drained soils having good tilth. Puddling reduces the infiltration of water but destroys soil structure (Sharma & De Datta, 1985). For the healthy growth of wheat, farmers generally do six to eight preparatory plowings when drying rice soils to achieve good seedbeds. Excessive tillage results in late planting of wheat and reduced yield (Harrington et al., 1993b; Hobbs & Gupta, 2003, Chapter 7 of this publication). The turnaround time between the rice and wheat phase is 2–6 wk. To plant wheat in time, many farmers harvest the rice crop using combines and burn loose residues, which helps farmers to clear the field faster than with natural decomposition. Such burning releases greenhouse gases and particulate matter in large quantities in sudden spurts. This deteriorates the air quality, results in significant losses of nutrients from the soil (Sharma & Mishra, 2001), and has adverse effects on human health. Besides late planting, key concerns of the rice–wheat system ecology include declining factor productivity (efficiency of specific input factor) and yields below the potential productivity of the system, nutrient mining, and the emergence of multiple nutrient deficiencies (Abrol et al., 2000; Byerlee et al., 2003, Chapter 4 of this publication; Ladha et al., 2003, Chapter 3 of this publication; Pathak et al., 2002).

In the semiarid northwest, excessive groundwater development for irrigation has led to receding water tables (Gupta & Abrol, 1990). In the eastern IGP, the delayed onset of rains during the monsoon season delays rice nursery and transplanting operations, leading to late planting of succeeding crops. Some of these sustainability concerns are depicted in Fig. 1–2. Some of these concerns of the rice–wheat system can be addressed by timely planting. If this can be combined with good crop and residue management that avoids burning and improves the use efficiency of costly external inputs, the benefits will be even greater.

Tillage and agronomic practices for the establishment of the rice–wheat system in the northwest differ considerably from those followed by farmers in the eastern IGP. The ways farmers establish crops in various transects of the IGP are best decided by socioeconomic conditions, seasonality of labor, water demand and supply, the availability of inputs, and local weather conditions. Numerous combinations of these conditions dictate how farmers plant. Therefore, a generalized package of agronomic and crop management recommendations is unlikely to work in all situations within and between transects. Farmers need options for addressing complex site-specific natural resource management problems. This implies that there is a need to fine-tune existing technologies and produce new information that provides options for tackling emerging concerns of post-GR agriculture and achieving the stated sustainability goal.

Commonly, rice is established through manual transplanting of seedlings in puddled soils. A few farmers establish rice with a mechanical rice transplanter. In India and China, rice is also grown in a limited area using a "double-transplanting rice culture" option (Zheng, 2000). In double transplanting, rice is seeded at 500 g m^{-2} in a small nursery and then 8- to 10-d-old seedlings are transplanted to another nursery bed for the next 40 to 50 d before shifting them to the main field. Double transplanting overcomes the problems associated with delays in water supplies because of canal operation or the late onset of monsoons, overaged seedlings, weak seedlings, and late availability of the main fields. The double-transplanting method improves grain yield by 13.8% over that of once-transplanted seedlings (Zheng,

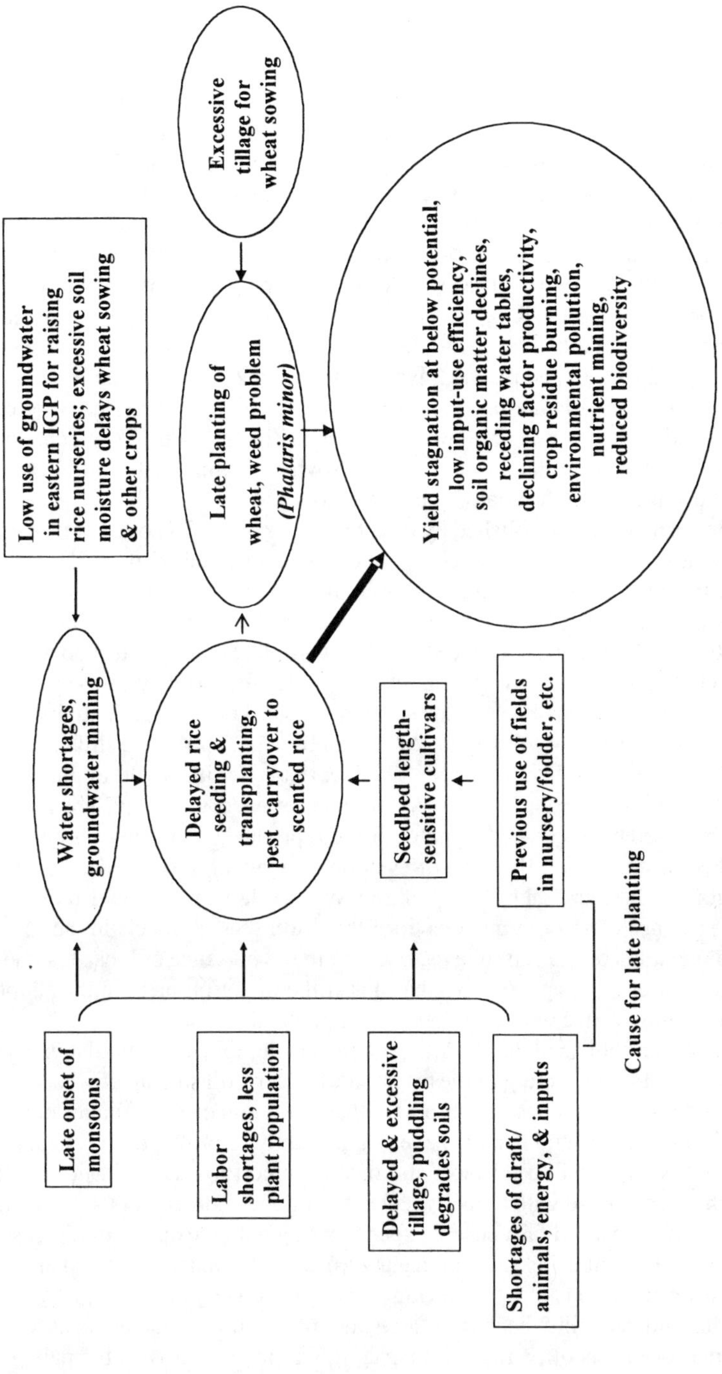

Fig. 1–2. Sustainability dimensions of the rice–wheat system in the Indo-Gangetic Plains.

2000). Double transplanting is becoming popular in the intensively cultivated areas of West Bengal (India) and Sichuan Province (China). Rice is mostly transplanted manually. Rice can also be transplanted into dry soils and irrigated. To do this, seedlings are placed in narrow slits opened with point-openers of a zero-till drill. In dry or moist permanent beds free of weeds, rice can be transplanted in "opened slits" (Balasubramanian et al., 2003, Chapter 6 of this publication). This practice is still in the experimental stage but can save water and avoid puddling.

In permanent raised-bed planting systems, a planter prepares beds of variable width. In Australia, depending on tractor wheel spacing, 130- to 160-cm-wide beds are prepared in finer textured soils with a narrow furrow for irrigation. In coarser soils of the IGP, narrower 60- to 70-cm-wide beds are made with a machine (Gupta & Rickman, 2002a). Two or three rows of rice can be placed on the bed by transplanting or direct seeding. Rice can also be transplanted by broadcasting seedlings onto a previously puddled or unpuddled wetted soil. In this method, 1 or 2 seeds are sown in bubble-cupped plastic trays having 1.5 to 2.5 cm^3 soil pellets per cup. Seedlings 20 to 25 d old are taken to the fields and broadcast onto the puddled or unpuddled soils using a power blower or by hand (Zheng, 2000). This technique is being practiced in several areas of Sichuan Province (China) but is still in the experimental stage in South Asia (Gupta et al., 2000).

Rice can also be established by direct dry seeding (DSR) or by wet seeding. Direct seeding of rice is an attractive alternative to puddled transplanted rice (TPRS). In India, Pakistan, and Nepal, rice direct-seeded into unpuddled soils is being evaluated in combination with zero- and reduce-till wheat (Gupta et al., 2002). Results show that, if rice weeds are not a problem or are controlled with herbicides, DSR is as good as TPRS. Direct-seeded rice followed by zero- or conventionally tilled wheat has been reported to improve the total productivity of the rice–wheat system (Hobbs & Gupta, 2000; Hobbs et al., 2001). We still need more research on stand establishment, seed rate interactions with N and blight infestation, and varieties developed that do better under direct dry seeding. The rice crop can also be raised by direct seeding of the primed sprouted seeds using a drum seeder on puddled or unpuddled flat soils or even on permanent raised beds. The use of this option in the permanent bed planting system, however, would require adjustable spacing for smooth movement of the drum seeder wheels in the furrows and seed openings on the raised beds. Seeding in the presence of loose residues of the previous crop (tonnage, 5–7 Mg ha^{-1}) is still a major problem in the adoption of direct drilling of rice with drum seeders or zero-till drills.

It is well established that puddling compacts the soil. The cumulative effects of puddling and freewheeling of the tractor during zero-till sowing of wheat add to compaction problems. It has been reported that minimum or zero-tillage combined with controlled traffic and residue cover significantly improve soil physical properties as reflected in reduced runoff and increased performance of wheat, maize, and sorghum compared with a free-wheeled treatment (Tullberg, 2001; Tullberg et al., 2001). Scientists and farmers are experimenting with the option of direct seeding of rice and wheat using the Pantnagar zero-till drills in free-wheeled and controlled traffic conditions in western Uttar Pradesh and Haryana, India. Direct dry-seeded rice and zero-till wheat crops are generally seeded in equally spaced rows 20 cm apart or in sets of paired rows (spacing within and between the pairs is 15

Table 1–2. Effect of paired-row planting, controlled traffic, and free wheeling of tractor on the yield of zero-till (ZT) wheat and irrigation water savings in farmer participatory trials, Ghaziabad, India.

Parameter	ZT-paired planting†	ZT-controlled traffic‡	Zero-tillage (free-wheeled)	Conventional§ (free-wheeled)
Water saving,%¶	26.2	30.8	35.4	--
Yield, Mg ha^{-1}	6.5	5.8	5.8	5.2

† Spacing between intra- and interpaired row is 13 to 15 and 25 to 30 cm, respectively. Traffic was controlled by removing tines immediately behind the tractor wheels.
‡ Two rows behind the tractor wheels were not sown.
§ Late-planted wheat.
¶ Percentage of the total irrigation time taken in respective treatments compared with conventional practices of farmers.

and 30 cm, respectively). Results of farmer field trials indicated that the productivity of wheat was in the order: paired row > controlled traffic with zero-till > conventional tillage. Thus, paired-row planting with zero-tillage improves yield more than the controlled traffic or free-wheeled zero-till treatments with equal row spacing (Table 1–2).

It emerges that farmers have the option of growing rice by direct seeding or by transplanting in puddled or unpuddled conditions on flat or raised beds. Wheat can be grown on flat or raised beds using minimum (surface seeding and zero-till) and reduced tillage options (Table 1–3). The remaining issue is where these combinations are best applied and how can the relevant technologies be disseminated for the maximum effect on the livelihoods of farmers? This assessment will be qualitative and subjective in the absence of georeferenced data. Table 1–4 assesses the place for combination tillage options. This option matrix provides an idea of the relative potential of crop establishment techniques in different transects.

It is common knowledge that most upland crops suffer in the IGP because of temporary water stagnation during the monsoon season and also in the winter during irrigation cycles (Yadav et al., 1998). Crops planted on raised beds do not suffer as much temporary waterlogging and hence improve the prospects for diversi-

Table 1–3. Matrix of options for establishment of rice and wheat crops in the system.

Planting methods†	Rice			Wheat	
	Puddled (flat surface)	Unpuddled (flat surface)	Raised beds	Flat surface	Raised beds
Direct-seeded					
Zero-till, controlled traffic, paired row	NA‡	A	A	A	A
Reduce till	NA	A	A	A	A
Sprouted seed (drum seeder), surface seeding	A	A	A	A	A
Transplanted					
Manual	A	A	A	NA	NA
Mechanical transplanter	A	A	DBT	NA	NA
Seeding broadcast	A	A	A	NA	NA

† Overarching concerns of cultivars, weed, pest, and crop residue management and machinery.
‡ NA, not applicable; A, applicable; DBT, doubtful.

Table 1–4. Matrix for crop establishment techniques and their potential in different Indo-Gangetic Plains transacts.

				Reduce tillage		
Wheat–rice†	Conventional tillage‡	Zero-till	Surface seeding	2-wheeled hand tractor	4-wheeled tractor	Bed planting
PTR	General control	TUM§	ML¶	ML§	TU§	Not applicable
PWSR	TU¶	TU§M¶	ML¶	ML§	TU§	Not applicable
NPTR	TUM#	TUM¶	TUML¶	ML¶	TU#	TUM§L¶††
NPDSR	TUML#	TUM§	TU#ML§	ML§	TU§	TUML§††

† PTR, puddled and transplanted rice; PWSR, puddled wet-seeded rice; NPTR, nonpuddled transplanted rice; NPDSR, nonpuddled direct dry-seeded rice.
‡ T, Trans-Gangetic plains; U, upper Gangetic plains; M, middle Gangetic plains; L, lower Gangetic plains.
§ Most important for a large area.
¶ Important for many areas.
# May be important in some areas.
†† Bed-planting system offers opportunities for diversification of rice–wheat.

fication of the rice–wheat system in the IGP. It appears that puddled and unpuddled transplanted rice or direct-seeded rice combine well with a zero-till option of wheat establishment in many areas in the trans-, upper, and middle IGP. Surface seeding or paracropping of wheat or other crops has large potential in the middle and lower IGP, where soils remain wet for long periods and rice vacates the fields late in the season. In weedy situations, a rotavator seeder and reduced-till wheat combined go well with various rice establishment options and are suited to the middle and lower IGP. The combination of options for rice–wheat is expected to have differential requirements for agronomic and crop management practices in different locations. Some discussion on the interaction effects of puddling on reducing tillage crop residues, nutrient dynamics (particularly N), and water management is warranted. In the ensuing sections, these issues are discussed as strategic entry points for further research.

How Does Puddling Affect Rice–Wheat System Productivity?

In the IGP, 5- to 6-wk-old seedlings are transplanted into puddled fields to grow the rice crop. Puddling is achieved through plowing and leveling the field, followed by a soaking irrigation and stirring or plowing of the surface layer while saturated. Puddling reduces soil water permeability, controls weeds, improves nutrient availability, and preserves the anaerobic conditions suitable for the growth of wetland rice (Sanchez, 1973). The presence of an impermeable puddled layer reduces water percolation, increases resistance to root penetration, and the puddled layers remain wet for longer periods because of reduced evaporation, with resulting delays in tillage operations, which adversely affect growth of the following crops (Sur et al., 1981; Greenland, 1997) until the soil is tilled again. Puddling alters the physical and chemical environment in the plow layer (Sharma & De Datta, 1985; Kar et al., 1997) and has implications for preparatory tillage for the succeeding wheat crop and to some extent its irrigation water requirement (Hobbs et al., 2001). Unless finer textured soils have the right moisture regime at plowing, large clods are formed that are difficult to pulverize for good seed–soil contact and crop germination (Hobbs et al., 2001). Hobbs et al. (2001) further reported that puddling had an adverse effect on the following wheat crop on finer textured but not on coarser textured soils. But puddling in the latter soils would reduce groundwater recharge during the monsoon season and promote runoff of rainwater, thus only adding to the problem of the receding water table in the northwestern IGP. No significant differences in yield between rice crops grown on puddled or nonpuddled soils were observed in long-term rice–wheat experiments at Pantnagar (India) and Bhairahawa (Nepal). Hobbs et al. (2001) observed higher numbers of tillers and panicles per square meter in the direct-seeded rice plots, but total yields were similar between direct-seeded and puddled transplanted plots. This was due to more filled grains per panicle in the transplanted crop and a higher incidence of insect pests, especially brown planthopper [*Nilaparvata lugens* (StååI); BPH] and disease such as blast (*Pyricularia grisea* Sacc.), in the direct-seeded plots. This was also possibly due to higher seed rates in DSR plots.

For South Asian conditions, we have observed that the effects of puddling are generally short-lived, as poor-quality irrigation systems and shortages of power and diesel fuel do not allow farmers to maintain the submerged soil conditions. Most puddled soils show cracks varying in size and depth depending on the soil texture within 10 to 15 d of rice transplanting when they are allowed to dry. Once the puddled soils crack, it takes more irrigation time to maintain the ponded condition. So the issue is, Can we grow rice (transplanted or direct-seeded) without puddling?

Can Puddling Be Avoided for Rice Growing?

The power and labor requirements of puddling soils and transplanting the rice crop are high, and peak seasonality of labor scarcity is a major problem. Pandey and Velasco (1999) suggested that dry seeding of rice is an appropriate alternative to puddled transplanted rice for South Asia. Direct dry-seeded rice planted in coarser textured soils commonly suffers from Fe chlorosis, which disappears at the time of flooding. We have observed that rice cultivars differ significantly in early growth, vigor, and susceptibility of rice seedlings to Fe chlorosis in DSR in flat and permanent bed planting systems. Hobbs et al. (2001) reported improvement in overall rice–wheat system productivity in nonpuddled soils. Thus, additional work is needed for screening and breeding of cultivars to realize the full potential of the dry-seeded rice system in flat and permanent bed planting systems. Interestingly, deepwater rice, an important rice crop in many low-lying areas of Bangladesh, eastern India, and Nepal, is dry-seeded onto nonpuddled soils. In recent trials, rice and wheat crops planted on a permanent system of raised beds have shown great promise in terms of water saving, efficient use of inputs, and higher yields (Table 1–5). Low yield in direct-seeded rice on raised beds was primarily due to Fe chlorosis manifested early in the season. Flooding the beds ameliorated the chlorosis. The use of seed priming also enhances early season growth of seedlings and may be an agronomic practice that would be useful for dry-seeded rice (Witcombe & Harris, 1997). Thus, it seems that farmers have options for doing away with the practice of puddling in many situations.

Table 1–5. Effect of tillage options on total irrigation time, yield attributes, and grain yield of rice.

Tillage	Total exp. area‡	No. of plants m^{-2}	Tillers/ plant	Total irrigation time§	Productive tillers/ plant	Spikelet length	Grains/ panicle	Grain yield
	ha			h ha^{-1}		cm		Mg ha^{-1}
DSRB	14 (22)	34	24	152 (39)	15	23	165	5.0
TRB	12 (20)	35	24	146 (41)	19	23	173	5.6
ZTR	12 (10)	56	16	205 (18)	13	22	163	5.7
RTR	1.6 (7)	32	13	216 (13)	13	23	169	5.2
C T¶	14 (35)	27	16	250	12	22	163	5.3

† DSRB, direct-seeded rice on beds; TRB, transplanted rice on beds; ZTR, zero-tilled rice on unpuddled flat soils; RTR, reduced-tilled transplanted rice on flat surface; CT, conventional tillage.
‡ Values in parentheses are numbers of farmers participating in the trials.
§ Values in parenthesses are percentages water savings in terms of irrigation time in relation to conventional practices of farmers.
¶ Severe Fe chlorosis was observed in initial crop stages.

Can the Mechanical Rice Transplanter Substitute for Manual Transplanting in the IGP?

Traditionally, rice is transplanted manually into puddled soils. In this case, the issue is to identify management practices that minimize the disadvantages of puddling and remove the problems associated with transplanting. Manual transplanting has several problems, such as seasonal labor shortages, low plant density, and random and deep placement of seedlings. In eastern India and Bangladesh, contractual farm workers generally transplant up to 10 seedlings per hill and place the seedlings 7 to 10 cm deep into puddled soils, with a consequent need for higher seed rates (Gupta et al., 2002b). Planting deep into puddled layers and more seedlings per hill adversely affect tillering and productivity of rice. In northwest India and Pakistan, scientists are experimenting with mechanical rice transplanters to overcome these problems. For mechanical transplanting of rice, seedlings are grown onto a thin soil layer of 3- to 5-cm thickness in wooden frames, placed on a raised bed covered with a plastic sheet. Rice seedlings, about 20 d old, make a thick mat of seedlings with good development of the root system. These seedling mats are sized and mounted onto the trays of the mechanical rice transplanter. Raising mat-type nurseries requires skill and training because operation of the rice transplanter is needed. Another major problem in the operation of the transplanter is "mud flows." During transplanter operations, the float that rests the transplanter on puddled soils pushes the mud sideways, burying one row of the newly planted seedlings. Variations of settling time for texturally different puddled soils and unevenly leveled fields result in variable depth of standing water and settling times. Uneven leveling of rice fields is the rule rather than the exception. These factors make it difficult to control mud flows during transplanter operations. To minimize mud flow, farmers generally puddle the fields a day ahead, allowing water to infiltrate and soil particles to settle ahead of transplanter operations, but this increases transplanting time. Laser-assisted precision land leveling would be an obvious help in overcoming the problem associated with smooth operation of the mechanical rice transplanter. Additional problems include machine access at reasonable costs and the small amount of time available for a service provider to make money renting out the machinery to farmers. Since seedling density is higher in this system, increased nutrient applications may be needed to ensure high yields. Use of the mechanical transplanter on zero-tilled soils may be an option, since settling time would not be an issue.

The above description of problems associated with the mechanical rice transplanter is further complicated by the need for puddling soils for smooth operation of the mechanical rice transplanter. Puddling adversely affects the deep percolation of water, thus reducing aquifer recharge in many areas where water tables are receding, particularly in the northwest parts of the IGP. For these reasons, in our opinion, it is unlikely that the mechanical rice transplanter will find favor with IGP farmers and replace manual rice transplanting operations. Instead, direct dry seeding of rice, an appropriate alternative to puddled transplanted rice, will probably receive a major boost in the coming years for the reasons explained above. An alternative option to transplanting is broadcasting of young rice seedlings in fields flooded with water. This method, described above, requires labor for some realignment of seedlings and gap filling but does provide farmers an option for overcoming the drudgery of transplanting.

Is Conventional Tillage Essential for Wheat?

Traditionally, the establishment of the wheat crop includes repeated plowing, cultivating, planking, and pulverizing the topsoil. Repeated tillage operations delay planting, escalate costs, and reduce farmers' profit margins (Hobbs et al., 1992; Harrington et al., 1993a,b). In the IGP, wheat planting after the first fortnight of November reduces wheat yields (Ortiz-Monasterio et al., 1994; Hobbs & Gupta, 2003, Chapter 7 of this publication). In the eastern IGP, the optimal date is the last week of November. It is generally conceived that, if tillage operations prior to wheat sowing were not performed, crop yield would suffer because of poor crop stand and also because of poor soil aeration following an irrigation or rainfall. However, many researchers (Aslam et al., 1993; Singh, 1995; Malik, 1996; Hobbs et al., 1997; Abrol et al., 2000; Gupta et al., 2000; Mehla et al., 2000; Tullberg et al., 2001) have shown that zero-till wheat after rice gives yields that are equal to or even higher than those of the conventional crop. Mehla et al. (2000) showed that a timely sown zero-till wheat crop yielded 6% more than the timely sown conventionally tilled wheat crop because of more efficient fertilizer use and less germination of grassy weeds. On a regional level, they reported a 15% improvement in the productivity of wheat planted with the zero-till method compared with conventionally tilled late-planted wheat. The gains in yield from timely planting with the zero-till drill were 9%. There is a need to adapt the equipment available for four-wheel tractors to smaller two-wheel and animal-power systems to make this technology more available and useful on the small fields of resource-poor farmers in the eastern IGP.

Surface seeding of wheat in the excessively moist, finer textured soils commonly found in low-lying areas of the eastern IGP is another option to completely eliminate tillage in wheat. The wheat seeds are broadcast onto the saturated soil surface either before or after the rice harvest. Seeds germinate and roots follow the saturation fringe as water recedes. Surface seeding options have been evaluated in longer-term studies in Nepal and Pantnagar (Hobbs et al., 2001) and Bihar, India

(Verma, 1999). Farmers in Kustia District in Bangladesh are also using this system (Ahmed & Meisner, 1996). Surface seeding of wheat is also practiced in large areas in China (Zheng, 2000). These and other studies suggest that land preparation is not needed for wheat following rice under a diverse set of soil conditions. This reduces production costs and allows less weed growth (including *Phalaris minor*), less use of fuel and steel for tractor parts, earlier planting, and improvements in the efficiency of water and fertilizer, thus resulting in higher yields. Surface seeding of lentil, grass pea, and wheat after or before the rice harvest (relay cropping) has great potential for increasing cropping intensity and avoids rice fallows (Subbarao et al., 2001) in eastern Uttar Pradesh and Bihar, India; Nepal; and Bangladesh. We estimate that IGP farmers have already reduced tillage in wheat on more than 250 000 ha where there are niches for zero-tillage, the rotavator seeder, and surface seeding.

The benefits of planting rice–wheat on a permanent raised-bed planting system have been discussed. This system is getting increased attention, promotes crop diversification, and provides an alternative source of productivity growth for wheat if planted with sugarcane in western Uttar Pradesh and parts of Haryana, Punjab, and the Nepal Terai. In Bangladesh, wheat–sugarcane planted with the raised bed–furrow system can prove helpful in sustaining the sugarcane crop (Gupta et al., 2002).

Can Crop Residues Be Managed Without Burning?

Farmers traditionally remove most of the rice and wheat straw during the crop harvest. With the introduction of combines for harvesting, more than 75% of the cropped area is now combine-harvested in northwest parts of the IGP. In most of these combine-harvested places, farmers use the wheat straw to feed the animals. It is the 5 to 7 Mg ha^{-1} of rice straw left unused that farmers generally burn to enable early wheat sowing. Combine harvesting of rice leaves behind a swath of loose straw, which interferes with operations of the seed drill used for planting wheat. Many farmers in Punjab (India and Pakistan), chop the rice stubbles with a stubble shaver, dry them, and burn them completely to facilitate the timely planting of wheat. As an alternative to complete burning and overcoming problems associated with the raking of loose straw during seed-drilling operations, many farmers opt for partial burning, wherein only the loose dry straw is burned. The remaining anchored green rice stubbles do not interfere with planting operations. Burning crop residue results in air pollution because of the release of CO_2 and particulate matter and in a substantial loss of nutrients and potential soil organic matter. Indeed, the loss of soil organic matter as a result of burning is one of the recognized threats to rice–wheat sustainability (Abrol et al., 2000; Timsina & Connor, 2001). There is an urgent need to find solutions to help farmers avoid burning.

To facilitate the drilling operation in combine-harvested fields, the first step will be to attach flappers or swappers to combines for some shredding of rice residues and to spread them more uniformly. Leaving a mulch on the surface improves soil moisture and thermal regimes, and improves the early vigor of wheat seedlings and subsequently system productivity (Tang et al., 2000; Zheng, 2000). Another way is to modify the zero-till drill machines available in the region by increasing tine lengths and spacing between tines for greater clearance of residues. Replacement of point-openers with pre-opener discs together with double-disc openers or the use of star-wheel-type seed placement devices used in Zimbabwe (K. Sayers, personal communication, 2000) can establish crops into loose residues without tillage (Gupta & Rickman, 2002a). In monocropped areas of Australia, residues of the previous rice crop are left on the surface to decompose until the planting of wheat in the next season. This residue management practice in Australia and elsewhere is hard to adopt in the IGP because the short turnaround time for planting wheat and the moisture and strength patterns of rice residue do not favor quick decomposition by microbes. Temporary immobilization of soil fertilizer N would also render N unavailable to the following wheat crop. However, N immobilization is not a serious problem and can be handled with a starter N dose or increasing N in the basal application. At this time, only a few farmers have opted to incorporate rice residues into the soil before planting wheat. The raised-bed planting system offers avenues for incorporating crop residues by attaching the shovel-type openers in front of the seed-cum-fertilizer openers and bed-shaper. The shovel shapes the old beds and covers the crop residues with soil taken from the furrows. Incorporation of residues into the beds avoids burning. Farmers did not observe any adverse effect on crop growth due to temporary immobilization of N, but farmers are still experimenting with this new option for residue management. The description above points to the need for improvement in machine design to suit different soil, moisture, and residue conditions.

Can Tillage Options Promote Intensification and Diversification of the Rice–Wheat System in the IGP?

Small and marginal farmers doing subsistence farming need assistance to make their agriculture profitable so they can improve their livelihoods and eventually help themselves escape the ill effects of the poverty spiral. The solution seems to lie in intensification and diversification of the rice–wheat system. Crop diversification ameliorates the adverse effects of seasonality on family income and peak labor demands, reduces risk from fluctuating monsoonal patterns, conserves rainwater and saves irrigation water by incorporating crops with low water requirements, facilitates easier weed and N management, reduces waterlogging, and often results in better yield. Intensification of the rice–wheat system is a very distinct possibility in eastern parts of the IGP where rice fallows are a common practice in different rice ecologies (upland, medium land, and lowland ecologies). Subbarao et al. (2001) indicated that many farmers in the east have to leave their fields fallow (rice fallows) because of excessive wetness of the soil. According to an estimate, nearly 14.3 million ha of 50.4 million ha planted to the main-season rice crop remain fallow. These soils having adequate residual soil moisture can be planted to legumes (lentil, chickpea, pea) and other crops through paracropping and surface-seeding practices. Farmers who wait for conventional tillage for establishment of a succeeding crop after rice often end up using fallows because of the very short winter window in the east. The prospects for introducing sugarcane, mint (*Mentha aravensis* L.), and vegetable crops as intercrops with wheat and winter maize provide good

Table 1–6. Prospects for crop diversification in the rice–wheat system in the Indo-Gangetic Plains.

	India							
Crop†	Uttar Pradesh	Haryana	Punjab	Bihar	West Bengal	Bangladesh	Nepal	Pakistan
Black gram	M				M			
Chickpea		M				G	G	G
Green ram		B	M	G				
Groundnut	S				G	M		G
Lentil	M			G			M	G
Peas and beans	B		M				M	
Pigeonpea	S	B		G			G	
Sugarcane‡	B	M	M			B		B
Soybean	M						G	

† B, bright; M, moderate; G, good; and S, stable.
‡ Intercrops with wheat and gram.

avenues for further intensification and diversification of the rice–wheat system. Table 1–6 gives a qualitative assessment of the prospects for crop diversification across the Consortium countries. While emphasis on grain legumes, fodder legumes, and green manure crops would continue, vegetables with sufficient potential in many peri-urban areas of the IGP could also be considered. Thus, a wide range of crops and cropping system options could be evaluated for their potential contributions to the income of farm households and for their effects on sustainability of the rice–wheat system. Of particular importance would be to elicit farmer preferences of different options for greater inclusion of legumes and other crops in rice- or wheat-based cropping systems.

Early harvest of short-duration rice provides an opportunity for diversification and intensification of the rice–wheat system with oilseeds, potato, pea, and other vegetables grown after the rice harvest and before wheat planting. The permanent system of planting crops on raised beds also promotes diversification of the rice–wheat system, leading to benefits in system ecology. The bed planting system opens opportunities for diversification of the rice–wheat system with pigeonpea, sugarcane, and maize (monsoon and winter season crops) in the northwest, upper, and middle parts of the IGP. To promote diversification, there is a need to develop technological options (genetic and management) to alleviate the major biotic and abiotic constraints to the adoption of these other crops.

Can We Be More Water-Wise in Direct-Seeded Rice and Puddled Transplanted Rice Systems (TPRS)?

Guerra et al. (1998) showed that rice grown with traditional practices in the tropics and subtropics requires 7000 to 15 000 m^3 of water. This consists of 150 to 250 mm for land preparation, 50 mm for growing rice seedlings, and 500 to 1200 mm to meet evapotranspiration (ET) demand and seepage losses, excluding rainfall. The difference between infiltration rates and between puddled and unpuddled soils may be only 250 to 400 mm, but not all of this water is lost from the system, since a high proportion will help to recharge the aquifer (Aslam, 1998). The extra water in the nonpuddled rice system would recharge the profile and aquifer and save water for the next wheat crop and could eliminate at least one irrigation of up to 1000 m^3.

Water management is a major problem for consideration in direct-seeded rice technology in spite of the rice crop maturing 10 to 15 d earlier than the transplanted rice, as it remains longer in the main field. This is not a problem in South Asia, where only a single crop of rice and wheat are grown, but would be a problem in eastern areas of the IGP where two rice crops and a wheat crop are sometimes grown. Several earlier reports showed that puddling significantly reduced percolation losses and the irrigation water requirement of transplanted rice (Sanchez, 1973; De Datta, 1981). However, the water requirement can be reduced through efficient scheduling and delaying irrigation for variable time periods depending on soil conditions (Hatta, 1967; Prihar & Grewal, 1985). Sandhu et al. (1980) reported that delaying irrigation for 1 to 5 d after the disappearance of ponded water in a sandy loam soil brought about no significant reduction in rice grain yield in an experiment conducted for 4 yr at Ludhiana, India. Thus, increasing the drainage period between irrigations can easily offset the effect of increased percolation rates in unpuddled soil. Direct-seeded or transplanted rice on a raised-bed system can also result in water savings through both lower water application rates and increased irrigation intervals. Guerra et al. (1998) suggested that a shift from TPRS to a more water-efficient direct-seeding rice technique is one of several other options to improve water use efficiency of rice at the farm level. In farmers' fields, farmers have observed relatively little soil cracking in bed-planted rice fields compared with puddled ones. Bhuiyan (1992) reported a 40% savings in water without yield losses by replacing shallow-depth water regimes with a saturated soil regime in puddled transplanted rice. He reported that, under high weed pressure, shallow flooding during the first 45 d after transplanting followed by maintenance of a saturated regime for the rest of the season resulted in similar yields and still saved 30% water over the flooded regime. Results of numerous farmer participatory field trials showed that initial flooding of DSR or transplanted rice in unpuddled flat land or bed-planted rice helped save total irrigation time without compromising on grain yield (Table 1–5). It can be mentioned here that lower yields obtained in DSR fields were mainly caused by Fe chlorosis. Some farmers who experimented with DSR and flooded their fields initially observed no Fe chlorosis and had good yields. Unfortunately, we could not monitor these fields from the beginning, and hence the data are not included here.

Inefficiency of the irrigation system and untimely delivery of canal water continue to be the major constraints to agriculture in South Asia as competition with other sectors of national economies (domestic and industry) makes water less available. The ultimate goal of research is to promote water-saving management techniques at the system and farm levels that sustain the environment and allow crop production to be maintained or increased in the face of growing demands for competing uses of water. To improve water use efficiency, the strategy often used in the past, in India and elsewhere in the region, has been to schedule irrigation to match crop water demands. Implicit in the present efficiency of the irrigation water delivery system is to supply sufficient water to meet crop demands in time. But this strategy of irrigation scheduling has not worked well. Farmers in the region continue to use crop phenology and visual symptoms to plan irrigation. Land leveling is also known to improve water use efficiency through uniformity in water application, improvement in crop stand, and improved nutrient and water interactions. Land leveling also facilitates the use of bed planting, zero-till machines, and the mechanical rice transplanter. In the eastern parts of the IGP, the strategy of raising the rice nursery with groundwater and transplanting with the onset of monsoons has been shown to enhance the productivity of not only the main-season (aman) rice crop but also of the rice–wheat system as a whole (S.R. Singh, personal communication, 2001).

What Are the Options for Weed Management in the Rice–Wheat System?

Although flooding helps control weeds in rice (Catling et al., 1983), weed control in general is more difficult in nonpuddled situations. Hobbs et al. (2001) observed mainly sedges in puddled soils and grassy and broadleaf weeds in nonpuddled situations. The use of the stale seedbed system may also be advantageous. In this system, the first flush of weeds is encouraged to germinate before killing them with nonselective herbicides or plowing. For weed control, varieties are needed that emerge quickly, cover the ground well, and are competitive against weeds. The use of rotations or alternative crops such as mustard, berseem, or sugarcane in the rotation also helps control weeds. The development of new, improved herbicides for dry-seeded rice is also needed. In China, some farmers are even growing zero-till rice. Weeds were a problem in the first years of the no-tillage system but were controlled with herbicides. In later years, weeds became less of a problem as the weed bank was exhausted (Zheng, 2000). Rice breeders will need to screen germplasm for the direct-seeding technology (on both flat surfaces and raised beds) to be able to identify suitable varieties for direct seeding and for resistance against planthopper and blast.

The continued use of isoproturon for years in India has resulted in *Phalaris minor* becoming resistant to this herbicide (Malik, 1996). Excessive preparatory tillage has been observed to further aggravate the situation. Farmers in such situations tend to rotate wheat with other crops.

What Are the Options for Efficient Nitrogen Management in the Rice–Wheat System of the IGP?

In the puddled rice system, any N left from the previous wheat crop would be in the NO_3 form. This NO_3–N would be lost through leaching or denitrification when the soil is flooded and puddled. In dry-seeded rice fields, this N leaching is likely to be much less in the absence of any standing water. Deep placement of urea supergranule (USG) tablets or briquettes, or even prilled urea below the soil surface (8–10 cm depth) 7 to 10 d after transplanting, is known to improve fertilizer N uptake by rice and wheat and reduce N losses through NH_3 volatilization in puddled soils (Ladha et al., 2000). With deep placement, steady and adequate supplies of N are made available to the crop for a relatively longer period, thus increasing grain yield and N use efficiency, and can reduce fertilizer requirements by 30 to 50% in clay, clay-loam, and loamy soils with low percolation rates (Mohanty et al., 1999).

This technique is especially effective for rice environments with the variable moisture conditions encountered in most fields. The deep placement of N in direct-seeded rice is as effective as in puddled conditions (Ladha et al., 2000). A major bottleneck in the use of the deep placement technique had been the lack of a reliable mechanical applicator that could apply prilled urea, USG, and other urea formulations such as the neem admixtures under ponded water and dry soil conditions at the required depth. Using the zero-till drill, it is now possible to simultaneously deep-place prilled urea or USG and plant rice or wheat in dry or relatively moist soils. With a modified fertilizer application system in the zero-till or bed-planting machine, USG can also be placed at any depth. However, as of now, there is no reliable mechanical applicator for the deep N placement in puddled transplanted rice. As an alternative to deep placement of N, urea can be applied before the presowing irrigation in direct-seeded rice and wheat to increase N use efficiency and yield. The application of urea just before presowing irrigation in wheat transports the N to a depth that reduces NH_3 volatilization losses and thereby increases wheat yield (Aggarwal & Sidhu, 1992; Sidhu et al., 1994). Thus, deep placement of fertilizer could be a good strategy for dry-seeded rice since the N would be placed in the wetter lower depths and would not be leached and volatilized, thus making the N available for early rice establishment. If N is also to be applied as a split application, topdressing of a second dose of N after watering is better than before irrigation (Verma & Srivastava, 1995).

SUMMARY AND CONCLUSIONS

Natural resource management problems of the rice–wheat system ecology are complex and differ greatly among farms within and among transects of the IGP. Problems of the rice–wheat system ecology have to be resolved through innovative technological options that are appropriate to the resource endowments of farmers and meet the local bioclimatic and soil conditions. For resource-conserving technologies to expand and provide affordable food in sufficient quantities and help improve farmer livelihoods, it is prudent for farmers to have more options to establish the rice and wheat crops and improve input efficiencies and to use natural resources more efficiently. Farmers have the breadth of knowledge about the nature and behavior of their land, water, and plant resources and their interactive effects on crop communities. To harness the benefits of modern agriculture, it is crucial that efforts be made to increase the depth of farmers' knowledge through the farmer-participatory mode of research for the speedy adoption of resource-conserving technologies. Crop establishment options such as direct dry seeding of rice or wheat on the flat-surface or bed-planting configuration are more water-wise agrotechniques that also address the problems of receding water tables, intensification, and diversification and result in increased sustainability of the rice–wheat system. In eastern areas of the IGP, surface seeding has the potential to increase cropping intensity and reduce rice fallows. Development of a permanent raised-bed planting system facilitates intercropping of wheat or chickpea with sugarcane and of maize with potato. Thus bed planting has the potential to provide for alternative sources of productivity growth for wheat, pulses, and maize. Efficient use of water and nutrients in the bed planting system enhances the factor productivity of these key inputs in rice. Farmer adoption of these technologies will require a concerted effort on the part of all stakeholders in the agricultural system: farmers, researchers, extension personnel, local manufacturers, nongovernment agencies, and the private sector. This will have to be coupled with proper policies that promote adoption by farmers and institutional changes that promote interactions, flexibility, and decentralized decision making to motivate national program staff to seek and extend these technologies. In this way, an immediate impact can be made on the productivity and sustainability of agricultural systems in the IGP, while at the same time livelihoods improve, leading to eventual poverty alleviation and improvements in environmental quality and the natural resource base on which agriculture depends.

ACKNOWLEDGMENTS

We thank the reviewers for technical help on earlier versions of the manuscript. Thanks also to S.V.R.K. Prabhakar (CIMMYT, New Delhi, India) for his help in formatting the manuscript.

REFERENCES

Abrol, I.P., K.F. Bronson, J.M. Duxbury, and R.K. Gupta. 2000. Long-term soil fertility experiments in rice–wheat cropping systems. RWC Paper Ser. 6. Rice–Wheat Consortium for the Indo-Gangetic Plains, New Delhi, India.

Aggarwal, G.C., and A.S. Sidhu. 1992. Efficiency of urea application with pre-sowing irrigation in wheat on sandy loam and loamy sand soils. J. Indian Soc. Soil Sci. 40:44–48.

Ahmed, S.M., and C. Meisner. 1996. Wheat research and development in Bangladesh. Bangladesh-Australia Wheat Improvement Project and CIMMYT. CIMMYT and Aus AID, Bangladesh. CIMMYT, Bangladesh.

Asia Pacific Association of Agricultural Research Institutions. 2000. Vision 2025: Agricultural research for development. APAARI, Bangkok, Thailand.

Aslam, M. 1998. Waterlogging and salinity management in Sindh province, Pakistan. Supplement I-A: Improved water management practices for the rice–wheat systems in Sindh province, Pakistan. Rep. No. R-70. 1a. International Irrigation Management Institute, Lahore, Pakistan.

Aslam, M., A. Majid, N.I. Hashmi, and P.R. Hobbs. 1993. Improving wheat yield in the rice–wheat cropping system of the Punjab through zero tillage. Pakistan J. Agric. Res. 14:8–11.

Balasubramanian, V., J.K. Ladha, R.K. Gupta, R.K. Naresh, R.S. Mehla, Bijay-Singh, and Yadvinder-Singh. 2003. Technology options for rice in the rice–wheat system in South Asia. p. 115–148. *In* J.K. Ladha et al. (ed.) Improving the productivity and sustainability of rice–wheat systems: Issues and impact. ASA Spec. Publ. 65. ASA, CSSA, SSSA, Madison, WI.

Bhuiyan, S.I. 1992. Water management in relation to crop production: Case study in rice. Outlook Agric. 21:293–299.

Byerlee, D., M. Ali, and A. Siddiq. 2003. Sustainability of the rice–wheat system in Pakistan's Punjab: How large is the problem. p. 77–96. *In* J.K. Ladha et al. (ed.) Improving the productivity and sustainability of rice–wheat systems: Issues and impact. ASA Spec. Publ.65. ASA, CSSA, SSSA, Madison, WI.

Catling, H.D., P.R. Hobbs, Z. Islam, and B. Alam. 1983. Agronomic practices and yield assessments of deepwater rice in Bangladesh. Field Crops Res. 6:109–132.

De Datta, S.K. 1981. Principles and practices of rice production. John Wiley, New York, NY.

Evans, L.T. 1993. Crop evaluation, adaptation and yield. Cambridge Univ. Press, Cambridge, UK.

Food and Agricultural Organization. 1999. FAO production book, Vol. 53. FAO, Rome, Italy.

Greenland, D.J. 1997. The sustainability of rice farming. IRRI, Philippines, and CAB Int., Wallingford, UK.

Guerra, R.P., S.I. Bhuiyan, T.P. Tuong, and R. Barker. 1998. Producing more rice with less water from irrigated systems. IRRI, Manila, Philippines.

Gupta, R.K., and I.P. Abrol. 1990. Salt-affected soils and their reclamation and management for crop production. Adv. Soil Sci. 11:223–288.

Gupta, R.K., P.R. Hobbs, and J.K. Ladha. 2000. From issue to action. p. 1–7. *In* Proc. 6th Meeting of the Regional Steering Committee. Islamabad, Pakistan. 7–8 Mar. 2000. Rice–Wheat Consortium for the Indo-Gangetic Plains, New Delhi, India.

Gupta, R.K., and J. Rickman. 2002a. Design improvements in existing zero-till machines for residue conditions. RWC Traveling Seminar Report Ser. 3. Rice–Wheat Consortium for the Indo-Gangetic Plains, New Delhi, India.

Gupta, R.K., A.K. Shukla, M. Ashraf, Z.U. Ahmed, R.K.P. Sinha, and P.R. Hobbs. 2002b. Options for establishment of rice and issues constraining its productivity and sustainability in eastern Gangetic Plains of Bihar, Nepal and Bangladesh. RWC Traveling Seminar Report Ser. 4. Rice–Wheat Consortium for the Indo-Gangetic Plains, New Delhi, India.

Gupta, R.K., N.T. Singh, and M. Sethi. 1994. Groundwater quality for irrigation in India. CSSRI Tech. Bull. 19. Central Soil Salinity Research Institute, Karnal, India.

Harrington, L.W., S. Fujisaka, P.R. Hobbs, C. Adhikary, G.S. Giri, and K. Cassaday. 1993b. Rice–wheat cropping systems in Rupandehi district of the Nepal terai: Diagnostic surveys of farmers' practices and problems, and needs for further research. CIMMYT, Mexico, D.F.

Harrington, L.W., S. Fujisaka, M.L. Morris, P.R. Hobbs, H.C. Sharma, R.P. Singh, M.K. Chaudhary, and S.D. Dhiman. 1993a. Wheat and rice in Karnal and Kurukshetra districts, Haryana, India: Farmers' practices, problems and an agenda for action. CIMMYT, Mexico, D.F.

Hatta, S. 1967. Water consumption in paddy field and water-saving rice culture in the tropical zone. Jpn. Trop. Agric. 11:106–112.

Hobbs, P.R., G.S. Giri, and P. Grace. 1997. Reduced and zero tillage options for the establishment of wheat after rice South Asia. RWC-IGP Paper No. 2. Rice–Wheat Consortium for the Indo-Gangetic Plains, New Delhi, and CIMMYT, Mexico, D.F.

Hobbs, P.R., and R.K. Gupta. 2000. Sustainable resource management in intensively cultivated irrigated rice–wheat cropping systems of the Indo-Gangetic Plains of South Asia: Strategies and options. p. 157. *In* A.K. Singh (ed.) Proc. Int. Conf. Managing Natural Res. for Sustainable Agric. Prod. in the 21st Century. New Delhi, India. 14–18 Feb. 2000. Indian Soc. Soil Sci., New Delhi, India.

Hobbs, P.R., G.P. Hettel, R.K. Singh, R.P Singh, L.W. Harrington, V.P. Singh, and K.G. Pillai (ed.). 1992. Rice–wheat cropping systems in Faizabad district of Uttar Pradesh, India: Exploratory surveys of farmers' practices and problems and needs for further research. CIMMYT, Mexico, D.F.

Hobbs, P.R., and M.L. Morris. 1996. Meeting South Asia's future food requirements from rice–wheat cropping systems: Priority issues facing researchers in the post-Green Revolution era. NRG Paper 96–01. CIMMYT, Mexico, D.F.

Hobbs, P.R., A. Razzaq, N.I. Hashmi, M. Munir, and B.R. Khan. 1985. Effect of mustard grown as a mixed or intercrop on the yield of wheat. Pakistan J. Agric. Res. 6:241–247.

Hobbs, P.R., Y. Singh, G.S. Giri, J.G. Lauren, and J.M. Duxbury. 2001. Direct-seeding and reduced-tillage options in the rice–wheat systems of the Indo-Gangetic Plains of South Asia. p. 201–205. *In* S. Pandey et al. (ed.) Direct seeding: Research issues and opportunities. Proc. of the Workshop on Direct Seeding in Asian Rice Systems: Strategic Research Issues and Opportunities. 25–28 Jan. 2000. Bangkok, Thailand. IRRI, Manila, Philippines.

Hobbs, P.R., and R.K. Gupta. 2003. Resource-conserving technologies for wheat in the rice–wheat system. p. 149–172. *In* J.K. Ladha et al. (ed.) Improving the productivity and sustainability of rice–wheat systems: Issues and impacts. ASA Spec. Publ. 65. ASA, CSSA, SSSA, Madison, WI.

Huke, R., E. Huke, and T. Woodhead. 1993a. Rice–wheat atlas of Nepal. IRRI, CIMMYT and NARC publication. IRRI, Manila, Philippines.

Huke, E., R. Huke, and T. Woodhead. 1993b. Rice–wheat atlas of Bangladesh. IRRI, CIMMYT and BARC publication. IRRI, Manila, Philippines.

Huke, R., E. Huke, T. Woodhead, and J. Huang. 1993c. Rice–wheat atlas of China. IRRI, CIMMYT and CNRRI publication. IRRI, Manila, Philippines.

Kar, S., C.L. Acharya, and S.S. Parihar. 1997. Soil management effects on physical edaphic environment and sustainability of rice–wheat system. p. 7–19. Tech. Bull. 18. Indian Soc. Soil Sci., New Delhi, India.

Ladha, J.K., H. Pathak, A.T. Padre, D. Dawe, and R.K. Gupta. 2003. Productivity trends in intensive rice–wheat cropping systems in Asia. p. 45–76. *In* J.K. Ladha et al. (ed.) Improving the productivity and sustainability of rice–wheat systems: Issues and impacts. ASA Spec. Publ. 65. ASA, CSSA, SSSA, Madison, WI.

Ladha, J.K., K.S. Fischer, M. Hossain, P.R. Hobbs, and B. Hardy (ed.) 2000. Improving the productivity and sustainability of rice–wheat systems of the Indo-Gangetic Plains: A synthesis of NARS–IRRI partnership research. IRRI Discussion Paper Ser. 40. IRRI, Los Baños, Philippines.

Lianzheng, W., and Yixian, G. 1994. Rice–wheat systems and their development in China. p. 160–171. *In* R.S. Paroda (ed.) Sustainability of rice–wheat production systems in Asia. FAO, Bangkok, Thailand.

Malik, R.K. 1996. Herbicide resistant weed problems in the developing world and methods to overcome them. p. 664–673. *In* Proc. 2nd Int. Weed Sci. Congr. Copenhagen, Denmark. FAO, Rome, Italy.

Mehla, R.S., J.K. Verma, R.K. Gupta, and P.R. Hobbs. 2000. Stagnation in the productivity of wheat in the Indo-Gangetic plains: Zero-till-seed-cum-fertilizer drill as an integrated solution. RWC Paper Ser. 8. Rice–Wheat Consortium for the Indo-Gangetic Plains, New Delhi, India.

Mohanty, S.K., U. Singh, V. Balasubramanian, and K.P. Jha. 1999. Nitrogen deep placement technologies for productivity, profitability, and environmental quality of rainfed lowland rice systems. Nutr. Cycl. Agroecosyst. 53:43–57.

Ortiz-Monasterio, J.I., S.S. Dhillon, and R.A. Fischer. 1994. Date of sowing effects on grain yield and yield components of irrigated spring wheat cultivars and relationships with radiation and temperature in Ludhiana, India. Field Crops Res. 37:169–184.

Pandey, S., and L. Velasco. 1999. Economics of direct seeding in Asia: Patterns of adoption and research priorities. Int. Rice Res. Notes 24:2.

Paroda, R.S., T. Woodhead, and R.B. Singh. 1994. Sustainability of rice–wheat production systems in Asia. RAPA Pub. 1994/11. FAO, Bangkok, Thailand.

Pathak, H., J.K. Ladha, P.K. Aggarwal, S. Peng, S. Das, Y. Singh, B. Singh, S.K. Kamra, B. Misra, A.S.R.A.S. Shastri, H.P. Aggarwal, D.K. Das, and R.K. Gupta. 2002. Climatic potential and on-farm yield trends of rice and wheat in the Indo-Gangetic plains. Field Crops Res. (In press).

Pingali, P.L. (ed.) 1999. Sustaining the rice–wheat production systems: Socioeconomic and policy issues. RWC Paper Ser. 5. Rice–Wheat Consortium for the Indo-Gangetic Plains, New Delhi, India.

Prihar, S.S., and S.S. Grewal. 1985. Improving irrigation water use efficiency in rice–wheat cropping sequence: Technical and policy issues. *In* National seminar on water management: The key to developing agriculture. Agricole Publishing Academy, New Delhi, India.

Razzaque, M.A., M. Badaruddin, and C.A. Meisner (ed.) 1995. Sustainability of rice–wheat systems in Bangladesh. Proc. Natl Workshop. Dhaka, Bangladesh. 14–15 Nov. 1994. Bangladesh and Australia Wheat Research Centre, Dhaka, Bangladesh.

Sanchez, P.A. 1973. Puddling tropical rice soils. 2. Effects on water losses. Soil Sci. 115:303–308.

Sandhu, B.S., K.I. Khera, S.S. Prihar, and B. Singh. 1980. Irrigation need and yield of rice on a sandy-loam soil as affected by continuous and intermittent submergence. Indian J. Agric. Sci. 50:492–496.

Sharma, P.K., and S.K. De Datta. 1985. Physical properties and processes of puddled rice soils. Adv. Soil Sci. 5:139–178.

Sharma, P.K., and B. Mishra. 2001. Effect of burning rice and wheat crop residues: Loss of N, P, K and S from soil and changes in the nutrient availability. J. Indian Soc. Soil Sci. 49:425–429.

Sidhu, A.S., H.S. Sur, and G.C. Aggarwal. 1994. Effect of methods of urea application on yield of wheat, nitrogen fertilizer and water use efficiency. J. Indian Soc. Soil Sci. 42:10–14.

Singh, Y. 1995. Rice–wheat cropping systems. NARP Basic Res. Project. Prog. Rep. 1994–95. Govind Ballabh Pant University of Agriculture and Technology, Pantnagar, India.

Subbarao, G.V., J.D.V.K. Kumar Rao, C. Johansen, U.K. Deb, I. Ahmed, M.V. Krishana Rao, L. Venkataratanam, K.R. Hebber, M.V.S.R. Sai, and D. Harris. 2001. Spatial distribution and quantification of rice-fallows in South Asia: Potential for legumes. ICRISAT, Patancheru, India.

Sur, H.S., S.S. Parihar, and S.K. Jalota. 1981. Effect of rice–wheat rotations on water transmission and wheat development in sandy loam soil of Punjab, India. Soil Tillage Res. 1:361–371.

Tang, Y., H. Gang, Y. Yao, and Y. Lixun. 2000. High-yielding cultivation techniques for wheat under the rice–wheat cropping system in Sichuan province of China. p. 11–23. *In* P.R. Hobbs and R.K. Gupta (ed.) Soil and crop management practices for enhanced productivity of rice–wheat cropping system in the Sichuan province of China. RWC Paper Ser. 9. Rice–Wheat Consortium for the Indo-Gangetic Plains, New Delhi, India.

Timsina, J., and D.J. Connor. 2001. The productivity and sustainability of rice–wheat cropping systems: Issues and challenges. Field Crops Res. J. 69:93–132.

Tran, D.V., and J.P. Marathée. 1994. Major issues in Asian rice–wheat production systems: Sustainability of rice–wheat production systems in Asia. RAPA Pub. 1994/11:61–67. FAO, Bangkok, Thailand.

Tullberg, J.N. 2001. Controlled traffic for sustainable cropping. Proc. of the 10th Australian Agron. Conf. Hobart, Australia. 16 May 2001. Austr. Soc. of Agron., Australia.

Tullberg, J.N., P.J. Ziebarth, and Y. Li. 2001. Tillage and traffic effects on runoff. Aust. J. Soil Res. 39:249–257.

Velayutham, M., D.K. Mandal, C. Mandal, and J. Sehgal. 1999. Agro-ecological sub-regions of India for planning and development. NBBS Pub. 35. National Bureau of Soil Survey and Land Use Planning, Nagpur, India.

Verma, U.N. 1999. Surface seeded wheat production system after transplanted rice. Indian Farm. 49:4–6.

Verma, U.N., and V.C. Srivastava. 1995. Balanced fertilization with NPK for increasing wheat production. p. 22–34. *In* G. Dev (ed.) Balanced fertilizer use for increasing food grain production in eastern India. Potash and Phosphate Institute of Canada and India, Gurgaon, India.

Witcombe, J.R., and D. Harris. 1997. Impact of the DFID Plant Sciences Research Program. A presentation for the Natural Resource Advisors' Conf. Bangor, UK. 6–11 June 1997. Centre for Arid Zone Studies. University of Wales, Bangor, UK.

Woodhead, T., R. Huke, and E. Huke. 1993. Rice–wheat atlas of Pakistan. IRRI, CIMMYT, and PARC publication. IRRI, Manila, Philippines.

Woodhead, T., R. Huke, E. Huke, and L. Balababa. 1994. Rice–wheat atlas of India. IRRI, CIMMYT, and ICAR publication. IRRI, Manila, Philippines.

Yadav, R.L., K. Prasad, and K.S. Gangwar. 1998. Analysis of eco-regional production constraints in rice–wheat cropping system. PDCSR Bull. 98–2. Project Directorate on Cropping Systems Research, Modipuram, India.

Zheng, J. 2000. Rice–wheat cropping system in China. p. 1–10. *In* P.R. Hobbs and R.K. Gupta (ed.) Soil and crop management practices for enhanced productivity of rice–wheat cropping system in Sichuan province of China. RWC Paper Ser. 9. Rice–Wheat Consortium for the Indo-Gangetic Plains, New Delhi, India.

2 Long-Term Sustainability of the Tropical and Subtropical Rice–Wheat System: An Environmental Perspective

Peter R. Grace and Larry Harrington

International Center for Wheat and Maize Improvement
El Batan, Mexico

Mahesh. C. Jain

Indian Agricultural Research Institute
New Delhi, India

G. Philip Robertson

Michigan State University
East Lansing, Michigan, USA

ABSTRACT

Arable lands in the Indo-Gangetic Plains are already intensively cropped with little scope for expansion because of the competing end uses of land for urbanization and industry. Evidence from long-term experiments in the region indicates that cereal yields are declining, which is in stark contrast to the needed increases in production to meet population demand in the future. The intensification of rice–wheat rotations has resulted in a heavy reliance on irrigation, increased fertilizer usage, and crop residue burning, which all have a direct effect on the variable that most affects global climate change—emissions of greenhouse gases. We estimate that the CO_2 equivalent emissions from a high-input conventionally tilled cropping system with residue burning and organic amendments would equal 8 Mg C or 29 Mg CO_2 yr^{-1} if applied to 1 million hectares of the Indo-Gangetic Plains. In a no-till, residue-retained system, with 50% of the recommended NPK application, the total emissions would equal 3.7 Mg C, or 14 Mg CO_2 yr^{-1}, an effective halving of emissions as we move from a high- to low-input system with improved nutrient use and environmental efficiency. The transition to intensified no-tillage systems, with recommended fertilizer levels, can be both productive and environmentally sound in a world that is rapidly becoming aware of the significant effects of global climate change in both the short and long term.

The need to increase production in rice- and wheat-growing regions of the tropics and subtropics is well recognized. Expected population growth will put enormous

 Improving the Productivity and Sustainability of Rice–Wheat Systems: Issues and Impacts. ASA Special Publication 65.

stress on existing production systems and is likely to lead to pressure to expand cropping systems into areas that are now either agronomically marginal or protected for environmental reasons. The long-term sustainability of these systems—their ability to provide food and fiber in a manner that is economically and socially viable without environmental harm (Robertson & Harwood, 2001)—will be stretched very thin.

Environmental problems are primarily associated with damage to the natural capital of the land, which often is not recognized until well after very serious and, in many cases, irreversible degradation of the natural resource base has occurred. Feedback in the system can either decelerate or accelerate this process, with the direction of this change depending on the type of agronomic management used in the system. For example, losses in soil organic matter under conventional tillage and residue removal may reduce yields, which can lead to further environmental damage as the land is farmed more intensively in an effort to recover yield that has been lost. If crop residues are retained and tillage is reduced in concert with legume rotations, stores of organic matter can be maintained or even increased under intensive agriculture.

Examples of this pattern are beginning to emerge across wheat–rice regions of the tropics and subtropics. A good example of this trend is in the Indo-Gangetic Plains (IGP), where serious questions about both the sustainability and potential environmental consequences of current and future production systems have been raised.

Concerns about the impact of current agronomic practices have been heightened by the evidence from long-term experiments that cereal yields in the region are generally declining (Ladha et al., 2002). A detailed assessment of 11 long-term rice–wheat experiments (ranging from 7 to 25 yr in duration) from the region (Duxbury et al., 2000) (Table 2–1) indicates a marked yield decline of up to 500 kg ha^{-1} yr^{-1} in rice in nine of the experiments (Fig. 2–1), while wheat yields have tended to stabilize (Fig. 2–2). There is evidence that decreases in solar radiation over the past decade across the IGP (Pathak et al., 2002) may have played a role in the overall decline. Higher genetic improvement in wheat relative to rice may also partly explain the anomaly between the trends in rice and wheat yields (Ladha et al., 2003, Chapter 3 of this publication). Strong evidence suggests that the main reasons for this overall decline are soil-based and include the ineffective use of N inputs; soil organic matter decline; soil structural decline; micronutrient deficiencies, particularly S, K, and Zn; and the buildup of pests and diseases.

The soil organic matter decline is of particular importance because it affects both nutrient availability and soil structural stability. In Fig. 2–3, rapid declines of 200 to 624 kg ha^{-1} yr^{-1} are evident in representative growing areas, even with recommended NPK levels, which indicates that tillage practices need to be reviewed.

The addition of organic inputs (both farmyard and legume green manures) has been promoted as a possible solution to the yield decline, but no clear evidence exists from the long-term experiments that the sustainability of crop yields is increased. The addition of 15 Mg ha^{-1} yr^{-1} of farmyard manure (FYM) before wheat in conjunction with NPK fertilizer consistently increased yields of rice in a 20-yr experiment in Uttar Pradesh (Ram, 2000) compared with fertilizer alone, but rice yields declined regardless. Only small effects of FYM have been observed in ex-

Table 2–1. Researchers and location of long-term rice–wheat experiments in South Asia (Duxbury et al., 2000).

Researchers	Location	Approx. duration
		yr
Prasad and Sinha	Samastipur, Bihar	7
Sakal	Pusa, Bihar	9
P. Singh and Khan	Jabulpar, Madya Pradesh	9
Kundu and Samui	Mohanpur Nadia, West Bengal	9
Yadvinder Singh et al.	Ludhiana, Punjab	10
Y. Singh et al.	Pantnagar, Uttar Pradesh	14
Regmi et al.	Bhairawaha, Nepal	18
Chhabra and Thakur	Karnal, Haryana	19
Ram	Pantnagar, UP	20
K. Singh and Swarup	Karnal, Haryana	20
Saha et al.	Barrackpore, West Bengal	25

periments in Bangladesh and West Bengal. It is possible that more emphasis needs to be given to the supply of nutrients such as K and S.

ENVIRONMENTAL CONSEQUENCES OF INTENSIFICATION

From a biophysical perspective, increases in production in the IGP have come from intensification (expressed as increasing production per hectare). This has been achieved by using several management options either singly or in combination. The major options lie in the degree of land preparation (tillage), the choice of germplasm (crop species or cultivar), the time of sowing, the use of appropriate nutrients to enhance growth, the use of irrigation where water is available, and the method of pest and weed control.

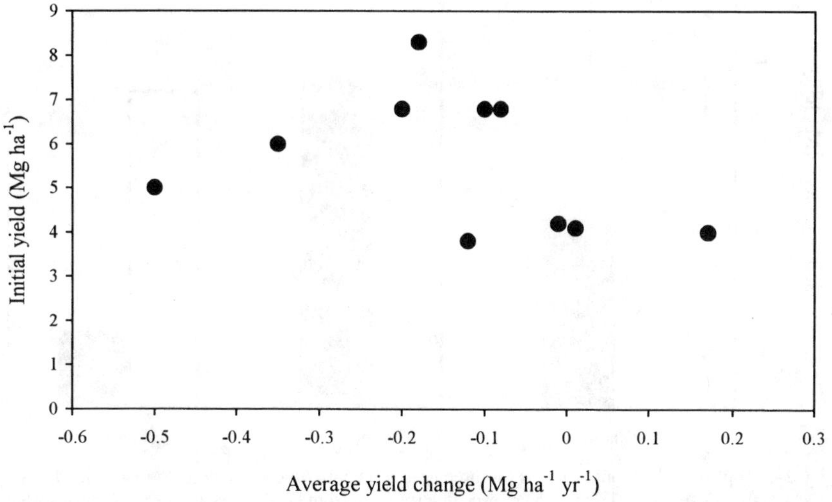

Fig. 2–1. Average annual yield change in rice in long-term experiments in the Indo-Gangetic Plains.

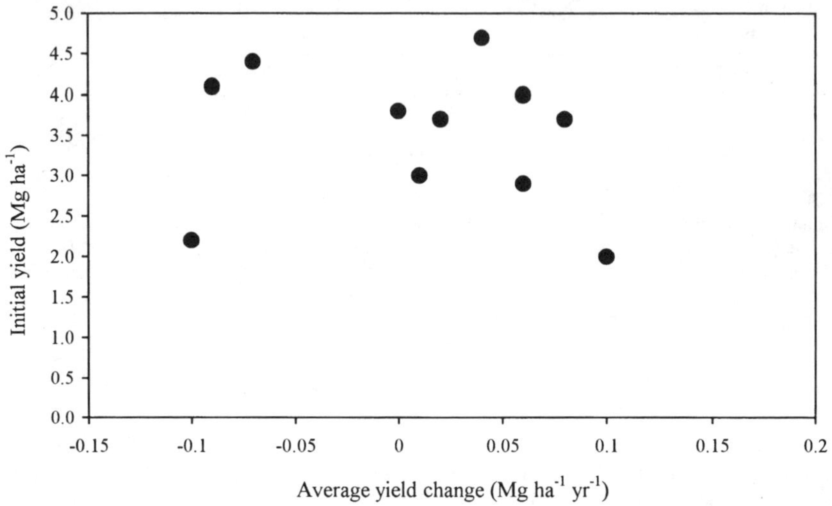

Fig. 2–2. Average annual yield change in wheat in long-term experiments in the Indo-Gangetic Plains.

Current production systems now occupy a continuum in a spectrum of intensification, governed, for a given climate and soils, by the level of inputs and crop management. For convenience, three types of intensification can be specified (Gregory et al., 2002), each with characteristic features (Table 2–2). Productivity may vary widely, however, within a given type.

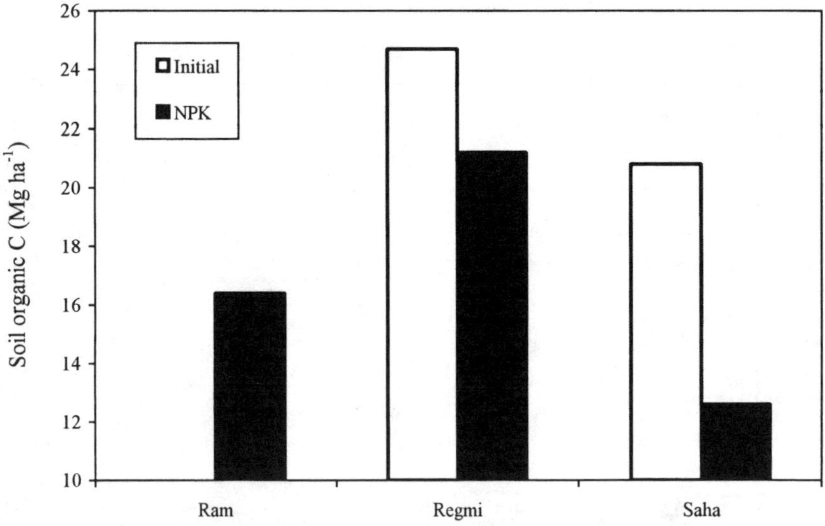

Fig. 2–3. Topsoil soil organic C levels for three long-term experiments in the Indo-Gangetic Plains by stated authors (after 20, 18, and 25 yr, respectively). Data from Duxbury et al. (2000). Assumes a bulk density of 1.3 g cm^{-3}.

Table 2–2. Categories of farming systems in the rice–wheat region of the Indo-Gangetic Plains and their characteristics.†

	Farming system		
Characteristic	Type I Low external inputs Pre-Green Revolution	Type II Conventional Green Revolution	Type III Alternative Doubly Green Revolution
Main objective	Minimizing food shortage	Maximizing food production	Maximizing profit and other land functions
Rural population density	Increasing	High	Reduced
Access to market and technology	Low	High	High
Environmental concerns	Medium	Low	High
Land efficiency	Increasing/declining	High	Reduced
Labor/energy efficiency	High or declining	Low	Reduced
Capital efficiency	No capital	Low or medium	High
Fallow time	Declining	Zero	Increasing
Technological package	Zero or limited	High	Reduced (minimum tillage, IPM)
Credit, land tenure	Zero	High	High
"Mining" agriculture	Yes or no	Yes	No
Example	Sahel	Indo-Gangetic Plains	South American "success" stories

† Reprinted from Gregory et al. (2002).

Type I intensification occurs when available management inputs are limited (pre-Green Revolution). Type II intensification is largely dominated by the technological features of the Green Revolution and has been widely adopted from about 1960. It involves the introduction of new cultivars coupled with large increases in the use of fertilizers, herbicides, pesticides, irrigation, and mechanization. The adoption of this technology successfully achieved its primary aim of substantially increasing food production in the IGP, but concern about the environment was, and still remains, low in this type. Type III intensification ("doubly" Green Revolution) necessitates a production system that is both high-yielding and more environmentally benign than Type II systems. Affluent populations are pressing for this type of system but demonstrated examples are rare, leading to the absence of comparative data sets in terms of production and environmental externalities.

All three intensification strategies result in detrimental environmental consequences to some extent, but their nature and magnitude differ markedly among types and particularly in whether the major consequences are on- or off-site. Type I is viewed as the most environmentally benign, while Type II has been the subject of most recent concern. This is because of its on-site effects in relation to sustainability and soil degradation and its off-site effects in relation to water pollution and climate forcing (mainly via greenhouse gas emissions to the atmosphere), leading to regional climate change. Any change in farming system results in changed environmental consequences, but the difficulty lies in assessing which changes are nonreversible and have the most deleterious effect. The advent of genetically modified technologies is becoming important in the development of Type III systems, bringing both potential environmental benefits and problems.

In the IGP, the intensification of rice–wheat rotations has resulted in the following changes in production systems: seasonal wet and dry crop cycles, a heavy reliance on irrigation, increased fertilizer usage, crop residue burning, asymmetry of planting schedule, and a uniformity in cultivar selection. Of these activities, irrigation, fertilizer application, and residue burning have a direct effect on the variables that most affect global climate change—emissions of the greenhouse gases (GHGs) CO_2, CH_4, and N_2O. While agricultural enterprises are normally local in scale (i.e., at the farm level), widespread changes in production practices in large high-input regions such as the IGP are of global consequence.

In a recent summary of the effects of climate change on wheat and rice production systems of the Indo-Gangetic region, Gregory et al. (1999) concluded:

1. Under optimal field conditions, wheat yields are unlikely to increase by more than about 10% for double preindustrial CO_2; a 5 to 7% increase is more likely for average management conditions. This estimate is lower than previously suggested by the Intergovernmental Panel on Climate Change (1996) because it allows for decreased crop duration (and hence yield) as a consequence of warming and the many factors that reduce yield from potential to actual levels.
2. Major rice models are consistent across a wide range of potential yields and indicate a reduction in yield of about 5% per °C rise above 32°C. This would largely offset any increase in yield as a consequence of increased CO_2. Rice yields are reduced when temperatures exceed 32°C at flowering because of spikelet sterility. This finding is unaffected by elevated CO_2 but might differ among genotypes.
3. Elevated CO_2 is likely to enhance yields of rice more than wheat because rice is usually grown intensively with inputs of water and nutrients, whereas wheat is grown extensively in drier areas.
4. For mid-latitude regions such as the IGP, an average warming of 1 to 3.5°C over the next 100 yr would be equivalent to a poleward shift of the present geographic bands of similar temperatures (or isotherms) of approximately 150 to 550 km, or an altitude shift of about 150 to 550 m.

SOURCES OF GREENHOUSE GASES IN THE RICE–WHEAT SYSTEM

The direct cause of climate change is primarily the enhanced greenhouse effect. Different gases persist for different lengths of time in the atmosphere, and they also trap heat to different degrees. The global warming potential (GWP) is used to compare the effectiveness of each greenhouse gas to trap heat in the atmosphere relative to some standard gas, by convention CO_2. The GWP for CH_4 (based on a 100-yr time horizon) is 21, while that for N_2O is 310. Although present in lower concentrations than either CO_2 or CH_4, N_2O is a very potent greenhouse gas, accounting for about 6% of the enhanced greenhouse effect.

By using GWP values for CH_4 and N_2O emissions as stated above (21 and 310, respectively), we can convert all our emission estimates to either a CO_2 or carbon equivalent (CE) basis. Carbon dioxide equivalents can be calculated by multiplying the CE values by 3.7.

Carbon Dioxide

The main source of CO_2 to the atmosphere in the rice–wheat system is through tillage. Two on-site tillage sources exist: the biological decomposition of soil organic matter and the production of CO_2 as a byproduct of machinery fuel usage. The burning of crop residues is not considered a CO_2 source to the atmosphere because an amount of CO_2 equivalent to the amount released will be taken up by the following crop; in other words, on an annual basis, there will be no change in standing C stock. Carbon dioxide from this source is therefore not counted in the accounting procedures adopted by the Intergovernmental Panel on Climate Change (IPCC).

During tillage, soil aggregates are broken, thus increasing oxygen supply and surface area exposure of organic material. This promotes the decomposition of organic matter. An example of the effect of three to four cultivations per year on soil organic C (SOC) content can be drawn from CIMMYT's long-term wheat–maize trial at El Batán in central Mexico (P. Grace, unpublished data, 2000). Initial levels of SOC in the top 20 cm were 1.37%. After 8 yr of continuous wheat, with residue retention and conventional tillage, SOC had declined to 1.12%, a total decline of nearly 6.5 Mg C ha^{-1}, or 800 kg ha^{-1} yr^{-1}. In the adjacent no-till treatment, where yields were similar, the total decline was only 3.6 Mg C ha^{-1}, or 450 kg ha^{-1} yr^{-1}, a savings of 350 kg ha^{-1} yr^{-1}. The rate of decline (or accumulation) of SOC therefore depends on several interacting factors: the amount of residue retained, the quality of residue retained, soil temperature and moisture interactions, tillage, and a reduction in fallow periods.

For residue quality, high-quality (i.e., high N) residues from grain legumes or fodders can accelerate the decomposition of soil organic matter because of the additional N being available to the microbial population. Usually, the additional N being released will also be taken up by the following crop, thereby enhancing yields and biomass production; however, where high temperatures coincide with high water-holding capacity, the conditions are ideal for the promotion of SOC decomposition and the emission of CO_2. Soils therefore are generally considered to be a source of CO_2, unless the inputs outweigh the outputs, which is not usually the case in conventionally tilled low- to medium-input production systems.

Heavy use of diesel fuel occurs with current land preparation practices. For every liter of diesel fuel consumed, 2.6 kg of CO_2 are released to the atmosphere. Assuming that 150 L ha^{-1} yr^{-1} of fuel are used for tractor usage and irrigation pumping in conventional systems, this would amount to nearly 400 kg of CO_2 being emitted. For the entire 12 million hectares under rice–wheat in the IGP, this would amount to 4.8 Mg CO_2 yr^{-1} or 1.3 million metric tonnes C equivalents (MMTCE). To put this in perspective, this value is approximately one-third of the CH_4 emissions (expressed as MMTCE) from rice paddy fields in the entire IGP (see section on methane below); therefore, diesel use is a greatly underestimated source of GHGs.

An off-site source that is often neglected is the production of CO_2 in the manufacture of fertilizers (Schlesinger, 2000). Although this is not relevant to our discussions of on-farm environmental effects, 0.58 mol of C is produced per mole of N fixed in fertilizer production, resulting in 1.8 kg CO_2 produced per kilogram of N fertilizer manufactured. The full cost of accounting (including transportation and application) would raise this value to 1.4 mol. In a no-till cropping system, this cost can significantly negate the mitigation gain by soil C sequestration.

A further source of CO_2 usually neglected is agricultural lime (usually calcium carbonate or dolomite) added to noncalcareous soils to counteract the acidity produced by N fertilizer use and by base cation removal during harvest (Robertson et al., 2000). Between lime applications, the lime previously applied dissolves to HCO_3^- or CO_2 depending on the acid–base balance of the soil solution; any CO_2 formed will be released to the atmosphere. For N fertilizer manufacture, the CO_2 cost of lime can significantly negate any mitigation achieved by soil C storage.

Methane

Methane is produced in soil during microbial decomposition of organic matter and reduction of CO_2 under strictly anaerobic conditions. Methane emissions are controlled by two microbial processes: CH_4 production by methanogenic bacteria and CH_4 oxidation by methanotrophic bacteria. Continuously flooded rice fields are a source of CH_4 because anoxic conditions that favor methanogenesis predominate, and rice plants serve as conduits for its release to the atmosphere once it is formed in sediments (Neue et al., 1997).

Under the UNDP-IRRI Inter-Regional Programme, CH_4 emissions from rice fields were measured at eight locations in five rice-producing countries covering the main rice ecosystems—irrigated, rainfed, and deepwater systems (Wassmann et al., 2000a,b). Irrigated rice fields had the highest emission rates. Seasonal emissions varied from 1 to 50 g m^{-2}, with continuously flooded fields emitting the most (Wassmann et al., 2000a). Depending upon local precipitation, emissions from rainfed rice fields may be less than one-half of the emissions from irrigated fields and may be less than 5 g m^{-2}.

On the basis of results obtained by various rice-growing countries, the IPCC has suggested that emission reduction factors under single and multiple aeration are 0.5 and 0.2 of those in continuously flooded fields. Using the IPCC accounting methodology and the country default value of 10 g m^{-2}, CH_4 emissions from rice cultivation in India (45 million ha) should not exceed 2.5 Mg yr^{-1} (or 14.3 MMTCE). In the 12 million ha of rice–wheat in the IGP, this would equal about 0.7 Mg CH_4 yr^{-1} (or 4 MMTCE yr^{-1}). The main reason for low CH_4 emissions from rice fields in India is that the soils of major portions of rice-growing areas have very low organic C or receive very little in the way of organic amendments (Jain et al., 2000). Incubation studies have shown that large differences in CH_4 production potential are related to organic C content (Majumdar et al., 1998). The use of organic manuring is also not very common in India, although where it occurs it can affect methane emissions (Debnath et al., 1996).

The burning of crop residues also contributes to the global CH_4 budget. For each tonne of crop residue burned, 2.3 kg CH_4 are emitted, equivalent to 48.3 kg CO_2 (using a GWP of 21). If we assume average annual residue production for rice and wheat on the IGP to be 10 Mg ha^{-1}, then, if one-half of the 12 million ha under rice–wheat is burned, a total flux of 0.14 Mg CH_4 is emitted. This is equivalent to 20% of the total CH_4 emitted from paddy rice fields in the IGP (total 0.7 Mg CH_4 as outlined above).

Nitrous Oxide

Both fertilized and unfertilized soils emit N_2O. Although nitrogenous fertilizer is a source in the case of fertilized soils, the built-in N of the soil itself also contributes to the release of this GHG. It has long been recognized that both soil water content and the availability of C enhance the production of N_2O provided a suitable NO_3 source is available. The total fluxes of N_2O have been reported to be many times higher in organic soils and mineral soils with higher contents of soil organic matter (Terry et al., 1981). The primary effect of water on N_2O production in aerobic and partially aerobic soils is the restriction of oxygen levels, which produces anaerobic conditions. Because of the relatively slow diffusion of oxygen through water (10 000 times slower than through air), denitrification in soil is highly affected by the relative amount of air- and water-filled porosity (Robertson, 2000; Burton & Beauchamp, 1985).

Generally, an increase in denitrification and potential N_2O losses is observed following irrigation and precipitation. However, rice paddies are not considered an important source of atmospheric N_2O because N_2O, as the intermediary product of denitrification, would be further reduced to N_2 under strong anaerobic conditions (Granli & Bockman, 1994). The exception is in irrigated rice grown in upland conditions with frequent natural drainage of standing water, in which N_2O production is possible. Nitrous oxide is also not usually detectable during the normal growing season of flooded rice (Bronson et al., 1997a,b) but is significant after N fertilizer applications. During the dry season, Bronson et al. (1997a,b) reported N_2O fluxes that were 2.5 times higher with ammonium sulfate than with urea. The N_2O flux also increased sharply during the drainage period at mid-tillering until reflooding, when it dropped back to near zero. Experiments in rice fields conducted at the Indian Agricultural Research Institute (IARI) have shown that, as the redox potential becomes positive, CH_4 emissions decrease, but N_2O emissions increase. It has also been shown that the use of the nitrification inhibitor DCD along with urea and ammonium sulfate reduces both CH_4 and N_2O emissions (Ghosh, 1998). There is much uncertainty about the amount of N_2O released; however, it is suspected that 0.1 to 6% of applied N is lost as N_2O. Intergovernmental Panel on Climate Change (1994) methodology assumes that 1.25% of fertilizer N is lost as N_2O. Crops such as wheat and maize, which receive nitrogenous fertilizers, are expected to release more N_2O than crops grown in flooded environments.

The burning of crop residues also contributes to the global N_2O budget. For each tonne of crop residue that is burned, 40 g N_2O are emitted, equivalent to 12.4 kg CO_2 (using a GWP of 310). If we assume that the average annual residue (rice and wheat) production on the IGP is 10 Mg ha^{-1}, then, if one-half of the 12 million ha under rice–wheat is burned, a total flux of 2400 tonnes of N_2O is possible; that is, about 0.74 Mt CO_2 (equivalent) is emitted.

MITIGATION MECHANISMS

Carbon Dioxide

Current land preparation practices for wheat after rice involve as many as 12 tractor passes. Changing to a zero-till system on 1 ha of land would save 98 L of diesel and approximately 1 million L of irrigation water; this represents about a quarter tonne less emission per hectare of CO_2, the principal contributor to global warming. These benefits increase dramatically if extended across even a portion of the region's 12 million ha. Adoption of zero-till on only 5 million ha would represent a savings of 5 billion m^3 of water each year. That would fill a lake 10 km long, 5 km wide, and 100 m deep. In addition, annual diesel fuel savings would come to 0.5 billion L, equivalent to a reduction of nearly 1.3 million Mg in CO_2 emissions each year.

Methane

Experiments at IARI during 1994 to 1998 have shown that intermittent irrigation, which is the usual practice, reduces CH_4 emissions up to about 28% (Jain et al., 2000). At Pantnagar, the reduction in CH_4 emissions because of intermittent irrigation was 30% compared with continuously flooded conditions. Experiments have also shown that the addition of organic manures (legume green manure, FYM, biogas slurry, and rice straw) enhances methane emissions from rice fields. The experiments conducted at IARI have also shown wide variations in CH_4 emissions because of the variety of organic amendments. However, it is apparent that CH_4 emissions from rice fields amended with biogas slurry are significantly less than with other organic amendments (Debnath et al., 1996). Experiments have also shown that there is no significant increase in CH_4 emissions if neem-coated urea is used in place of urea, whereas grain yields increase significantly (Sarkar, 1997). Methane emissions have been reported to depend on the cultivar sown (Mitra et al., 1999), and efforts are needed to identify a few suitable cultivars in different regions with high yield potential but low CH_4 emission potential.

In summary, methane emissions can be reduced significantly by adopting the following mitigation practices: water management through intermittent irrigation or drainage, the use of digested manure instead of fresh manure, and the selection of suitable cultivars that emit less methane. Varietal differences have been shown to affect CH_4 emissions (Neue et al., 1994; Wang et al., 1997), with differences usually becoming evident in the middle and late growth stages (Wang & Adachi, 2000). Indications are that roots and their relative size play a significant role in CH_4 production as they provide a source of C through exudation and upon their death.

Nitrous Oxide

Strategies that increase N-fertilizer efficiency (Robertson, 1997) reduce emissions of N_2O (Mosier et al., 1996; Majumdar et al., 2001). Plant uptake of N can be improved and total N losses can be reduced by deep fertilizer placement (Youngdahl et al., 1986) and by banding fertilizer within the crop rows. Nitrogen use efficiency can also be improved by applying fertilizer close to the time that it is needed by the crop, and by using soil and plant tests before fertilization to tailor the amount of fertilizer applied to that needed by the plant. Cover crops can also keep available soil N away from the microbes responsible for N_2O production.

Nitrous oxide emissions from soil can also be reduced by nitrification inhibitors (Mosier et al., 1998; Majumdar et al., 2000; Pathak & Nedwell, 2001; Kumar et al., 2000). Compounds such as nitrapyrin, $HgCl_2$, toluene, CS_2, DCD, thiosulfate, and acetylene retard the conversion of NH_4^+–N into NO_3^-–N. This results in the reduction of direct N_2O emissions from nitrifiers and reduces the availability of NO_3^- necessary for denitrification. Low emissions of N_2O have been observed through the use of used encapsulated calcium carbide (a source of acetylene) and nitrapyrin (Bronson et al., 1992). These authors reported losses of 1.5 to 3.1 kg N_2O ha^{-1} from urea alone compared with 0.87 to 1.0 kg N_2O ha^{-1} from urea plus nitrapyrin. Other biocidal inhibitors, such as neem and karanja seed extracts, have been found to retard nitrification by 60 to 70% (Sahrawat, 1996); however, the general lack of availability of inhibitors in the developing world and cost-ineffectiveness in broad-area situations usually prohibit their use.

We therefore propose three feasible, cost-effective agronomic interventions that would have an immediate effect by reducing GHG production in the IGP: (i) a reduction in residue burning, (ii) a reduction in flood irrigation frequency for rice, and (iii) the use of minimum or no-tillage for wheat.

Based on the calculations in the sections above, we can make an estimate of the total savings in CO_2 equivalent emissions (kg ha^{-1} yr^{-1}) if we

Reduce burning	60
Reduce irrigation frequency	1000
No-tillage (diesel savings)	260
No-tillage (soil C), based on 100 kg C ha^{-1} increase	360
Total CO_2 saved (kg ha^{-1})	1680

These estimates also assume no change in basic nutrient management, only tillage and reduced irrigation. If these measures were in place on one-half of the IGP, a conservative estimate would be a savings of 10 Mg CO_2. To put this in perspective, during the 30 yr of the Green Revolution, 45 million ha of land were cleared in all of Asia, resulting in 360 Mg being lost to the atmosphere as CO_2 on average every year (Grace et al., 2000).

COMPARATIVE GREENHOUSE GAS BUDGETS FOR FARMING SYSTEMS

The management of the rice–wheat system in the IGP is extremely diverse in terms of the amount of N applied and the rotation and tillage options that exist. It is therefore instructive to look at the GHG emission potential of some parts of this system and, in line with the need for poverty alleviation, assess their potential in terms of overall productivity in relation to the amount of GHG emissions from each system. Emissions usually depend on both inputs (inorganic N, organic manure) and tillage. To make this comparison, a carbon/productivity ratio (CPR)

Table 2–3. Agronomic details of 20-yr irrigated rice–wheat–cowpea treatments and average yields at Pantnagar, Uttar Pradesh (Ram, 2000).

System	Rice	Wheat	Fodder	N fert. per crop	Manure	Soil organic C
	Mg ha^{-1}			kg ha^{-1}	Mg ha^{-1}	%
Control	3.74	1.71	1.86	0	0	0.49
Trt 1: 50%	5.02	3.13	1.93	60	0	0.64†
Trt 2: 100%	5.67	3.97	2.36	120	0	0.84
Trt 3: 150%	5.92	4.38	2.32	180	0	1.06†
Trt 8: FYM	6.41	4.6	2.47	120	15	1.48

† Estimated for 0–15 cm based on published measurements in other treatments.

(Grace et al., 2000) is used to assess how environmentally inefficient various production systems are with respect to GHG emissions and how much food is actually placed on the table. Systems with higher CPRs are more inefficient at producing food with respect to the damage to the global environment. The CPR for Latin America, Africa, and Asia for the 30 yr of the Green Revolution was 0.52 (Grace et al., 2000).

To make this comparison more realistic, we have chosen data from a long-term trial in northern India at Pantnagar (Ram, 2000). This trial started in 1971 and its objective was to assess the effects of nutrient management (NPK) on productivity and soil fertility. The rice–wheat–cowpea trial consisted of 12 treatments with 50, 100, or 150% of the recommended dose of NPK fertilizer. The recommended dose of N was 120 kg ha^{-1} for each crop of rice and wheat. In our assessment, we constructed annual GHG budgets for individual gases (CO_2, CH_4, N_2O) for 5 of the 12 treatments (Table 2–3) on the basis of the activities outlined in the earlier part of this chapter. We included one treatment that incorporated farmyard manure. In all budgets, we then converted this information to C equivalents to give an overall assessment of C lost to the atmosphere in relation to total edible product.

Table 2–4 outlines the conventional tillage budget with all crop residues being retained, which we assume represents the experiment most closely. The highest total emissions in terms of C equivalents were found in Treatment 8 (7 Mg C ha^{-1}), which included FYM. Treatment 2, which received the recommended dosage of fertilizer only, had emissions of 4.7 Mg C ha^{-1}, but it had the lowest CPR, which meant that it was the most efficient of all the nutrient management systems in this analysis of conventional tillage with residue retention. However, Treatments 1 and 3, which supplied 50 and 150% of the recommended dosage, respectively, had similar CPRs. In Treatment 3, the higher fertilizer N input may also result in leaching of excess fertilizer, so this treatment is not recommended if we are considering other environmental problems other than GHG emissions. In terms of recommendations, obviously Treatment 1, with the reduced N input (and hence reduced cost), would be the most suitable, particularly if we consider the off-site environmental costs (i.e., fertilizer production) discussed earlier.

Tables 2–5 and 2–6 outline the respective budgets where burning of all residues and no-tillage are alternative managements. The no-till scenario assumes complete retention of all crop residues and a hypothetical two-thirds reduction in on-site fuel use (cultivation only). Both scenarios exhibit the same trend as the earlier conventional residue-retained scenario in the final treatment analysis. The

Table 2–4. Estimated annual (on-site) emissions of greenhouse gases (GHG) from the irrigated rice–wheat system in the Indo-Gangetic Plains based on 20-yr long-term trial data of Ram (2000). This example assumes conventional tillage with all crop residues being retained.

	GHG emissions				
GHG source	Control	Trt. 1	Trt. 2	Trt. 3	Trt. 8
	kg ha^{-1}				
CO_2: Soil tillage	3539	3003	2288	1502	0
CO_2: Diesel†	260	260	260	260	260
N_2O: Burning residues	0	0	0	0	0
N_2O: Manure application‡	0	0	0	0	3
N_2O: N fertilizer application‡	0	2.4	4.7	7.1	4.7
N_2O: Cereal residues retained§	0.1	0.1	0.1	0.1	0.1
N_2O: N-fixing crops§	1.1	1.1	1.4	1.4	1.5
CH_4: Burning residues	0	0	0	0	0
CH_4: Rice cultivation	100	100	100	100	200
kg CE ha^{-1}#	3496	4103	4721	5232	7137
CPR††	0.64	0.50	0.49	0.51	0.64

† 2.6 kg CO_2 L^{-1} of diesel consumed and 100 L yr^{-1}.
‡ 1.25% of applied N.
§ 1.25% of fixed or residue N.
¶ 10 g CH_4 m^{-2}, doubled if manure added.
# Carbon equivalents, global warming potentials (GWPs) of 21 and 310 used for CH_4 and N_2O, respectively.
†† Carbon/productivity ratio, kg C emitted (equivalents) per kg of edible dry product (cereals).

Table 2–5. Estimated annual (on-site) emissions of greenhouse gases (GHG) from the irrigated rice–wheat system in the Indo-Gangetic Plains based on 20-yr long-term trial data of Ram (2000). This example assumes conventional tillage with all crop residues being burned.

	GHG emissions				
GHG source	Control	Trt. 1	Trt. 2	Trt. 3	Trt. 8
	kg ha^{-1}				
CO_2: Soil tillage	3539	3003	2288	1502	0
CO_2: Diesel†	260	260	260	260	260
N_2O: Burning residues	0.4	0.6	0.6	0.7	0.7
N_2O: Manure application‡	0	0	0	0	3
N_2O: N fertilizer application‡	0	2.4	4.7	7.1	4.7
N_2O: Cereal residues retained§	0	0	0	0	0
N_2O: N-fixing crops§	1.1	1.1	1.4	1.4	1.5
CH_4: Burning residues	17	25.3	29.9	31.1	34.1
CH_4: Rice cultivation¶	100	100	100	100	200
kg CE ha^{-1}#	3953	4774	5510	6086	8032
CPR††	0.73	0.59	0.57	0.59	0.73

† 2.6 kg CO_2 L^{-1} of diesel consumed and 100 L yr^{-1}.
‡ 1.25% of applied N.
§ 1.25% of fixed or residue N.
¶ 10 g CH_4 m^{-2}, doubled if manure added.
# Carbon equivalents, global warming potentials (GWPs) of 21 and 310 used for CH_4 and N_2O, respectively.
†† Carbon/productivity ratio, kg C emitted (equivalents) per kg of edible dry product (cereals).

Table 2–6. Estimated annual (on-site) emissions of greenhouse gases (GHG) from the irrigated rice–wheat systems in the Indo-Gangetic Plains based on 20-yr long-term trial data of Ram (2000). Assuming no-tillage† and all crop residues being retained in this example.

	GHG emissions				
GHG source	Control	Trt. 1	Trt. 2	Trt. 3	Trt. 8
	kg ha^{-1}				
CO_2: Soil tillage	1769	1501	1144	751	+366‡‡‡
CO_2: Diesel‡	86	86	86	86	86
N_2O: Burning residues	0	0	0	0	0
N_2O: Manure application§	0	0	0	0	3
N_2O: N fertilizer application§	0	2.4	4.7	7.1	4.7
N_2O: Cereal residues retained¶	0.1	0.1	0.1	0.1	0.1
N_2O: N-fixing crops¶	1.1	1.1	1.4	1.4	1.5
CH_4: Burning residues	0	0	0	0	0
CH_4: Rice cultivation#	100	100	100	100	200
kg CE ha^{-1}††	2966	3646	4362	4981	6724
CPR‡‡	0.54	0.45	0.45	0.48	0.61

† Assuming 66% fuel savings and soil organic C levels decrease only by 50% compared with conventional till.
‡ 2.6 kg CO_2 L^{-1} of diesel consumed and 33 L yr^{-1}.
§ 1.25% of applied N.
¶ 1.25% of fixed or residue N.
# 10 g CH_4 m^{-2}, doubled if manure added.
†† Carbon equivalents, global warming potentials (GWPs) of 21 and 310 used for CH_4 and N_2O, respectively.
‡‡ Carbon/productivity ratio, kg C emitted (equivalents) per kg of edible dry product (cereals).
‡‡‡ Assuming a C accumulation of 100 kg yr^{-1} in the topsoil.

seemingly small amount of N_2O and CH_4 being emitted from residue burning in fact contributed an additional 15% to the overall C emission budgets of all treatments (ranging from 457 to 895 kg C ha^{-1} yr^{-1}). These increased emissions with residue burning exceed any benefit of moving from conventional to no-tillage, which showed only a 9% reduction in the C emission budget. In other words, no-till must be complemented by residue retention. The use of FYM is not recommended as it increases GHG emissions, and Treatments 1 through 3 all showed similar CPRs. Again, the tendency to use a less N-intensive system would be desirable, but it is not essential provided N losses through leaching can be minimized, perhaps through better timing and placement of fertilizers.

Many experiments exist in the literatures that demonstrate that the judicial use of fertilizers can produce results similar to the application of greater amounts. Also, crop simulation models such as CERES-Wheat and Rice can mimic these systems very well in terms of water, N movement, and yield responses. A version of CERES-Rice also gives CH_4 emissions (Matthews et al., 2000) and has shown conclusively the significant effect of irrigation frequency on reducing emissions.

CONCLUSIONS

The summary in Table 2–7 compares the total annual emissions in terms of CEs and CPRs of three different systems. The most inefficient systems, expressed

Table 2–7. Estimated total annual (on-site) emissions of greenhouse gases (GHG) from the irrigated rice–wheat system in the Indo-Gangetic Plains in terms of C equivalents (CE) and the C/productivity ratio (CPR).

GHG source	GHG emissions				
	Control	Trt. 1	Trt. 2	Trt. 3	Trt. 8
	kg ha^{-1}				
	CE (kg C emitted)				
Conventional/retain residues	3496	4103	4721	5232	7137
Conventional/burn residues	3953	4774	5510	6086	8032
No-till/retain residues	2966	3646	4362	4981	6724
	CPR (kg C emitted per kg grain yield)				
Conventional/retain residues	0.64	0.50	0.49	0.51	0.64
Conventional/burn residues	0.73	0.59	0.57	0.59	0.73
No-till/retain residues	0.54	0.45	0.45	0.48	0.61

in terms of the CPR, were the control and FYM treatments in conventionally tilled systems with residue burning. Even with conventional tillage and residue retention, both these treatments had high CPRs.

If we applied these scenarios to 1 million ha of the IGP, the CE emissions from a high-input conventionally tilled system with residue burning with FYM (the lowest environmental efficiency, i.e., high CPR) would be equivalent to 8 Mt C or 29 Mt CO_2 yr^{-1}. In a no-till, residue-retained system with 50% NPK (the highest environmental efficiency, i.e., low CPR), the total CE emissions would be equivalent to 3.7 Mt C or 14 Mt CO_2 yr^{-1}, an effective halving of emissions as we move from a high- to low-input system with improved nutrient-use and environmental efficiency.

We must recognize that these values are for reasonably high production systems within the IGP; the overall magnitude may be reduced in terms of C loss from soil in the sandier soils of the Punjab. Nevertheless, it is obvious that, even though a vast combination of management practices exists on the IGP, the transition to intensified no-tillage systems, with recommended fertilizer levels, can be both productive and environmentally sound in a world that is rapidly becoming aware of the significant effects of global climate change in both the short and long term.

REFERENCES

Bronson, K.F., A.R. Mosier, and S.R Bishnoi. 1992. Nitrous oxide emission in irrigated corn as affected by nitrification inhibitors. Soil Sci. Soc. Am. J. 56:161–165.

Bronson, K.F., H.U. Neue, U. Singh, and E.B. Abao, Jr. 1997a. Automated chamber measurements of methane and nitrous oxide flux in a flooded rice. Soil. I. Residue, nitrogen, and water management. Soil Sci. Soc. Am. J. 61:981–987.

Bronson, K.F., U. Singh, H.U. Neue, and E.B. Abao, Jr. 1997b. Automated chamber measurement of methane and nitrous oxide flux in a flooded rice soil. II. Fallow period emission. Soil. Sci. Soc. Am. J. 61:988–993.

Burton, D.L., and E.G. Beauchamp. 1985. Denitrification rate relationships with soil parameters in the field. Commun. Soil Sci. Plant Anal. 16:539–549.

Debnath, G., M.C. Jain, S. Kumar, K. Sarkar, and S.K. Sinha. 1996. Methane emission from rice fields amended with biogas slurry and farmyard manure. Clim. Change 33:97–109.

Duxbury, J.M., I.P. Abrol, R.K. Gupta, and K.F. Bronson. 2000. Analysis of long-term soil fertility experiments with rice–wheat rotations in South Asia. p. vii–xxii. *In* I.P. Abrol et al. (ed.) Long-term soil fertility experiments in rice–wheat cropping systems. Rice–Wheat Consortium for the Indo-Gangetic Plains, New Delhi, India.

Ghosh, S. 1998. Ph.D. Thesis. IARI, New Delhi.

Grace, P.R., P. Sanchez, J. Ingram, C. Palm, R. Wassman, M. Fisher, R. Thomas, F. Chandler, W. Bowen, R. Reid, M. Woperies, and S. Waddington. 2000. The consequences of international agricultural research on greenhouse gas emissions and global climate change. USAID, Washington, DC.

Granli, T., and O.C. Bockman. 1994. Nitrous oxide from agriculture. Nor. J. Agric. Sci. (Supple. 12) Agricultural University of Norway, Advisory Service, Ås, Norway.

Gregory, P.J., J.S. Ingram, R. Anderson, R.A. Betts, V. Brovkin, T. N. Chase, P.R. Grace, A.J. Gray, N. Hamilton, T.B. Hardy, S.M Howden, A. Jenkins, M. Maybeck, M. Olsson, I. Ortiz-Monasterio, C.A. Palm, T. Payne, M. Rummukainen, R.E Schulze, M. Thiem, C. Valentin, and M.J. Wilkinson. 2002. Environmental consequences of alternative practices for intensifying crop production. Agric. Ecosyst. Environ. 88:279–290.

Gregory, P.J., J.S. Ingram, B. Campbell, J. Goudriaan, A. Hunt, J. Landsberg, S. Linder, M. Stafford-Smith, W. Sutherst, and C. Valentin. 1999. Managed production systems. p. 229–270. *In* B. Walker et al. (ed.) The terrestrial biosphere and global change implications for natural and managed ecosystems. Synthesis Volume. Book Ser. 4. International Geosphere-Biosphere Program, Cambridge, UK.

Intergovernmental Panel on Climate Change. 1994. Climate change: Radiative forcing and climate change and an evaluation of the IPCC IS92 second assessment report on methane emission from rice cultivation. *In* J.T. Houghton et al. (ed.) Cambridge University Press, Cambridge, UK.

Intergovernmental Panel on Climate Change. 1996. Climate change 1995: Impacts, adaptations and mitigation of climate change. Contribution of Working Group II to the Second Assessment Report of the IPCC. Cambridge University Press, Cambridge, UK.

Jain, M.C., S. Kumar, R. Wassman, S. Mitra, S.D. Singh, J.P. Singh, R. Singh, A.K. Yadav, and S. Gupta. 2000. Methane emissions from irrigated rice fields in northern India (New Delhi). Nutr. Cycl. Agroecosyst. 58:75–83.

Kumar, U., M.C. Jain, H. Pathak, S. Kumar, and D. Majumdar. 2000. Nitrous oxide emissions from different fertilizers and its mitigation by nitrification inhibitors in irrigated rice. Biol. Fertil. Soils 32:474–478.

Ladha, J.K., D. Dawe, H. Pathak, A.T. Padre, R.L. Yadav, Bijay Singh, Yadvinder Singh, Y. Singh, P. Singh, A.L. Kundu, R. Sakal, N. Ram, A.P. Regmi, S.K. Gami, A.L. Bhandari, R. Amin, C.R. Yadav, E.M. Bhattarai, S. Das, H.P. Aggarwal, R.K. Gupta, and P.R. Hobbs. 2002a. How extensive are yield declines in long-term rice–wheat experiments in Asia? Field Crops Res. (in press).

Ladha, J.K., H. Pathak, A.T. Padre, D. Dawe, and R.K. Gupta. 2003. Productivity trends in intensive rice–wheat cropping systems in Asia. p. 45–76. *In* J.K. Ladha et al. (ed.) Improving the productivity and sustainability of rice–wheat systems: Issues and impacts. ASA Spec. Publ. 65. ASA, CSSA, and SSSA, Madison, WI.

Majumdar, D., S. Kumar, and M.C. Jain. 1998. Methane entrapment in different rice soils of India. Curr. Sci. 75:951–955.

Majumdar, D., S. Kumar, H. Pathak, M.C. Jain, and U. Kumar. 2000. Reducing nitrous oxide emission from rice field with nitrification inhibitors. Agric. Ecosyst. Environ. 81:163–169.

Majumdar, D., M. Rastogi, S. Kumar, H. Pathak, M.C. Jain, and U. Kumar. 2001. Nitrous oxide emissions from an alluvial soil with different nitrogenous fertilizers and nitrogen levels. J. Indian Soc. Soil Sci. 48:732–741.

Matthews, R.B., R. Wassmann, and J. Arah. 2000. Using a crop/soil simulation model and GIS techniques to assess methane emissions from rice fields in Asia. I. Model development. Nutr. Cycl. Agroecosyst. 58:141–159.

Mitra, S., M.C. Jain, S. Kumar, S.K. Bandhopadhyay, and N. Kalra. 1999. Effect of rice cultivars on methane emission. Agric. Ecosyst. Environ. 73:177–183.

Mosier, A.R., J.M. Duxbury, J.R. Freney, O. Heinemeyer, and K. Minami. 1996. Nitrous oxide emissions from agricultural fields: Assessment, measurement and mitigation. Plant Soil 181:95–108.

Mosier, A.R., J.M. Duxbury, J.R. Freney, O. Heinemeyer, and K. Minami. 1998. Assessing and mitigating N_2O emissions from agricultural soils. Clim. Change 40:7–38.

Neue, H.U., R. Wassmann, H.K. Kludze, B. Wang, and R.L Lantin. 1997. Factors and processes controlling methane emissions from rice fields. Nutr. Cycl. Agroecosyst. 49:111–117.

Neue, H.U., R. Wassman, R.S. Lantin, M.C.R. Alberto, and J.B. Aduna. 1994. Effect of rice cultivars on methane emissions. Int. Rice Res. Notes 19:32.

Pathak, H., J.K. Ladha, P.K. Aggarwal, S. Peng, S. Das, R.L. Yadav, Yadvinder Singh, Bijay Singh, S.K. Kamra, P. Singh, B. Mishra, A.S.R.A.S. Sastri, H.P. Aggarwal, D.K. Das, and R.K. Gupta. 2002.

Climatic potential and on-farm yield trends of rice and wheat in the Indo-Gangetic Plain. Field Crops Res. (in press).

Pathak, H., and D.B. Nedwell. 2001. Strategies to reduce nitrous oxide emission from soil with fertilizer selection and nitrification inhibitor. Water Air Soil Pollut. 129:217–228.

Ram, N. 2000. Long-term effects of fertilizers on rice–wheat–cowpea productivity and soil properties in mollisols. p. 50–55. *In* I.P. Abrol et al. (ed.) Long-term soil fertility experiments in rice–wheat cropping systems. Rice–Wheat Consortium for the Indo-Gangetic Plains, New Delhi, India.

Robertson, G.P. 1997. Nitrogen use efficiency in row-crop agriculture: Crop nitrogen use and soil nitrogen loss. p. 347–365. *In* L. Jackson (ed.) Ecology in agriculture. Academic Press, New York, NY.

Robertson, G.P. 2000. Denitrification. p. 181–190. *In* M.E. Sumner (ed.) Handbook of soil science. CRC Press, Boca Raton, FL.

Robertson, G.P., and R. R. Harwood. 2001. Sustainable agriculture. p. 99–108. *In* S.A. Levin (ed.) Encyclopedia of biodiversity. Academic Press, New York, NY.

Robertson, G.P., E.A. Paul, and R.R. Harwood. 2000. Greenhouse gases in intensive agriculture: Contributions of individual gases to the radioactive forcing of the atmosphere. Science 289:1922–1925.

Sahrawat, K.L. 1996. Nitrogen inhibitors, with emphasis on natural products and the persistence of fertilizer nitrogen in soil. p. 379–388. *In* N. Ahmad (ed.) Nitrogen economy in tropical soils. Kluwer Academic Publ., Dordrecht, the Netherlands.

Sarkar, K. 1997. Ph.D. thesis. IARI, New Delhi.

Schlesinger, W.H. 2000. Carbon sequestration in soils: Some caution amongst optimism. Agric. Ecosyst. Environ. 82:121–127.

Terry, R.E., R.L. Tate (III), and J.M. Duxbury. 1981. Nitrous oxide emissions form drained cultivated organic soils of South Florida. Am. Pollut. Contr. Assoc. J. 31:1173–1176.

Wang, B., and K. Adachi. 2000. Differences among rice cultivars in root exudation, methane oxidation and populations of methanogenic and methanotrophic bacteria in relation to methane emission. Nutr. Cycl. Agroecosyst. 58:349–356.

Wang, B., H.U. Neue, and H.P. Samonte. 1997. Effect of cultivar difference ('IR72', 'IR65598' and 'Dular') on methane emission. Agric. Ecosyst. Environ. 62:31–40.

Wassmann, R., H.U. Neue, R.S. Lantin, L.V. Buendia, and H. Rennenberg. 2000a. Characterization of methane emissions from rice fields in Asia. I. Comparison among field sites in five countries. Nutr. Cycl. Agroecosyst. 58:1–12.

Wassmann, R., H.U. Neue, R.S. Lantin, K. Makarim, N. Chareonsilp, L.V. Buendia, and H.R. Rennenberg. 2000b. Characterization of methane emissions from rice fields in Asia. II. Differences among irrigated, rainfed, and deepwater rice. Nutr. Cycl. Agroecosyst. 58:13–22.

Youngdahl, L.J., M.S. Lupin, and E.T. Crasswell. 1986. New developments in nitrogen fertilizers for rice. Fert. Res. 9:149–160.

3 Productivity Trends in Intensive Rice–Wheat Cropping Systems in Asia

Jagdish K. Ladha, Himanshu Pathak, Agnes Tirol-Padre, and David Dawe

International Rice Research Institute
Manila, Philippines

Raj K. Gupta

Rice–Wheat Consortium for Indo-Gangetic Plains
New Delhi, India

ABSTRACT

The rice–wheat system occupies 24 million ha in South Asia and China and is important for the food security of the region. Recently, concerns have been raised regarding the stagnation and, in some cases, decline in the productivity and sustainability of this system. To understand the productivity trends of the rice–wheat cropping system, we reviewed (i) yield trends in long-term experiments (LTE), (ii) trends in climatic constraints to the potential yield, and (iii) trends of on-farm yields. Gaps between climatic potential yields and on-station and on-farm yields were also estimated. Data showed that, in treatments in which the recommended rates of N, P, and K were applied, yields of rice and wheat declined in several LTE. In the rice–wheat system, particularly in the Indo-Gangetic Plains (IGP), rice yields tended to have more of a declining trend than wheat yields. There was also a negative trend in climatic potential yields of rice across a west–east transect of the IGP. On-farm yields of rice also showed mixed trends, with trends in some districts negative but positive in others. At the provincial level, yields have stagnated since 1990 in both Punjab and Haryana. However, the wheat yield trends were mostly positive. More detailed data collection and archives of soil and plant samples in the LTE and regular on-farm monitoring of crops and climatic factors are required to identify the specific cause of the yield decline and to develop suitable agronomic and soil management interventions for improving and sustaining productivity.

Rice and wheat have been grown as food crops in Asia for more than 6000 yr. The rice–wheat cropping system has been practiced for 1000 yr (Tran & Marathée, 1994), but the intensification that exists today has evolved rapidly since the 1960s after the introduction of modern high-yielding varieties, access to irrigation, fertilizer, and better infrastructure. This system provides a staple grain supply for more than 400

 Improving the Productivity and Sustainability of Rice–Wheat Systems: Issues and Impacts. ASA Special Publication 65.

million people, or about 8% of the world's population, and it is critical for the food security of Asia. The rice–wheat cropping system is practiced over a variety of soil and climatic conditions with a wide range of agronomic conditions. To ensure food security for an increasing population, the challenge is to increase production while maintaining and enhancing the quality of the land resource. During the Green Revolution, production increases resulted from increases in both the area under production and system productivity. However, little additional land is now available, and traditional farmlands are increasingly lost to urbanization. In addition, since most of the land is already double- and even triple-cropped, further intensification is not a viable option with current resources and management. Therefore, future demands will have to be met primarily through increases in yield per unit land area. An agricultural system is regarded as sustainable if biophysical, environmental, economic, and social objectives are met. Therefore, it is essential that the performance of the system be continuously monitored for productivity, soil nutrient stocks, and nutrient-supplying capacity to ensure and improve sustainability (Powlson et al., 1998).

Another issue of concern is the yield gap between the genetic potential of the region and the yields obtained at experimental stations and in farmers' fields. It is argued that the stagnating or declining yields in both research stations and farmers' fields, declining factor productivity (decreased return per unit input), and degrading soil and water resources are threatening the sustainability of this important agricultural production system (Hobbs & Morris, 1996; Byerlee & Siddiq, 1994; Sinha et al., 1998; Ali & Byerlee, 2000; Duxbury et al., 2000; Kumar & Yadav, 2001; Ladha et al., 2003). Though there is variability in yield gaps among regions, the gap seems to be large even in the most productive regions (Sinha et al., 1998; Singh, 2000; Pathak et al., 2002b). It is important to quantify the yield gaps in different agroecological regions, and to identify and address opportunities to narrow the gap to achieve an increased food supply.

In this review, our goal is to characterize the yield trends in the rice–wheat system in Asia and to explain these trends on the basis of the changes in biophysical factors at climatic potential, on-station, and on-farm levels. The specific objectives are to (i) analyze the rice–wheat LTE conducted in South Asia and China to investigate the extent of the yield decline, (ii) review rice–wheat productivity trends in terms of climatic potential and on-farm yields, (iii) estimate the yield gaps in the IGP, and (iv) analyze the causes contributing to yield stagnation and decline and yield gaps.

DISTRIBUTION OF THE RICE–WHEAT SYSTEM

The rice–wheat rotation is one of the largest agricultural production systems in the world, occupying 24 million ha of the most productive land in the IGP and in China. In South Asia, this system occupies about 13.5 million ha (10 million in India, 2.2 million in Pakistan, 0.8 million in Bangladesh, and 0.5 million in Nepal), extending across the Indo-Gangetic floodplain into the Himalayan foothills (Ladha et al., 2000). It accounts for about one-third of the area of both rice and wheat grown in South Asia, where about 85% of the total rice–wheat system is practiced in the

IGP. Although the IGP agroecological region is a relatively homogeneous region, based on agroclimatic characteristics, soil type, and physiographic features, it is divided into five transects: (i) Trans-Gangetic Plains or IGP Transects 1 and 2 (areas in Pakistan and parts of Punjab and Haryana in India), (ii) Upper-Gangetic Plains or IGP Transect 3 (most of Uttar Pradesh and parts of Bihar, India, and parts of Nepal), (iii) Middle-Gangetic Plains or IGP Transect 4 (parts of Bihar, India, and parts of Nepal), and (iv) Lower-Gangetic Plains or Transect 5 (parts of Bihar and West Bengal, India, and parts of Bangladesh) (Narang & Virmani, 2001). In addition, a substantial area in rice–wheat exists outside the IGP, in China, India, and Nepal. In India, the system includes hilly parts of Himachal Pradesh and Uttranchal and parts of Jharkhand, Madhya Pradesh, and Rajasthan. A considerable amount of rice–wheat farming is also practiced in hilly parts of Nepal. In China, the rice–wheat rotation is common in the provinces of Jiangsu, Zhejiang, Hubei, Guizhou, Yunnan, Sichuan, and Anhui. Descriptions of areas under rice–wheat in the IGP and outside the IGP, including China, are available elsewhere (Zheng, 2000; Narang & Virmani, 2001; Gupta et al., 2003, Chapter 1 of this publication; Ladha et al., 2003).

The rice–wheat areas are located within subtropical to warm temperate climates characterized by cool and dry winters and warm and wet summers (Timsina & Connor, 2001; Ladha et al., 2003). Solar radiation decreases as we move from IGP Transect 1 to 5 in the rice season, while the trend is reversed in the wheat season (Pathak et al., 2002b). Minimum temperature in rice and wheat increases as you move from IGP Transects 1 to 5. This is also true for maximum temperature in the wheat season, but during the rice season, the average maximum temperature is similar across all of the IGP. Rainfall also follows a distinct pattern, increasing from Transects 1 to 5 of the IGP, with Transects 1 and 2 receiving only 650 mm of rainfall annually and Transect 5 receiving more than twice as much. Except for rainfall, the climatic conditions make the upper transects of the IGP more favorable for rice and wheat cultivation. Access to assured irrigation has alleviated the problem of low rainfall periods and made the zone (Transects 1 and 2) very productive. The less favorable climatic conditions and limited irrigation facilities are the major constraints to achieving higher yields in the lower transects (3, 4, and 5) of the IGP.

TRENDS OF RICE AND WHEAT YIELDS

Yields in Long-Term Experiments

Long-term experiments are valuable repositories of information regarding the sustainability of intensive agriculture. They provide opportunities for monitoring long-term changes in crop yields and nutrient balances, and identifying factors associated with such changes. They are of particular value in studying biophysical and environmental factors and provide data on which to base rational judgments about the biophysical aspects of sustainability (Powlson et al., 1986). Several LTE began in the 1970s and 1980s in double- and triple-cropped rice–wheat systems in the IGP of South Asia and in China to monitor yield trends and system sustainability. These LTE have a range of treatments representing different levels of mineral and organic fertilizer. However, for the study of sustainability, only data from the treatments with the recommended NPK rate were included. This is because treatments in which one or more of the nutrients are omitted will show declines in yield with time because of incorrect nutrient management and are not suitable for assessing the sustainability of a cropping system (Gami et al., 2001; Bhandari et al., 2002; Regmi et al., 2002a,b). Several researchers have studied long-term rice–wheat experiments in Asia (Nambiar, 1994; Cassman et al., 1995; Yadav et al., 1998; 2000; Dawe et al., 2000; Duxbury et al., 2000), but the analyses are mostly restricted to yield trends. The analyses of some of these experiments have shown declining rice and wheat yields (Nambiar, 1995; Brar et al., 1998; Yadav et al., 1998, 2000; Duxbury et al., 2000), whereas in others, yields either increased or were maintained (Aggarwal et al., 2000a; Yadav et al., 2000; Dawe et al., 2000). Dawe et al. (2000) analyzed yield trends from 30 long-term rice experiments within rice–rice or rice–wheat systems conducted in seven Asian countries (China, India, Indonesia, Bangladesh, Vietnam, the Philippines, and Malaysia), covering a wide variety of soil types. They suggested that yield declines are not very common, particularly at yields of 4 to 7 Mg ha^{-1} (Dawe et al., 2000). Of the seven long-term rice–wheat experiments examined, none had a significant yield decline in the wheat crop, but rice yields in the Pantnagar-LTFE (long-term fertility experiment), started in 1984 in India, declined at 2.3% yr^{-1}. In all six of the remaining rice–wheat LTE studied, rice yields with the highest fertilizer inputs remained unchanged for 10 to 25 yr. However, Duxbury et al. (2000) demonstrated that 8 of the 11 long-term rice–wheat experiments that had run for more than 8 yr exhibited a downward trend in rice yield over time. This is in contrast to only 3 of 11 showing a downward trend in wheat yield. Although statistical analyses were not conducted, rice yield appeared to decline substantially in several of the LTE.

Recently, analyses of 33 LTE conducted at 21 different sites in India, Nepal, China, and Bangladesh was completed to make global and regional interpretations using the most recent data and identical statistical procedures (Ladha et al., 2003) (Fig. 3–1, Table 3–1). This included the rice–wheat LTE analyzed by Dawe et al. (2000), Duxbury et al. (2000), and Yadav et al. (1998) and several new LTE. On a global basis (from 33 LTE), a negative average yield trend of –0.02 Mg ha^{-1} yr^{-1} (0.5% yr^{-1} of the average yield) for rice was obtained. Downward yield trends were observed in 55 and 48% of the LTE for rice and wheat, respectively. Region-wide averages were also negative for the IGP except for Transect 5, but no significant declines were observed at sites outside the IGP and in China (Table 3–2). Increasing average trends were obtained for wheat in IGP Transects 2 and 5 and in the non-IGP (Table 3–2). On a systems (rice + wheat) basis, the average yield trend was similar for rice. Globally, the system yield declined at –0.02 Mg ha^{-1} yr^{-1}, although statistically significant ($P < 0.05$) declining trends were observed at only six sites within the IGP. In the rice–wheat system, rice yields had more disturbing trends than wheat yields. Declining yield trends in rice were observed in many of the LTE in the IGP where the initial yields were >6 Mg ha^{-1}. The initial yield and yield change in the LTE in the IGP were significantly correlated, suggesting that the higher yielding sites were more prone to decline.

Data from 12 LTE conducted in India under the Indian Council of Agricultural Research (ICAR) Coordinated Project on Cropping Systems were analyzed

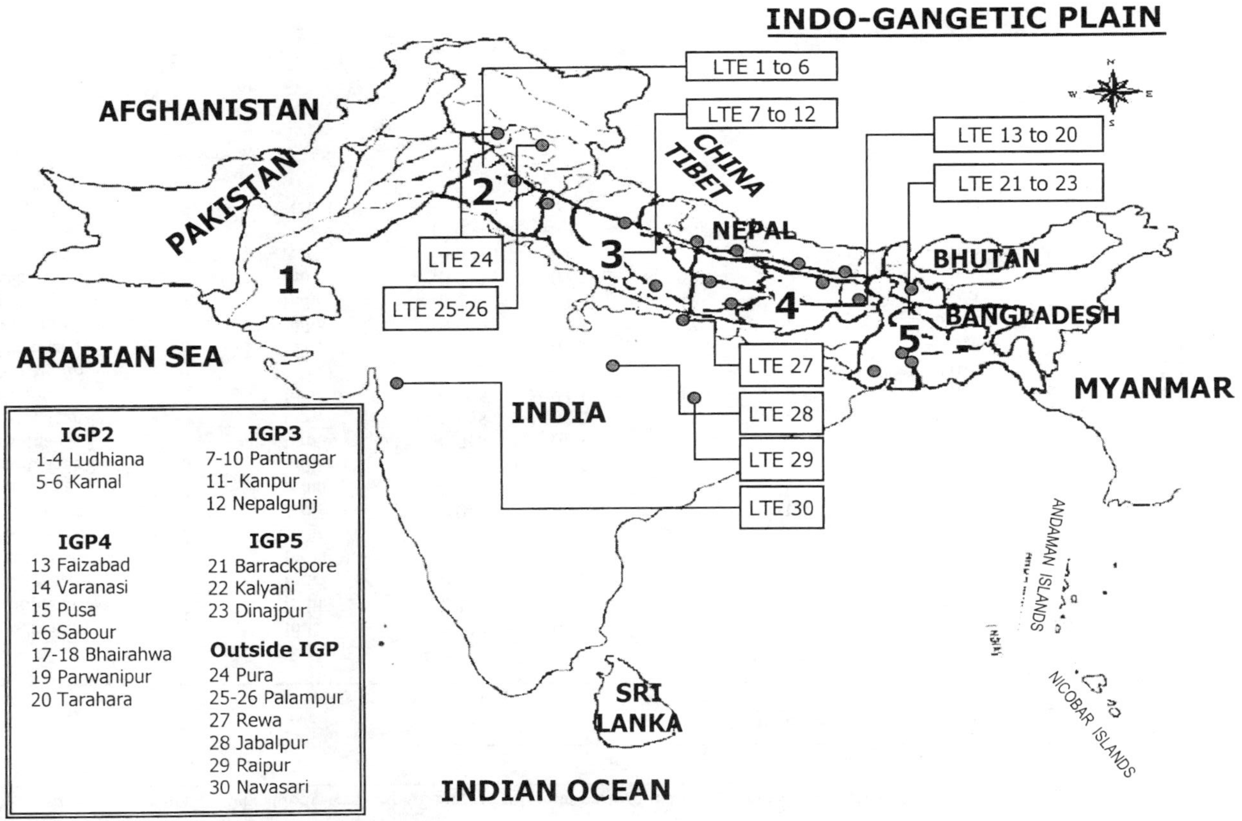

Fig. 3–1. Locations of long-term rice–wheat experiments in the Indo-Gangetic Plains of India, Nepal, and Bangladesh.

Table 3–1. Yield trends of rice in rice–wheat long-term experiments in different parts of the Indo-Gangetic plains (IGP), non-IGP in India, and China.

Site no.	Site	IGP	Country	Rice: Yield change	Rice: P value	Rice: Avg. yield	Rice: Potential yield†	Wheat: Yield change	Wheat: P value	Wheat: Avg. yield	Wheat: Potential yield†
				$Mg\ ha^{-1}\ yr^{-1}$		—— $Mg\ ha^{-1}$ ——		$Mg\ ha^{-1}\ yr^{-1}$		—— $Mg\ ha^{-1}$ ——	
1	Ludhiana 1	2	India	−0.113	0.00	6.03	10.7	0.019	0.45	4.62	7.9
2	Ludhiana 2	2	India	0.163	0.18	5.31	10.7	0.081	0.49	4.91	7.9
3	Ludhiana 3	2	India	0.029	0.78	5.50	10.7	0.207	0.00	5.17	7.9
4	Ludhiana 4	2	India	−0.010	0.90	5.74	10.7	0.065	0.13	4.38	7.9
5	Karnal	2	India	−0.052	0.01	6.67	10.4	0.008	0.74	4.68	7.3
6	Karnal 1	2	India	−0.047	0.53	7.02	10.4	0.151	0.06	4.45	7.3
7	Pantnagar 1	3	India	−0.171	0.00	6.42	9.0	0.040	0.22	4.58	6.5
8	Pantnagar 2	3	India	0.018	0.62	4.46	9.0	0.077	0.00	4.00	6.5
9	Pantnagar 3	3	India	−0.016	0.58	4.56	9.0	−0.060	0.14	3.87	6.5
10	Pantnagar 4	3	India	−0.145	0.00	6.45	9.0	−0.065	0.06	3.33	6.5
11	Kanpur	3	India	0.106	0.01	4.48	9.5	−0.020	0.22	4.55	7.0
12	Nepalgunj	3	Nepal	−0.058	0.28	3.64	na‡	0.019	0.57	2.56	Na
13	Faizabad	4	India	−0.046	0.35	4.17	9.1	−0.018	0.77	3.42	6.7
14	Varanasi	4	India	0.010	0.75	4.07	9.2	−0.029	0.60	3.80	7.0
15	Pusa	4	India	−0.237	0.00	3.47	9.8	−0.071	0.16	3.35	6.8
16	Sabour	4	India	0.012	0.65	4.06	9.8	0.040	0.13	3.16	6.8
17	Bhairahwa 1	4	Nepal	−0.097 (0.038)§	0.01 (0.12)	2.73 (3.06)	9.4	−0.045	0.24	2.32	6.7
18	Bhairahwa 2	4	Nepal	−0.107	0.02	4.82	9.4	−0.083	0.12	2.10	6.7

19	Parwanipur	4	Nepal	0.021	0.58	3.13	na	–0.025	0.39	1.92	na
20	Tarahara	4	Nepal	–0.100	0.01	3.73	na	–0.082	0.04	3.20	na
21	Barrackpore	5	India	–0.022	0.39	4.39	7.7	0.002	0.91	2.96	5.2
22	Kalyani	5	India	–0.015	0.64	3.35	7.7	0.016	0.59	2.70	5.2
23	Dinajpur	5	B'desh	0.052	0.53	1.77	8.6	0.114	0.09	3.75	5.4
				(0.011)§	(0.09)	(3.37)					
24	RS Pura	NIGP¶	India	0.049	0.24	4.85	na	–0.023	0.45	3.29	na
25	Palampur	NIGP	India	–0.033	0.42	2.86	10.0	0.124	0.01	2.70	7.5
26	Palampur 1	NIGP	India	–0.091	0.32	3.12	10.0	–0.02	0.62	2.48	7.5
27	Rewa	NIGP	India	–0.037	0.29	4.08	na	0.06	0.06	3.14	na
28	Jabalpur	NIGP	India	0.058	0.46	4.96	na	–0.056	0.22	2.53	na
29	Raipur	NIGP	India	0.169	0.02	5.40	8.2	0.052	0.59	2.77	6.2
30	Navsari	NIGP	India	0.041	0.23	3.47	na	–0.136	0.01	2.20	na
31	Nangong	China	China	0.048	0.47	6.80	na	–0.077	0.24	3.00	na
32	Sichuan	China	China	0.077	0.41	5.77	na	–0.063	0.25	3.57	na
				(–0.067)	(0.42)	(5.80)					
33	Jiangsu	China	China	–0.200	0.07	7.34	na	0.020	0.50	4.87	na

† Potential yields were calculated using CERES-RICE 3.5 (Singh et al., 1998) and GENERIC-CERES 3.5 (Ritchie et al., 1998) models for rice and wheat, respectively.
‡ Not available.
§ Values in parentheses are for the late rice crop.
¶ Non-IGP.

Table 3–2. Yield trends of rice and wheat in rice–wheat long-term experiments in different parts of the Indo-Gangetic Plains (IGP), non-IGP in India, and China.†

Crop	Location	No. of data set	Yield change	Avg. yield	No. of positive and negative trends‡	
					Positive	Negative
			Mg ha^{-1} yr^{-1}	Mg ha^{-1}		
Rice	IGP2	6	−0.002	6.04	2 (0)	4 (2)
	IGP3	6	−0.044	5.00	2 (1)	4 (2)
	IGP4	9	−0.056	3.69	4 (0)	5 (4)
	IGP5	4	0.007	3.22	2 (0)	2 (0)
	Non-IGP	7	0.022	4.11	4 (1)	3 (0)
	China	4	−0.036	6.43	2 (0)	2 (0)
	Total	36	−0.023	4.63	16 (2)	20 (8)
Wheat	IGP2	6	0.090	4.70	6 (1)	0
	IGP3	6	−0.002	3.82	3 (1)	3 (0)
	IGP4	8	−0.036	2.91	1 (0)	7 (1)
	IGP5	3	0.044	3.14	3 (0)	0
	Non-IGP	7	0.000	2.73	3 (1)	4 (1)
	China	3	−0.040	3.82	1 (1)	2 (0)
	Total	33	0.007	3.47	17 (4)	16 (2)
System§	IGP2	6	0.070	10.69	3 (1)	3 (1)
	IGP3	6	−0.042	8.81	2 (2)	4 (2)
	IGP4	8	−0.101	7.06	1 (0)	7 (3)
	IGP5	3	0.050	7.39	2 (0)	1 (0)
	Non-IGP	7	0.023	6.84	5 (0)	2 (0)
	China	3	−0.092	12.39	0	3 (0)
	Total	33	−0.019	8.50	13 (3)	20 (6)

† Source: Ladha et al. (2003).
‡ Numbers in parentheses indicate the number of long-term experiments with time trends that are statistically significant at $P = 0.05$.
§ Rice–rice–wheat systems are included in IGP 4, IGP 5, and China.

to evaluate the effect of different sources of organic matter (OM) (farmyard manure [FYM] and green manure [GM]) in combination with inorganic fertilizer on the productivity of the rice–wheat system (Yadav et al., 2000). The F test for homogeneity of regression coefficients, two-way analysis of variance (ANOVA) to test for treatment effects within sites, followed by Tukey's pairwise multiple comparison were done. The F test did not show significant ($P < 0.05$) differences in yield trends among the treatments within a site, in agreement with the analysis of Dawe (unpublished data, 2002). However, the two-way ANOVA and pairwise multiple comparison indicated that the average yield change in rice was significantly ($P < 0.05$) higher with the addition of organic manures when compared with the 100% NPK treatment. The fertilizer treatments showed no significant effect on the final yield, although the initial yield was significantly higher with 100% NPK than with FYM (Table 3–3). The average rice yield with 100% NPK was still significantly higher than with FYM. In wheat, average yields and yield trends were not significantly different among the fertilizer treatments. These results seem to indicate that organic manure (such as FYM) application can contribute to maintaining and enhancing the productivity of rice–wheat systems in the long run. This is consistent with the farmers' perception (unpublished). But the nonavailability of FYM because

Table 3–3. Comparison of yield and yield trends of rice among treatments with full NPK and NPK supplemented with farmyard manure (FYM), green manure (GM), and straw in 12 rice–wheat long-term experiments in India.

Crop	Treatment	Yield change	Initial yield†	Final yield†	Avg. yield	No. of positive and negative trends‡	
						Positive	Negative
		$Mg\ ha^{-1}\ yr^{-1}$		$Mg\ ha^{-1}$			
Rice	NPK	0.019b§	4.3a	4.3a	4.36a	7 (2)	5 (1)
	NPK + FYM	0.053a	3.9b	4.3a	4.17b	11 (3)	1 (0)
	NPK + GM	0.050a	4.1ab	4.5a	4.33ab	10 (5)	2 (0)
Wheat	NPK	–0.009a	3.3a	3.1a	3.29a	4 (1)	8 (1)
	NPK + FYM	–0.002a	3.4a	3.3a	3.42a	5 (2)	7 (2)
	NPK + GM	0.006a	3.2a	3.3a	3.32a	6 (2)	6 (1)

† Initial yield is the average of the first 3 yr and final yield is the average of the last 3 yr.

‡ Numbers in parentheses indicate the number of long-term experiments with time trends that are statistically significant at $P < 0.05$.

§ For each crop, values followed by a common letter in a column are not significantly different from each other $P < 0.05$ by the Tukey Test.

of burning of cow dung for fuel and the high labor cost of transporting the FYM to fields restrict its extensive and widespread application in agriculture.

A linear response of rice and wheat yields to increasing levels of NPK was observed in the majority of the 12 LTE during the initial and final 3 yr of the experiment (the averages of the initial and final 3 yr were used for regression analyses), indicating that both rice and wheat yields could still be increased with higher nutrient inputs (Fig. 3–2). On the basis of data sets of different LTE, Duxbury et al. (2000) also drew a similar conclusion. These results support the hypothesis that greater inputs of N, P, and K are needed to reach maximum yield. However, at a few sites (Navsari, Palampur, Varanasi, and Raipur), responses appeared to plateau. The increase in grain yield per unit of NPK applied (slope) was highest in Ludhiana, which also exhibited the highest cumulative yield with NPK.

Potential Yield

Potential yield is defined as the maximum yield of a variety restricted only by the specific climatic conditions of a particular season. Climatic components—temperature, solar radiation, rainfall, relative humidity, and wind velocity—independently or in combination influence plant life and can inhibit, stimulate, alter, or modify crop performance. The potential yields of different sites (Table 3–1) were estimated using CERES-RICE 3.5 (Singh et al., 1998) and GENERIC-CERES 3.5 (Richie et al., 1998) models for rice and wheat, respectively. The potential yields of rice and wheat in the IGP ranged from 7.7 to 10.7 $Mg\ ha^{-1}$ and 5.2 to 7.9 $Mg\ ha^{-1}$, respectively (Table 3–1). Yields of rice and wheat were highest in Ludhiana (IGP 2) and decreased toward the eastern region (IGP 5) of the IGP. Yield decreased by 28% for rice and 34% for wheat from Transect 2 (Ludhiana) to Transect 5 (24-Pargana). This is because of lower solar radiation and higher daily minimum temperature in the lower part of the IGP, resulting in decreased photosynthesis, increased respiration, and a shortened vegetative and grain-filling period (Yoshida & Parao,

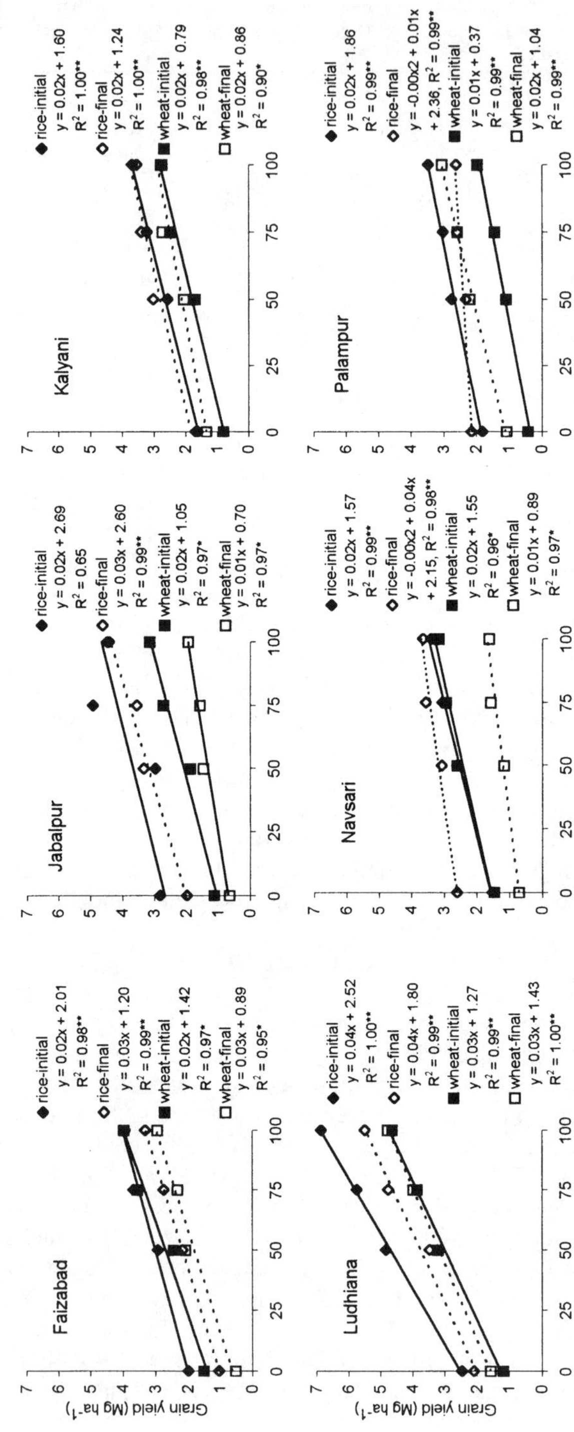
Grain yield (Mg ha-1)
0 1 2 3 4 5 6 7
0 25 50 75 100
Faizabad
rice-initial y = 0.02x + 2.01 R^2 = 0.98**
rice-final y = 0.03x + 1.20 R^2 = 0.99**
wheat-initial y = 0.02x + 1.42 R^2 = 0.97*
wheat-final y = 0.03x + 0.89 R^2 = 0.95*
Jabalpur
rice-initial y = 0.02x + 2.69 R^2 = 0.65
rice-final y = 0.03x + 2.60 R^2 = 0.99**
wheat-initial y = 0.02x + 1.05 R^2 = 0.97*
wheat-final y = 0.01x + 0.70 R^2 = 0.97*
Kalyani
rice-initial y = 0.02x + 1.60 R^2 = 1.00**
rice-final y = 0.02x + 1.24 R^2 = 1.00**
wheat-initial y = 0.02x + 0.79 R^2 = 0.98**
wheat-final y = 0.02x + 0.86 R^2 = 0.90*
Ludhiana
rice-initial y = 0.04x + 2.52 R^2 = 1.00**
rice-final y = 0.04x + 1.80 R^2 = 0.99**
wheat-initial y = 0.03x + 1.27 R^2 = 0.99**
wheat-final y = 0.03x + 1.43 R^2 = 1.00**
Navsari
rice-initial y = 0.02x + 1.57 R^2 = 0.99**
rice-final y = -0.00x2 + 0.04x + 2.15, R^2 = 0.98**
wheat-initial y = 0.02x + 1.55 R^2 = 0.96*
wheat-final y = 0.01x + 0.89 R^2 = 0.97*
Palampur
rice-initial y = 0.02x + 1.86 R^2 = 0.99**
rice-final y = -0.00x2 + 0.01x + 2.36, R^2 = 0.99**
wheat-initial y = 0.01x + 0.37 R^2 = 0.99**
wheat-final y = 0.02x + 1.04 R^2 = 0.99**

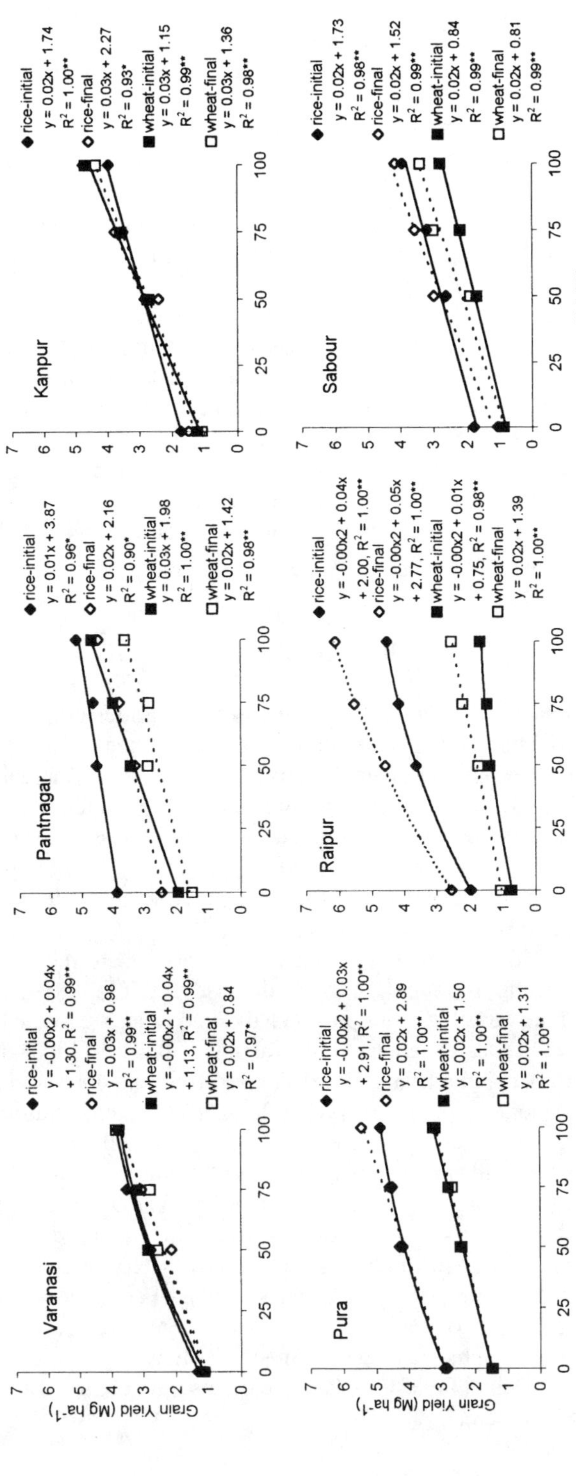

Fig. 3–2. Rice and wheat grain yield response to NPK applied during the initial and final 3 yr of 12 long-term experiments in India.

1976; Penning de Vries, 1993; Horie et al., 1995). Using daily weather data generated from monthly means by WGEN-weather, Aggarwal et al. (2000c) estimated similar potential yields of rice and wheat in the IGP and observed that Punjab and Haryana had 10 Mg ha^{-1} or more potential rice yield, which decreased in an eastward direction in the IGP. The potential yield of wheat was about two-thirds that of the rice crop and followed the same declining trend from northwest to east along the transect.

Pathak et al. (2003b) estimated the potential yield trend of rice and wheat at different sites in the IGP and outside the IGP in India. The potential yield trend of rice ranged from –0.12 Mg ha^{-1} yr^{-1} at Delhi to 0.05 Mg ha^{-1} yr^{-1} at Kanpur (Table 3–4). Negative yield trends were observed in six of the nine data sets, four of which were significantly different from 0 ($P < 0.05$). These declining trends were observed throughout the transects of the IGP, indicating that yield declines are not localized. On the other hand, positive trends were observed at three sites but none was statistically significant ($P < 0.05$). In wheat, the rate of the annual yield trend ranged from –0.07 Mg ha^{-1} yr^{-1} at Delhi to 0.04 Mg ha^{-1} yr^{-1} at Faizabad and Pantnagar (Table 3–4). Of the nine sites, six showed negative trends and three showed positive trends, but none was significantly different from 0. Thus, wheat yield appeared to be more stable than rice yield.

District Yields

The government of India reports grain yields of various crops obtained by farmers on a district basis every year. Studies have examined the trends in yields of rice and wheat at the district level in the northwestern states of Haryana and Punjab (Sinha et al., 1998). During the 1960s and 1970s, through the introduction of new high-yielding varieties, large, continuous, and widespread gains occurred in rice and wheat yields. All districts in these two states have shown a much lower growth rate in rice yields since the late 1980s compared with the 1960s and 1970s (Aggarwal et al., 2000a).

Recently, Pathak et al. (2003b) estimated on-farm yield trends of rice and wheat at several locations along the IGP for the last 15 yr (Table 3–4). Yields of rice and wheat for the various districts in which the study sites are located were collected from the Fertilizer Association of India Regional Statistics (1986–2000), New Delhi, India. Both rice and wheat yields decreased from the upper to lower transects of the IGP. Among other factors, favorable climate, better irrigation facilities and infrastructure, and better socioeconomic conditions of the farmers were responsible for greater yields in the upper transects (Narang & Virmani, 2001). The time trend of rice yield ranged from –0.15 Mg ha^{-1} yr^{-1} at Delhi to 0.13 Mg ha^{-1} yr^{-1} at Varanasi. Negative yield trends were observed in five of the nine data sets, but only one (Delhi) was significantly different. Positive yield trends were observed at four sites, with three being (Kanpur, Varanasi, and 24-Pargana) statistically significant. In wheat, the rate of annual yield change differed markedly from that of rice. The estimated yield change ranged from –0.02 Mg ha^{-1} yr^{-1} at 24-Pargana to 0.11 Mg ha^{-1} yr^{-1} at Delhi (Table 3–4). Positive trends were observed at eight sites, with seven being significant.

Table 3–4. Climatic potential and district average yields and yield changes over the years in the Indo-Gangetic plains.†

Site	Potential yield‡					District yield§				
		Rice		Wheat			Rice		Wheat	
	Year	Yield¶	Change	Yield§	Change	Year	Yield§	Change	Yield§	Change
		Mg ha^{-1}	Mg ha^{-1} yr^{-1}	Mg ha^{-1}	Mg ha^{-1} yr^{-1}		Mg ha^{-1}	Mg ha^{-1} yr^{-1}	Mg ha^{-1}	Mg ha^{-1} yr^{-1}
Ludhiana	1985–1999	10.7	–0.07*	7.9	–0.02	1985–98	5.6	–0.05	4.3	0.06*
Karnal	1985–1999	10.4	–0.08**	7.3	0.02	1985–98	3.8	–0.01	3.6	0.05**
Delhi	1985–1999	9.8	–0.12**	7.1	–0.07	1985–99	2.4	–0.15*	2.5	0.11*
Kanpur	1990–1999	9.5	0.05	7.0	–0.00	1985–98	2.8	0.07*	2.8	0.04**
Varanasi	1985–1999	9.2	–0.01	7.0	–0.01	1985–98	3.2	0.13**	2.2	0.06*
Faizabad	1985–1997	9.1	–0.00	6.7	0.04	1985–99	2.8	0.03	2.3	0.03*
24-Pargana	1985–1998	7.7	–0.10**	5.2	–0.07	1985–98	2.8	0.08*	2.1	–0.02
Raipur	1985–1999	8.2	0.03	6.2	–0.03	1987–98	2.1	–0.03	1.0	0.03
Pantnagar	1985–1999	9.0	0.00	6.5	0.04	1989–98	4.2	–0.03	2.7	0.07**

*,** Significant at $P = 0.05$ and $P = 0.01$, respectively.
† Source: Pathak et al. (2002b).
‡ Potential yields were calculated using CERES-RICE 3.5 (Singh et al., 1998) and GENERIC-CERES 3.5 (Ritchie et al., 1998) models for rice and wheat, respectively.
§ Source: Fertilizer Association of India Regional Statistics (1986–2000).
¶ Average yield.

Table 3–5. Possible changes in soil properties under rice based cropping resulting in yield decline.

Soil changes	References
Decline in soil C	Dawe et al. (2000), Ram (2000), Yadav et al. (2000), Ladha et al. (2003)
Decline in soil N	Saha et al. (2000), Dawe et al. (2000), Ram (2000), Yadav et al. (2000), Bhandari et al. (2002), Ladha et al. (2003)
Decline in soil P availability	Yadvinder Singh et al. (2000), Regmi et al. (2002a), Ladha et al. (2003)
Decline in soil K	Yadav et al. (1998, 2000), Bhandari et al. (2002), Regmi et al. (2002a, 2002b), Ladha et al. (2003)
Decline in available soil S	Yadav et al. (1998)
Decline in available soil Zn	Yadav et al. (1998), Ram (2000), Sakal (2000), Ladha et al. (2003)
Decline in B, Cu, Fe, or Mn	Yadav et al. (1998)
Delay in planting	Ladha et al. (2003), Regmi et al. (2002a), Gami et al. (2001)
Decrease in solar radiation	Ladha et al. (2003), Pathak et al. (2003b)
Increase in minimum temperature	Ladha et al. (2003), Pathak et al. (2003b)

POSSIBLE CAUSES OF YIELD STAGNATION OR DECLINE

The widespread stagnation and occasional decline in both the observed (on-station and on-farm) and climatic potential yield of rice and wheat on experimental farms are a matter of concern. The agricultural productivity of any region depends upon several biophysical and socioeconomic factors. The key biophysical factors are soil health, climate, crop variety, and pest and agronomic management. A yield decline or stagnation in a production system can occur because of a negative change in any of these factors. Various possible causes of a yield decline, summarized in Table 3–5, are discussed below.

Soil Carbon and System Productivity

Soil organic matter (SOM), although usually constituting <5% of soil weight, is one of the most important components of a field ecosystem. It serves as a soil conditioner, nutrient source, a substrate for microbial activity, and a major determinant for sustaining or increasing agricultural productivity (Schnitzer, 1991). The biological activity induced by microbial decomposition of SOM has the potential to antagonize root diseases. Soil OM induces changes in physical properties of soils as well. A marked improvement in the structure and aggregate stability of the soils is often detected (Tisdale & Oades, 1982), which is of prime importance in intensive agriculture. Soil aeration becomes a limiting factor under conditions of generous nutrient and water supply and can be ensured only by proper soil structure. Although soil physical properties and associated water relations are generally more important in wheat, they are critical for rice as well in light-textured soils. The fact that rice yield declines are more widespread in the IGP with light-textured soil seems to indicate that adverse soil physical properties are more detrimental to rice than to wheat or are equally detrimental.

Unfortunately, in most of the rice–wheat experiments, the SOM content appears to have declined with time (Table 3–6), and there is a corresponding yield de-

Table 3–6. Soil organic C levels with different treatments in some long-term experiments in rice-based cropping systems.†

Treatments	Pantnagar	Pantnagar	Ludhiana	Bhairahwa	Barrackpore	Barrackpore
System	RW‡	RWC	RW	RW	RW	RWJ
Initial	1.48	1.48	0.36	1.03	0.71	0.71
Unfertilized	0.49	0.60	0.34	0.73	0.40	0.42
NPK	0.84	0.90	na§	0.88	0.43	0.45

† Source: Nambiar (1994), Duxbury et al. (2000).
‡ RW, rice–wheat; RWC, rice–wheat–cowpea; RWJ, rice–wheat–jute.
§ Not available.

cline of rice and wheat in the recommended NPK treatments (Nambiar, 1994; Abrol et al., 2000; Duxbury et al., 2000; Yadav et al., 2000). In the major rice–wheat regions of northwestern India, the soil organic C (0–15 cm) has decreased from 0.5% in the 1960s to 0.2% at present (Sinha et al., 1998). A similar decline in SOM was evidenced in Pakistan's Punjab (Ali & Byerlee, 2000). During 1971 through 1974, SOM was 1.02%, decreasing to 0.72% in 1975 through 1984, and further declining to 0.59% in the post-Green Revolution period of 1985 through 1994. The lack of recycling of organic matter is the principal cause of such a decline. The majority of the total plant biomass is removed from the field in irrigated rice systems because rice is usually manually harvested in Asia. After the grain harvest, the remaining stubble is sometimes removed to be used as forage, fuel, or building material. Stubble is also burned to facilitate land preparation for a subsequent rice crop. In areas where harvesting is undertaken mechanically (e.g., northwestern India), rice straw is generally burned before planting the next crop (Samra et al., 2003, Chapter 8 of this publication). Only roots and a small amount of residual stubble are recycled.

Using the data of 33 LTE, Ladha et al. (2003) attempted to derive a relationship between soil organic C and average long-term yields or the magnitude of yield change. Although the yield parameters did not correlate with soil organic C, suggesting that it may not be linked to a decline in crop productivity, this sort of conclusion is not robust until it is ruled out whether factors other than SOM have limited productivity (Ladha et al., 2003). We examined two sets of LTE to further study linkages between soil C and productivity. In both cases, the Ludhiana (IGP Transect 2) site, which represents inherent low soil C, was compared with Bhairahwa (IGP Transect 4) and Pantnagar (IGP Transect 3), representing medium to high inherent soil C.

Ludhiana vs. Bhairahwa

The two LTE represented high (Ludhiana) and low (Bhairahwa) crop productivity, with an average rice–wheat system productivity of 10.6 Mg ha^{-1} at Ludhiana and 8.1 Mg ha^{-1} at Bhairahwa. The system yields declined by 0.04 to 0.22 Mg ha^{-1} in both the LTE, which had run for 16 to 20 yr. In both LTE, the system yield declines were dominated by the decline in rice yields. There was no clear effect of FYM on system yield when it supplemented or replaced inorganic fertilizer.

Total C was measured in soil samples archived at intervals during the two LTE. Unfortunately, soil samples in the early years of the experiments were not available, and soil sampling started at different years from the start of the experiments. To deal with these deficiencies, in 1999 the surface soil from uncultivated or unmanaged reference sites was sampled and compared with the surface soil in the long-term test plots. The reference sites were located in an uncleared and lightly grazed fence line or roadside within 50 m of the cropped site and away from animal trails, eroded areas, and trees. The soil samples from the reference sites probably represent C and nutrient status prior to the introduction of an agroecosystem. In other words, a reference sample can be used to determine the magnitude of changes in soil parameters that have taken place because of cultivation in an LTE. Total soil C appeared to have declined with continuous cultivation when the reference samples were used as the base. It declined in both the soils from 27 to 44%, and the decline was greater in the loamy sand (Ludhiana) than in the silt loam (Bhairahwa) soil (Fig. 3–3). However, during the later years of the experiment, total C remained unchanged and maintained an equilibrium. Although the analysis in the initial years is lacking, the trends in available data suggest that most of the loss or gain of C probably occurred immediately after the change in the system, followed by the establishment of an equilibrium. During the experiment, the C content in NPK treatments was similar to that of the control (no NPK), but it increased when FYM was added. The increase calculated by subtracting the C content of the available initial samples (Years 5 and 10 at Ludhiana and Bhairahwa sites, respectively) from that of the soil sample in the final year (1999) varied from 21 to 48%. The increase was proportional to the amount of FYM added. It was the largest (48%) at Bhairahwa, where 12 Mg ha^{-1} yr^{-1} of FYM was applied, and the lowest (21%) at Ludhiana, where 6 Mg ha^{-1} yr^{-1} of FYM was applied. The total soil N followed a trend similar to that of soil C (data not shown).

In summary, the soil C content remained unchanged in the NPK treatments and increased in the FYM treatment. Nevertheless, yields declined. The continuous addition of FYM also did not arrest the yield decline. Based on these data, it could possibly be concluded that soil C and crop productivity are not linked. However, in subsequent studies, detailed investigations revealed that a gradual depletion of soil K and its insufficient application may have caused the yield decline (Bhandari et al., 2002; Regmi et al., 2002a,b).

Ludhiana vs. Pantnagar

An interesting example of a lack of relationship between soil C and productivity is Ludhiana vs Pantnagar. Soil C in Ludhiana is known to be remarkably low (3–5 g kg^{-1}) compared with high C in Pantnagar (10–14 g kg^{-1}). This large difference in soil C is also reflected in higher system (rice + wheat) productivity in unfertilized long-term plots in Pantnagar than in Ludhiana. On average, system productivity without fertilizer in Pantnagar was twice that of Ludhiana (Fig. 3–4). However, the productivity and response to NPK at the Ludhiana site were much higher than in Pantnagar. To examine the reason for the higher productivity despite much lower soil C in Ludhiana than in Pantnagar, we simulated the long-term (1984–1995) climatic potential yield of rice and wheat and analyzed the weather data of these two sites. On average, the potential system yield of Ludhiana was 20% higher than that of Pantnagar, which could be attributed to increased amounts of solar radiation received at Ludhiana compared with Pantnagar. The difference in solar radiation received in Ludhiana and Pantnagar is greater during rice than in wheat (Pathak et al., 2003b). Moreover, the daily minimum temperature during wheat is lower at Ludhiana, thus favoring crop growth. During rice, the minimum temperature is 0.4°C higher in Ludhiana, but an increase in solar radiation results in higher yield. Thus, simulated potential yield partly explains the fact that, in spite of the greater fertility status in soil of Pantnagar than in Ludhiana, the actual yield in Ludhiana can be higher.

The comparison of results from the LTE highlights the complexity of the relationship between soil C and productivity. Undoubtedly, high SOM is critical for

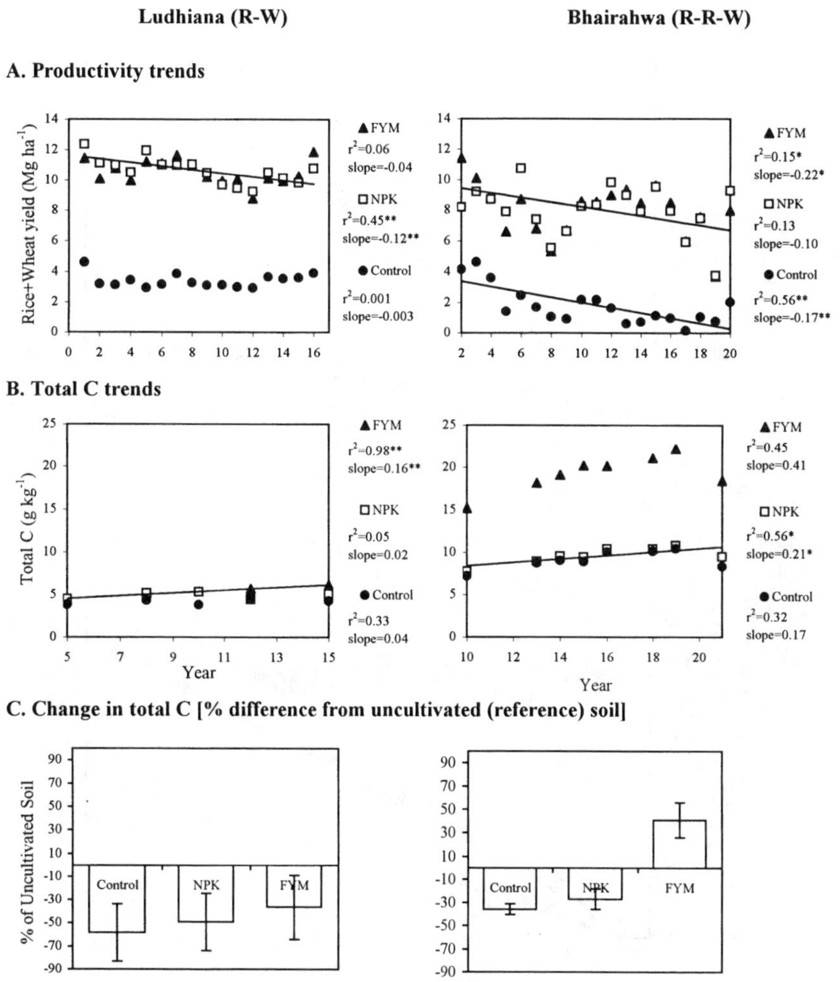

*, ** significant at $P = 0.05$ and $P = 0.01$ respectively.

Fig. 3–3. Changes in productivity and soil C content in unfertilized, NPK, and farmyard manure (FYM) treatments in long-term rice–wheat experiments at Ludhiana, India and Bhairahwa, Nepal.

attaining yields close to potential yields, but other factors should also be optimal (Ladha et al., 2003). Our understanding about the role of SOM in the rice–wheat system is still inadequate. There is a need to understand and quantify the role of soil C, particularly how it may change in alternating anaerobic and aerobic environments, and to develop suitable biochemical and physical indices to monitor changes in the quality of SOM in relation to the performance of a modern agroecosystem.

A. Ludhiana

B. Pantnagar

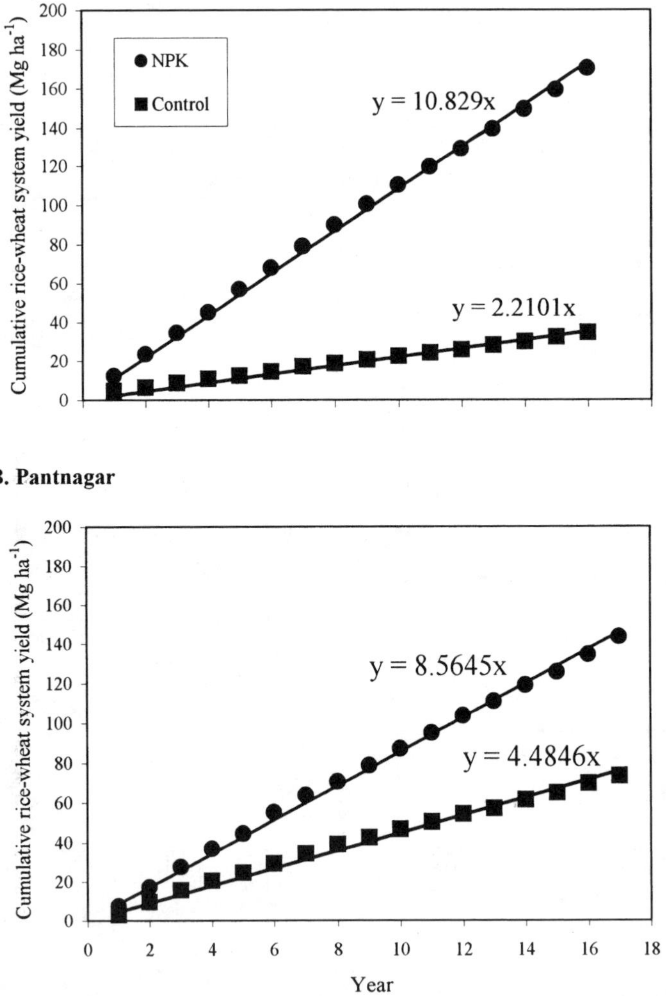

Fig. 3–4. Cumulative rice–wheat system yields in Ludhiana and Pantnagar, India.

Nutrient Depletion and Imbalances

Diagnostic surveys conducted by regional universities and international agencies indicated that in recent years higher amounts of fertilizer need to be applied to get the same yield as that achieved in the 1970s and early 1980s (Harrington et al., 1993; Hobbs et al., 1996). Crops remove substantial quantities of major nutrients from soil, particularly in high-output intensive rice–wheat systems. Sinha et al. (1998) indicated that with time the agriculture of the northwestern region of India has been removing more nutrients than the amount added externally through fertilizer. It is therefore not surprising that farmers would eventually need to apply higher amounts of nutrients per unit of yield.

There could also be a case of declining nutrient use efficiencies over time (Aggarwal et al., 2000a). Nutrient use efficiencies generally plateau and then decrease as the level of fertilizer application increases (Pathak et al., 2003a). For this reason, in farmers' fields where the rate of fertilizer application to rice and wheat crops is now close to 150 kg N ha^{-1} for each crop, it is not surprising to see a marked decline in N use efficiency.

Nutrient balance estimates indicate that the recommended rates of NPK either balanced or exceeded nutrient removal for P, but that large negative balances occurred with K (Ladha et al., 2003). While only 3% of the LTE had a negative P balance, others had either a neutral or positive balance (Fig. 3–5). In 91% of the LTE, applications of K fertilizer were not sufficient to sustain a neutral K input–output balance. All the LTE with a significant yield decline had large negative balances of K. The P and K balances were consistent with the changes in soil P and K status in Ludhiana and Bhairahwa. In these two places, soil K declined by 62 and 33%, respectively, after about 10 yr of cultivation (Bhandari et al., 2002; Regmi et al., 2002b). Large K deficits and declining soil K pools in all the experiments examined would indicate that K is a major contributing factor to a yield decline or stagnation. Duxbury et al. (2000) reported a large K deficit in several rice–wheat LTE that they analyzed, and they also reported a prevalent K deficiency in the farmers' fields in the region. In an LTE at Bhairahwa, a large response to K and a reversal of the yield decline, particularly of wheat, were apparent when additional K was applied to the main LTE plots (Regmi et al., 2002a). Regmi et al. (2002a) also reported a K deficiency enhancing the incidence of *Helminthosporium* leaf blight (*Helminthosporium* spp. *Cochliobolus sativus*) and yield losses of wheat.

In a LTE at Ludhiana (Ludhiana 1), a gradual depletion of soil N together with a decline in plant-available N also appear to be a cause of a yield decline of both rice and wheat (Bhandari et al., 2002). A contrasting trend of total soil C and N in this rice–wheat LTE seems to indicate unique C–N dynamics. Bhandari et al. (2002) hypothesized that the continuous addition of rice and wheat residue (root and stubble of high C–N) to the soil together with marked seasonal changes in water regime and temperature disturbed the C–N mineralization–immobilization balance. In a well-drained, sandy rice–wheat soil with high pH, where leaching and volatilization are major pathways (Bijay-Singh et al., 1991; Katyal et al., 1985), mineral N may be readily lost, but C is retained. The total soil N pool was maintained, however, in the treatment where FYM partially replaced inorganic NPK, suggesting that the application of organic matter, such as FYM, is critical for maintaining soil fertility in the sandy soils of Punjab, India.

Deficiency of S and micronutrients could also be a cause of a yield decline as there are reports of isolated micronutrient deficiencies in the IGP, particularly where the soils are calcareous (Duxbury et al., 2000). The majority of the LTE used single superphosphate as a source of P in the NPK treatment; therefore, the deficiency of S can be ruled out as the cause of a yield decline because single superphosphate contains a considerable amount of S. In some of these LTE, blanket applications of micronutrients were made either once in 3 yr or whenever the deficiency symptoms appeared (Bhandari et al., 2002). A declining trend in available Zn content of soils causing a significant decline in the yield of rice and wheat at

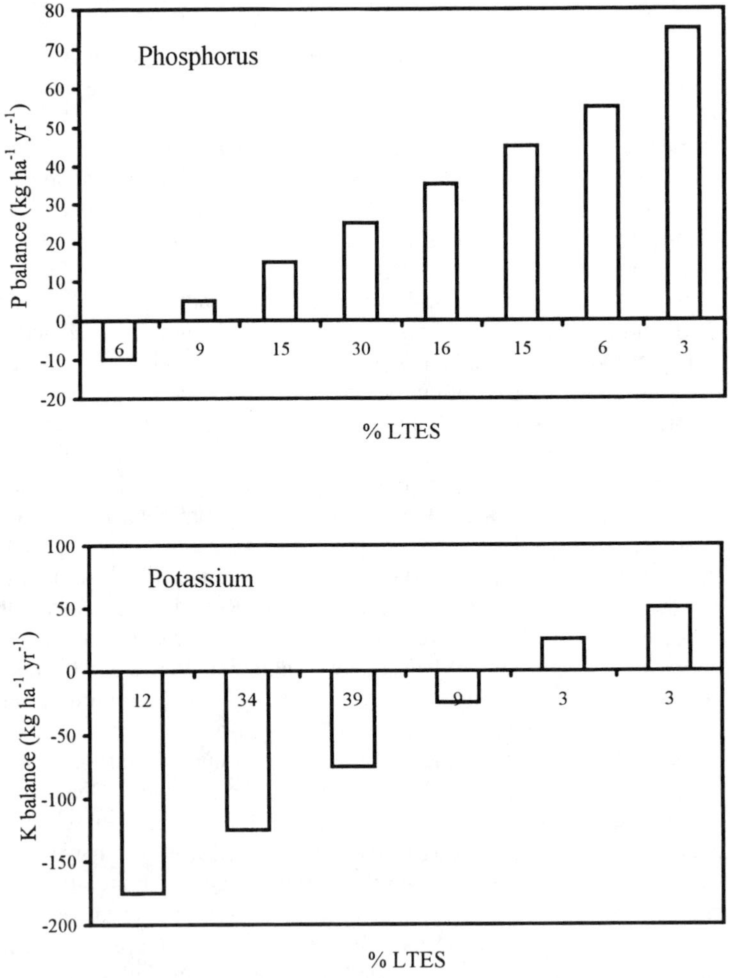

Fig. 3–5. Apparent P and K balances in 33 long-term experiments in Asia.

recommended levels of NPK was reported earlier from the IGP of India (Nambiar, 1988). Sakal (2000) reported that significant yield declines in the Pusa LTE were arrested when Zn was applied along with the NPK, indicating that Zn deficiency was the cause of the yield decline. The more stable yield with the FYM treatment compared with the NPK treatment, as discussed above, could partially be ascribed to the supply of micro- and secondary nutrients with FYM. However, actual measurements of micro- and secondary nutrients in soil and plant have been done in only a limited number of LTE, making it difficult to identify their deficiency as a cause of the yield decline (Duxbury et al., 2000). Although detailed soil and plant analyses will be required to further evaluate the role of these nutrients in yield stability, indirect evidence suggests that the deficiency of such elements was not widespread.

When explaining on-farm rice yields, Pathak et al. (2003b) highlighted soil nutrient depletion (particularly K), a decline in soil C, imbalanced fertilizer use, and a delay in planting as the primary reasons for a yield decline. Sinha et al. (1998) and Singh (2000) observed a marked decline in soil fertility in Punjab and Haryana, India, which are dominated by the rice–wheat system. Inherently, the soils were low in organic C and N, which declined further during the last two decades. For example, data from soil-testing laboratories in Haryana showed that only about 3% of the soil samples had a low Olsen P (<10 kg ha^{-1}) in 1980, which increased to 73% by 1995 (Sinha et al., 1998). A positive P balance is usually observed in LTE as well as in some farmers' fields in Punjab. This positive P balance in farmers' fields has been observed very recently. Likewise, 91% of the soil samples had a high available K content (>280 kg ha^{-1}) in 1980, which decreased to 61% in 1995. A decline in available soil P was also observed in Punjab, Pakistan (Ali & Byerlee, 2000). Ghuman and Narang (1998) reported widespread Zn deficiency in Punjab and Haryana, India. About 74% of the soils of Punjab are reported to be deficient in Zn, 9.5% in Fe, and 1% in Cu.

Soil Salinity and Other Changes in Soil Properties

The use of poor-quality water for irrigation and increasing soil salinity could also contribute to a yield decline. Ali and Byerlee (2000) observed an increase in total soluble salts and alkalinity in soils in the rice–wheat system of Punjab, Pakistan. Similar increases in salinity were reported for Punjab and Haryana in northwestern India because of the use of poor-quality (saline) water for irrigation (Sinha et al., 1998). Singh (1994) also observed that more areas are being salinized through the excessive use of canal water and irrigation with saline water in the states of Punjab, Haryana, and Uttar Pradesh in India. Presently, IGP farmers use poor-quality groundwater and run a high a risk of increasing soil salinity. However, in the LTE, either surface water or good-quality groundwater was used for irrigation, and hence salinity was not a reason for a yield decline.

Related changes in soil physical properties, the formation of hardpans in subsoil, the accumulation of toxic substances, and potential changes in soil microbial biomass have not been studied to a great extent; therefore, it is difficult to judge their contribution to a yield decline.

Variety

During the course of the LTE, rice and wheat varieties changed as a result of the availability of improved germplasm (Ladha et al., 2003). Enough evidence suggests that continuing efforts in selection and breeding have improved the adaptation of both rice and wheat varieties to changing biotic and abiotic stresses. In fact, there is also evidence of improvement in the genetic potential of wheat. A genetic gain of 16 kg Mg^{-1} yr^{-1} has been reported in yield since the mid 1960s (Morris et al., 1994; Nagarajan, 1998) largely because of the higher number of grains per spike and grains per square meter (Sankaran et al., 2000). The potential yield of rice has remained at the same level since the late 1960s, though disease and insect resistance and yield stability have improved (Aggarwal et al., 1997; Peng et al., 1999, 2000). More recently, varieties are also showing better adaptation to the biotic and abiotic constraints in intensive rice systems (Peng et al., 2000). This discussion basically rules out a possible breakdown in the genetic potential of a crop variety as a possible reason for a yield decline, and it could be argued that, if genetic improvement had not occurred, the yields of both rice and wheat would have declined even more.

Delay in Crop Sowing

Another possible cause of a yield decline or low yield, especially of wheat, is a delay in sowing. Normally, delays can be attributed to nontrafficability of fields at the time when sowing is considered to be optimal for maximum yield. It has been shown that a delay in sowing of wheat results in a yield loss because of a rise in temperature during grain filling. Although this problem is very common in farmers' fields throughout the IGP, it was also a constraint in some LTE where three crops were grown. At Bhairahwa, the yield of wheat declined because of a gradual delay in sowing. This yield loss was estimated to be 0.04 Mg ha^{-1} d^{-1} of delay (Hobbs et al., 1996; Regmi et al., 2002b). When actual yields were adjusted for the yield reduction caused by a delay in sowing, no yield decline occurred. It is suspected that the much lower wheat yields in the LTE at Parwanipur were also caused by a general delay in wheat sowing (Gami et al., 2001). Hobbs and Gupta (2003, Chapter 7 of this publication) reported that the introduction of no tillage in the rice–wheat system has a beneficial effect on soil structure and significantly reduces the likelihood that sowing will be delayed.

Climate Change

As yields in some of the most productive regions are approaching a plateau or even declining, the likely effect of climate change on crop production adds to the already complex problem. Concern is widespread about the possible climate change caused by the increase in the concentration of greenhouse gases such as CO_2, CH_4, and N_2O in the atmosphere. To study the effects of climate change on yield, weather data along the transects of the IGP were analyzed across several years (Pathak et al., 2003b). The significant decline in climatic potential rice yields at Ludhiana, Karnal, Delhi, and 24-Pargana (Table 3–4) was the result of a decrease in solar radiation, while at Ludhiana it was also partly the result of an increase in minimum temperature. The decrease in radiation and increase in minimum temperature influence rice yields through decreased photosynthesis, increased respiration, and a shortened vegetative and grain-filling period. (Yoshida & Parao, 1976; Penning de Vries, 1993; Horie et al., 1995).

Hundal and Kaur (1996) also observed that increasing temperature and decreasing radiation levels were the reasons for a decline in simulated potential yields of rice and wheat at Ludhiana. It is widely perceived that, in all the major cities of India, increased aerosol concentrations have contributed to decreased solar radiation and consequently increased minimum temperature (Hundal & Kaur, 1996; Sinha et al., 1998; Aggarwal et al., 2000c). Crop residue burning is widespread in the IGP (particularly in IGP Transects 2 and 3) and, while the CO_2 produced from burning is not considered to be a greenhouse gas contributor under these circumstances, other gases (e.g., N_2O and CH_4) are considered in greenhouse gas budgets and in turn can influence climate (Miura & Kanno, 1997; Kukla & Karl, 1993).

Pest Problems

A change in the spectrum of pests is possible with a change in agroecosystems such as that brought about by the intensification of the rice–wheat area of northwestern India. If new pests evolve or existing ones become more virulent, they may have a significant deleterious effect on yield. Aggarwal et al. (2000a) reported that shootfly[*Atherigona soccata* (Rondani)], brown mite [*Bryobia rubrioculus* (Scheuten)], aphids (*Diuraphis tritici*), and pink borer [*Sesamia nonagrioides* (Lefebvre)] were the most important insect pests of wheat in rice–wheat, and there has been a considerable rise in their mean incidence since the 1980s. Similarly, the incidence of pests such as leaffolder (*Cnaphalocerus medinalis*), whitebacked planthopper (*Sogatella furcifera*), green leafhopper (*Nephotettix* spp.), and Hispa [*Dicladispa armigera* (Olivier)], the important pests of rice, rose in the 1980s but declined in the 1990s. But no such changes were observed for the diseases of rice or wheat in the rice–wheat system. Sinha et al. (1998) also indicated that the incidence of diseases such as false smut [*Ustilaginoidea virens* (Cooke) Takah]], sheath rot [*Sarocladium oryzae* (Sawada) Gums and Hawksworth], and sheath blight (fungus—*Rhizoctonia solani* Kühn) has increased over the years. Among weeds, there has been a significant increase in the *Phalaris minor* population over time (Sinha et al., 1998). There is a growing recognition now that resistance to herbicides has appeared in large areas under the rice–wheat sequence and that this is a real threat to wheat production (Malik et al., 1998). These factors might have contributed to the on-farm yield decline, but not in the LTE, as pests and diseases were regularly controlled on the research farm.

Water Management

In the IGP, water availability for irrigation is decreasing because of (i) the increased area under rice, which reduces the availability per unit area (Sinha et al., 1998), (ii) a reduced water supply through canal irrigation, which is linked to less capacity of reservoirs because of increasing siltation, and (iii) increasing competition for water for domestic and industrial use (Hobbs & Gupta, 2003, Chapter 7 of this publication). Water will become a major constraint to agriculture in the future as demand from domestic and industrial use increases and new irrigation facilities are not created. Soils of the upper transect of the IGP are light- to medium-textured, and a large amount of water (>250 cm) is required for growing rice (Narang & Singh, 1988; Aggarwal et al., 2000c). Since the rainfall is not sufficient to meet the water requirement, frequent irrigation is needed. Canal water accounts for 35 to 40% of the total irrigation requirement, and the remaining is met from groundwater. This has resulted in a decline in the water tables and water quality in many regions (Sinha et al., 1998; Singh, 2000). Progressive farmers in this region have started planting summer rice (April–June) along with rice and wheat in the main season (Aggarwal et al., 2000b). The additional groundwater required for summer rice can further imbalance the groundwater budget. Pumping from deeper layers increases the total operational costs and thus decreases the profitability of growing rice. A long-term view on water policy is urgently needed for sustainable management of the rice–wheat system.

Another aspect of water management in the Trans-Gangetic Plains in India is waterlogging. An impressive canal network has made the intensive rice–wheat cropping system possible, but with water percolation from canals, distributaries, and water courses, the water table is rising in some areas. The use of excess amounts of irrigation water is another cause of water-table rise. Ghuman and Narang (1998) reported that, in Haryana, the water table is rising at 0.14 to 1.0 m yr^{-1}, and more than 0.4 million ha of land (of the total geographic area of about 4.0 million ha) has the water table within 3 m of the soil surface. Similarly, in the low-lying poorly drained areas of the Faridkot district of Punjab, India, a steady rise in the water table is occurring. Singh (1994) observed that the areas prone to waterlogging are increasing in Punjab, Haryana and Uttar Pradesh, India.

In summary, we argue that there is not a general cause for the yield decline across the IGP based on evidence from the LTE; rather, the causes are location-specific. Depletion of soil C, N, K, and Zn; the reduced availability of P; the delay in planting; decreases in solar radiation; and increases in minimum temperature are the primary influences on the yield decline in one location or another. Depletion of soil K, however, seems to be relatively more common. We consider the lack of historical soil and plant samples to be a major deterrent to identifying the specific cause of the yield decline. Therefore, more detailed data collection and archives of soil and plant samples should be maintained to unravel the cause–effect relationship.

YIELD GAP AND CAUSES OF YIELD GAP

The yields of both rice and wheat in most LTE as well as in farmers' fields are low (Tables 3–1 and 3–4). Although it is important to be concerned about the yield decline, it is equally or perhaps more important to be aware of yield gaps. Using the crop simulation models described above, we estimated yield gaps across the IGP and outside the IGP (except China). We estimated two types of yield gap. Yield Gap 1 is the difference between the climatic potential yield (Table 3–7) calculated by the model and the actual yield obtained at the experimental station. The average yields from the LTE were used as the actual yields on the experimental plot. The second gap (Yield Gap 2) is the difference between the climatic potential yield and the actual yield in farmers' fields.

Yield Gap 1 ranged from 34 to 63% for rice and from 31 to 55% for wheat across the IGP. Yield Gap 2 ranged from 48 to 74% for rice and from 35 to 66% for wheat (Table 3–7). While the principal causes of Yield Gap 1 are inadequate and inefficient crop and input management, various other factors (summarized in Table 3–8) probably cause what we consider to be Yield Gap 2.

In summary, the yield gap analysis suggests tremendous scope for improving yield by improved crop management. Although breeders must continue to focus on increasing potential yield to increase overall productivity, the most obvious route to rapid increases in production in the IGP is through research that concentrates efforts on overcoming soil biological and chemical constraints.

CONCLUSIONS

The yields of rice and wheat in the LTE of the IGP have remained either stagnant or to some extent have declined. If similar stagnation or declining yield trends occur in farmers' fields, the sustainability of this important cropping system will be threatened. A declining trend raises serious concerns about the region's food security because a "population plateau" is yet to be realized and may not be realized for many years. We have attempted to answer several questions pertaining to yield stagnation or decline in this region:

Is the present fertilizer application inadequate for higher yield? This appears to emerge as the most obvious cause of the yield decline and lower yield. However, this is not a matter of serious concern because this can be readily remedied. We argue that the present fertilizer recommendation needs revision for N and K and that N management needs improvement to achieve higher efficiency (Gami et al., 2001; Bhandari et al., 2002; Regmi et al., 2002a,b). Using N fertilizer efficiently reduces environmental problems and costs of pollution.

Is the decline in inherent SOM quantity, and perhaps quality, a reason for the yield decline and our inability to achieve yields close to the climatic potential yields? Many farmers and scientists believe that the high yields of rice and wheat they had at the start of the Green Revolution cannot be achieved with any amount of nutrient inputs. Unfortunately, because of the lack of systematic sampling and archiving of soil samples in the LTE, understanding of the role of SOM in the region is inadequate. The LTE data that do exist are consistent with the widespread observation that SOM will decline under conventional management and removal of aboveground biomass, but we do not know to what extent the quality of the SOM has declined. Soil organic matter quality does not refer to the absolute amount of nutrient being supplied per unit of SOM, but to changes in nutrient-supplying capacity of the soil, which is regulated by a combination of soil biological, chemical, and physical properties that are of crucial importance, particularly in light-textured soils. This may be a serious concern, as SOM and soil quality, once degraded, are difficult to restore.

Table 3–7. Potential, on-station, and on-farm yields and yield gaps of rice and wheat in different zones of the Indo-Gangetic Plains (IGP) and non-IGP.

		Rice					Wheat				
					Yield gap (%)					Yield gap (%)	
IGP zone	Site	Potential yield (*A*)	On-station† (*B*)	On-farm‡ (*C*)	1 (*A* – *B*)/*A*100	2 (*A* – *C*)/*A*100	Potential yield (*D*)	On-station† (*E*)	On-farm‡ (*F*)	1 (*D* – *E*)/*D*100	2 (*D* – *F*)/*D*100
2	Ludhiana	10.7	5.6	5.6	48	48	7.9	4.7	4.3	41	46
2	Karnal	10.4	6.8	3.8	35	63	7.3	4.6	3.6	37	51
3	Kanpur	9.5	4.5	2.8	53	71	7.0	4.6	2.8	35	60
3	Pantnagar	9.0	5.5	4.2	39	53	6.5	3.9	4.2	40	35
4	Varanasi	9.2	4.1	3.2	55	65	7.0	3.8	3.2	46	54
4	Faizabad	9.1	4.2	2.8	54	69	6.7	3.4	2.8	49	58
5	Dinajpur	8.6	3.2	3.1	63	64	5.4	3.8	2.3	31	57
5	24-Pargana	7.7	4.4	2.8	43	64	5.2	3.0	2.8	43	46
Non-IGP	Raipur	8.2	5.4	2.1	34	74	6.2	2.8	2.1	55	66

† Yield with the recommended level of NPK in the long-term experiments.
‡ Source: Fertilizer Association of India Regional Statistics (1986–2000), Statistical Yearbook of Bangladesh (2001).

Table 3–8. Causes of gap between potential and farmers' yields in the Indo-Gangetic Plains (IGP).

Factor	IGP 1 and 2	IGP 3	IGP 4	IGP 5
Water	Shortage of good quality irrigation water and low water use efficiency	Shortage of irrigation water	Shortage of irrigation water and occasional flood and drought	Water logging, flood, excessive soil moisture in wheat
Soil	Low soil organic matter, rising salinity	Low soil organic matter, rising salinity, alkalinity	Nutrient mining	Nutrient mining, rising salinity
Fertilizer	Inefficient N use, imbalanced fertilizer use	Inefficient N use, imbalanced fertilizer use	Low fertilizer use	Imbalanced fertilizer use
Crop management	Delay in sowing of wheat	Poor land leveling, low plant density in rice, late sowing of wheat	Nonadoption of high yielding varieties, poor land leveling, low mechanization	Low mechanization, poor land leveling
Climate	--	Rain during maturity of rice	Rise in temperature during grain filling of wheat, shorter wheat season, rain and storm during maturity of rice	Rain and storm during maturity in rice and wheat, shorter wheat season
Pest	Low	Moderate	Severe	Severe

Has the climate worsened and thus reduced yield? This is the most crucial among the issues of yield decline because it is beyond the immediate control of local farmers. It is more in the hands of international and national policymakers (e.g., the Kyoto Protocol). The analysis showed some disturbing trends of a decrease in solar radiation and an increase in minimum temperature at a few sites of the IGP for which the climate data were available. The short-term changes in radiation and temperature may be related to an increase in aerosol concentration in the atmosphere; however, the evidence in support of global climate changes and the effect on regional productivity is increasing. Nevertheless, there is an urgent need to further investigate the impact of these issues on production and how farming communities may adapt, and also to help mitigate climate change.

Has the genetic potential of the new cultivars remained the same as that of the cultivars of the 1960and 1970s or broken down despite the all-out efforts to break the yield barrier? Though the existing LTE and available data do not allow us to analyze this possibility, some published work suggests that the adaptation of new cultivars has improved. In such a case, the LTE yields will probably increase with current input use and crop and soil management. We argue that if that does not happen, yield would have further negative trends. This suggests the importance of maintenance breeding to improve the adaptation of varieties.

Are rice yields in the rice–wheat system more susceptible to a decline than wheat yields? We observed that, in the rice–wheat system, rice yields have more negative trends than wheat yields. In contrast to these results, it is often perceived that wheat yields suffer more after rice because of frequent puddling and the formation of a hard pan (Sharma & De Datta, 1986; Humphreys et al., 1994; Bhushan & Sharma, 1999). On the other hand, rice–rice systems are considered to be more sustainable (Greenland, 1997; Dawe et al., 2000). However, it is important to note that the soils in rice–rice generally are fine textured with a high SOM content, while alluvial soils of the IGP in rice–wheat are coarse to medium textured with a generally low SOM content and much lower nutrient supplying capacity. We hypothesize that these soils will be depleted of nutrients more rapidly. If nutrient application is insufficient, then rice, because of its shallower root system, would probably suffer more nutrient deficiency than wheat.

The present analyses suggest that the yield decline in the LTE is not widespread and therefore may not be of serious concern, but the widespread yield stagnation and large yield gap both at the experimental station and in farmers' fields is to be taken seriously. There is still scope for improving yield by suitable crop management practices. Research must now focus on increasing the potential yield of cultivars to increase overall productivity and develop tools to increase input use efficiency and close the yield gap.

REFERENCES

Abrol, I.P., K.F. Bronson, J.M. Duxbury, and R.K. Gupta. 2000. Long-term soil fertility experiments in rice–wheat cropping systems. Rice–Wheat Consortium Paper Series No. 6. Rice–Wheat Consortium for the Indo-Gangetic Plains, New Delhi, India.

Aggarwal, P.K., S.K. Bandyopadhyay, H. Pathak, N. Kalra, S. Chander, and S. Sujith Kumar. 2000a. Analyses of yield trends of the rice–wheat system in north-western India. Outlook Agric. 29:259–268.

Aggarwal, P.K., N. Kalra, S.S. Kumar, H. Pathak, S.K. Bandyopadhyaya, A.K. Vasisth, R.P. Roetter, and C.T. Hoanh. 2000b. Haryana State case study: Trade-off between cereal production and environmental impact. p. 11–18. *In* R.P. Roetter et al. (ed.) Synthesis of methodology development and case studies. SysNet Research Paper Series No. 3. IRRI, Manila, Philippines.

Aggarwal, P.K., M.J. Kropff, K.G. Cassman, and H.F.M. ten Berge. 1997. Simulating genetic strategies for increased yield potential in irrigated, tropical environments. Field Crops Res. 51:5–18.

Aggarwal, P.K., K.K. Talukdar, and R.K. Mall. 2000c. Potential yields of rice–wheat system in the Indo-Gangetic plains of India. Rice–Wheat Consortium Paper Series No.10. Rice–Wheat Consortium for the Indo-Gangetic Plain, New Delhi, India, and Indian Agricultural Research Institute, New Delhi, India.

Ali, M., and D. Byerlee. 2000. Productivity growth and resource degradation in Pakistan's Punjab: A decomposition analysis. Policy Research Working Paper 2480. World Bank, Washington, DC.

Bhandari, A.L., J.K. Ladha, H. Pathak, A.T. Padre, D. Dawe, and R.K. Gupta. 2002. Yield and soil nutrient changes in a long-term rice–wheat rotation in India. Soil Sci. Soc. Am. J. 66:167–170.

Bhushan, L., and P.K. Sharma. 1999. Effect of depth, bulk density and aeration status of root zone on productivity of wheat. J. Indian Soc. Soil Sci. 47:29–34.

Bijay-Singh, Y. Singh, C.S. Khind, and O.P. Meelu. 1991. Leaching losses of urea-N applied to permeable soils under lowland rice. Fert. Res. 28:179–184.

Brar, B.S., Y. Singh, N.S. Dhillon, and B. Singh. 1998. Long-term effects of inorganic fertilizers, organic manure and crop residues on the productivity and sustainability of a rice–wheat cropping system in northwest India. p. 335. *In* A. Swarup et al. (ed.) Long-term soil fertility management through integrated plant nutrient supply. Indian Institute of Soil Science, Bhopal, India.

Byerlee, D., and A.K. Siddiq. 1994. Has the green revolution been sustained: The quantitative impacts of the seed-fertilizer technology in Pakistan revisited. World Dev. 22:1345–1361.

Cassman, K.G., S.K. De Datta, D.C. Olk, M.J. Alcantara, M.I. Samson, J. Descalsota, and M.A. Dizon. 1995. Yield decline and the nitrogen economy of long-term experiments in continuous, irrigated rice–wheat systems in the tropics. p. 181–122. *In* R. Lal and B.A. Stewart (ed.) Soil management: Experimental basis for sustainability and environmental quality. Lewis Publ., Boca Raton, FL.

Cassman, K.G., and P.L. Pingali. 1995. Extrapolating trends from long-term experiments to farmers' fields: The case of irrigated rice systems in Asia. p. 63–86. *In* V. Barnett et al. (ed.) Agricultural sustainability in economic, environmental, and statistical terms. John Wiley & Sons, London, UK.

Dawe, D., A. Dobermann, P. Moya, S. Abdulrachman, P. Lal, S.Y. Li, B. Lin, G. Panaullah, O. Sariam, Y. Singh, A. Swarup, P.S. Tan, and Q.X. Zhen. 2000. How widespread are yield declines in long-term rice experiments in Asia? Field Crops Res. 66:175–193.

Duxbury, J.M., I.P. Abrol, R.K. Gupta, and K.F. Bronson. 2000. Analysis of long-term soil fertility experiments with rice–wheat rotations in South Asia. p. vii–xxii. *In* I.P. Abrol et al. (ed.) Long-term soil fertility experiments with rice–wheat rotations in South Asia. Rice–Wheat Consortium Paper Series No. 6. Rice–Wheat Consortium for the Indo-Gangetic Plains, New Delhi, India.

Fertilizer Association of India. 1986–2000. Fertilizer Association of India Regional Statistics. FAI, New Dehli, India.

Gami, S.K., J.K. Ladha, H. Pathak, E. Pasuquin, and D. Joshy. 2001. Long-term changes in yield and soil fertility in a twenty-year rice–wheat experiment in Nepal. Biol. Fertil. Soils 34:73–78.

Ghuman, B.S., and R.S. Narang. 1998. Sustainable management of soil and water resources in the upper and Trans-Gangetic plains of India. p. 140–154. *In* R. Lal (ed.) Soil quality and agricultural sustainability. Ann Arbor press, Chelsea, MI.

Greenland, D.J. 1997. The sustainability of rice farming. CAB Int. and IRRI, Manila, Philippines.

Gupta, R.K., R.K. Naresh, P.R. Hobbs, Z. Jiaguo, and J.K. Ladha. 2003. Sustainability of post-Green Revolution agriculture: The rice–wheat cropping systems of the Indo-Gangetic Plains and China. p. 1–26. *In* J.K. Ladha et al. (ed.) Improving the productivity and sustainability of rice–wheat systems: Issues and impacts. ASA Spec. Publ. 65. ASA, CSSA, SSSA, Madison, WI.

Harrington, L.W., S. Fujisaka, P.R. Hobbs, C. Adhikary, G.S. Giri, and K. Cassaday. 1993. Rice–wheat cropping systems in Rupandehi District of the Nepal Terai: Diagnostic survey of farmers' practices and problems, and needs for further research. Mimeograph. NARC/CIMMYT/IRRI, Mexico City, Mexico.

Hobbs, P.R., and R.K. Gupta. 2003. Resource-conserving technologies for wheat in the rice–wheat system. p. 149–172. *In* J.K. Ladha et al. (ed.) Improving the productivity and sustainability of rice–wheat systems: Issues and impacts. ASA Spec. Publ. 65. ASA, CSSA, and SSSA, Madison, WI.

Hobbs, P.R., L.W. Harrington, C. Adhikari, G.S. Giri, S.R. Upadhya, and B. Adhikari. 1996. Wheat and rice in the Nepal Tarai: Farm resources and production practices in Rupandehi District. NARC and CIMMYT, Mexico City, Mexico.

Hobbs, P.R., and M.L. Morris. 1996. Meeting South Asia's future food requirements from rice–wheat cropping systems: Priority issues facing researchers in the post green revolution era. p. 1–45. NRG Paper 96-01. CIMMYT, Mexico City, Mexico.

Horie, T., H. Nakagawa, H.G.S. Centeno, and M.J. Kropff. 1995. The rice crop simulation model SIMRIW and its testing. p. 51–66. *In* R.B. Matthews et al. (ed.) Modeling the impact of climate change on rice production in Asia. CAB International and IRRI, Manila, Philippines.

Humphreys, E., A. van der Lely, W.A. Muirhead, and D. Hoey. 1994. The development of on-farm restrictions to minimize recharge from rice in New South Wales. Aust. J. Soil Water Conserv. 7:11–20.

Hundal, S.S., and P. Kaur. 1996. Climatic change and its impact on crop productivity in Punjab, India. p. 377–393. *In* Y.P Abrol et al. (ed.) Climatic variability and agriculture. Narosa Publishing House, New Delhi.

Katyal, J.C., Bijay Singh, P.L.G. Vlek, and E.T. Crasswell. 1985. Fate and efficiency of nitrogen fertilizers applied to wetland rice. II. Punjab, India. Fert. Res. 6:279–290.

Kukla, G., and T.R. Karl. 1993. Nighttime warming and greenhouse effect. Environ. Sci. Technol. 27:1469–1474.

Kumar, A., and D.S. Yadav. 2001. Long-term effects of fertilizers on the soil fertility and productivity of a rice–wheat system. J. Agron. Crop Sci. 186:47–54.

Ladha, J.K, D. Dawe, H. Pathak, A.T. Padre, R.L. Yadav, Bijay-Singh, Yadvinder-Singh, Y. Singh, P. Singh, A.L. Kundu, R. Sakal, N. Ram, A.P. Regmi, S.K. Gami, A.L. Bhandari, R. Amin, C.R. Yadav, E.M. Bhattarai, S. Das, H.P. Aggarwal, R.K. Gupta, and P.R. Hobbs. 2003. How extensive are yield declines in long-term rice–wheat experiments in Asia? Field Crops Res. (In press).

Ladha, J.K., K.S. Fischer, M. Hossain, P.R. Hobbs, and B. Hardy. 2000. Improving the productivity and sustainability of rice–wheat systems of the Indo-Gangetic Plains: A synthesis of NARS–IRRI partnership research. p. 1–31. Discussion Paper No. 40. IRRI, Manila, Philippines.

Malik, R.K., G. Gill, and P.R. Hobbs. 1998. Herbicide resistance: A major issue for sustaining wheat productivity in rice–wheat cropping systems in the Indo-Gangetic Plains. Paper Series No. 3. Rice–Wheat Consortium, New Delhi, India.

Miura, Y., and T. Kanno. 1997. Emissions of trace gases (CO_2, CO, CH_4, and N_2O) resulting from rice straw burning. Soil Sci. Plant Nutr. 43:849–854.

Morris, M.L., H.J. Dubin, and T. Pokhrel. 1994. Returns to wheat breeding research in Nepal. Agric. Econ. 10:269–282.

Nagarajan, S. 1998. Perspectives on wheat demand and research needs. p. 1–12. *In* S. Nagarajan et al. (ed.) Wheat research needs beyond 2000 AD. Narosa, New Delhi, India.

Nambiar, K.K.M. 1988. Annual reports, long-term fertilizer experiments in India (1971–1988). IARI, New Delhi, India.

Nambiar, K.K.M. 1994. Soil fertility and crop productivity under long-term fertilizer use in India. Indian Council for Agricultural research, New Delhi, India.

Nambiar, K.K.M. 1995. Major cropping systems in India. p. 135–142. *In* V. Barnett et al. (ed.) Agricultural sustainability: Economic, environmental and statistical considerations. John Wiley and Sons, London, UK.

Narang, R.S., and N. Singh. 1988. Water requirement studies in rice. IRRI Newsl. Philippines 13:43.

Narang, R.S., and S.M. Virmani. 2001. Rice–wheat cropping systems of the Indo-Gangetic Plain of India. Rice–Wheat Consortium Paper Series 11. Rice–Wheat Consortium for the Indo-Gangetic Plains, New Delhi, and ICRISAT, Patancheru, India.

Pathak, H., P.K. Aggarwal, R. Roetter, N. Kalra, S.K. Bandyopadhaya, S. Prasad, and H. Van Keulen. 2002a. Modeling the quantitative evaluation of soil nutrient supply, nutrient use efficiency, and fertilizer requirements of wheat in India. Nutr. Cycl. Agroecosyst. (In Press).

Pathak, H., J.K. Ladha, P.K. Aggarwal, S. Peng, S. Das, Yadvinder Singh, Bijay-Singh, S.K. Kamra, B. Mishra, A.S.R.A.S. Sastri, H.P. Aggarwal, D.K. Das, and R.K. Gupta. 2003b. Climatic potential and on-farm yield trends of rice and wheat in the Indo-Gangetic plains. Field Crops. Res. (In Press).

Peng, S., K.G. Cassman, S.S. Virmani, J. Sheehy, and G.S. Khush. 1999. Yield potential trends of tropical rice since the release of IR8 and the challenge of increasing rice yield potential. Crop Sci. 39:1552–1559.

Peng, S., R.C. Laza, R.M. Visperas, A.L. Sanico, K.G. Cassman, and G.S. Khush. 2000. Grain yield of rice cultivars and lines developed in the Philippines since 1966. Crop Sci. 40:307–314.

Penning de Vries, F.W.T. 1993. Rice production and climate change. p. 175–189. *In* F.W.T. Penning de Vries et al. (ed.) Systems approaches for agricultural development. Kluwer Academic Publ., Dordrecht, the Netherlands.

Powlson, D.S., P.R. Poulton, and J.L. Gaunt. 1998. The role of long-term experiments in agricultural development. p. 1–15. *In* A. Swarup et al. (ed.) Proc. of a National Workshop on Long-Term Soil Fertility Management through Integrated Plant Nutrient Supply. Indian Institute of Soil Science, Bhopal, India.

Powlson, D.S., G. Pruden, A.E. Johnston, and D.S. Jenkinson. 1986. The nitrogen cycle in Broadbalk wheat experiment–recovery and losses of ^{15}N-labelled fertility applied in springs and inputs of N from the atmosphere. J. Agric. Sci. (Cambridge) 107:591–609.

Ram, N. 2000. Long-term effects of fertilizers on rice–wheat–cowpea productivity and soil properties in Mollisols. p. 50–55. *In* I.P. Abrol et al. (ed.) Long-term soil fertility experiments in rice–wheat cropping systems. Research Series No. 6. Rice–Wheat Consortium, New Delhi, India.

Regmi, A.P., J.K. Ladha, E. Pasuquin, H. Pathak, P.R. Hobbs, L.L. Shrestha, D.B. Gharti, E. Duveiller. 2002a. Potassium in sustaining yields in a long-term rice–wheat experiment in the Indo-Gangetic Plains of Nepal. Biol. Fert. Soils 36:240–247.

Regmi, A.P., J.K. Ladha, H. Pathak, E. Pasuquin, C. Bueno, D. Dawe, P.R. Hobbs, D. Joshy, S.L. Maskey, and S.P. Pandey. 2002b. Yield and soil fertility trends in a 20-year rice–rice–wheat experiment in Nepal. Soil Sci. Soc. Am. J. 66:857–867.

Ritchie, J.T., U. Singh, D.C. Godwin, W.T. Bowen, P.W. Wilkens, B. Baer, G. Hoogenboom, and L.A. Hunt. 1998. GENERIC-CERES V3.5. IFDC, Michigan State University, University of Georgia, USA.

Saha, M.N., A.R. Saha, B.C. Mandal, and P.K. Ray. 2000. Effects of long-term jute–rice–wheat cropping system on crop yields and soil fertility. p. 94–104. *In* I.P. Abrol et al. (ed.) Long-term soil fertility experiments in rice–wheat cropping systems. Rice–Wheat Consortium Paper Series No. 6. Rice–Wheat Consortium for the Indo-Gangetic Plains, New Delhi, India.

Sakal, R. 2000. Long-term effects of N, P and Zn management on crop yields and micronutrient availability in rice–wheat cropping system in calcareous soils. p. 73–82. *In* I.P. Abrol et al. (ed.) Long-term soil fertility experiments in rice–wheat cropping systems. Rice–Wheat Consortium Paper Series No. 6, Rice–Wheat Consortium for the Indo-Gangetic Plains, New Delhi, India.

Samra, J.S., Bijay-Singh, and K. Kumar. 2003. Managing crop residues in the rice–wheat system of the Indo-Gangetic Plain. p. 173–196. *In* J.K. Ladha et al. (ed.) Improving the productivity and sustainability of rice–wheat systems: Issues and impacts. ASA Spec. Publ. 65. ASA, CSSA, SSSA, Madison, WI.

Sankaran, V.M., P.K. Aggarwal, and S.K. Sinha. 2000. Improvement in wheat yields in northern India since 1965: Measured and simulated trends. Field Crops Res. 66:141–149.

Schnitzer, M. 1991. Soil organic matter—the next 75 years. Soil Sci. 151:41–58.

Sharma, P.K., and S.K. De Datta. 1986. Physical properties and processes of puddled rice soils. Adv. Soil Sci. 5:139–178.

Singh, N.T. 1994. Land and water management in north and northwestern India. p. 93–120. *In* Management of land and water resources for sustainable agriculture and environment. Indian Soc. Soil Sci., New Delhi, India

Singh, R.B. 2000. Environmental consequences of agricultural development: A case study from the green revolution state of Haryana. Agric. Ecosyst. Environ. 82:97–103.

Singh, U., D.C. Godwin, J.T. Ritchie, W.T. Bowen, P.W. Wilkens, B. Baer, G. Hoogenboom, and L.A. Hunt. 1998. CERES-RICE V3.5. IFDC, Michigan State University, University of Georgia, USA.

Sinha, S.K., G.B. Singh, and M. Rai. 1998. p. 89. *In* Decline in crop productivity in Haryana and Punjab: Myth or reality? Indian Council of Agricultural Research, New Delhi, India

Statistical Yearbook of Bangladesh. 2001. Statistical Division, Govt. of Bangladesh, Bangladesh.

Timsina, J., and D.J. Connor. 2001. Productivity and management of rice–wheat cropping systems: Issues and challenges. Field Crops Res. 69:93–132.

Tisdale, J.M., and J.M. Oades. 1982. Organic matter and water soluble aggregates in soils. J. Soil Sci. 33:141–163.

Tran, D.V. and J.P. Marathée. 1994. Major issues in Asian rice–wheat production systems: Sustainability of rice–wheat production systems in Asia. RAPA Publ. 11:61–67.

Yadav, R.L., B.S. Dwivedi, and P.S. Pandey. 2000. Rice–wheat cropping system: Assessment of sustainability under green manuring and chemical fertilizer inputs. Field Crops Res. 65:15–30.

Yadav, R.L., D.S. Yadav, O. Singh, and A. Kumar. 1998. Long term effects of inorganic fertilizer inputs on crop productivity in a rice–wheat cropping system. Nutr. Cycl. Agroecosyst. 51:193–200.

Yadvinder-Singh, Bijay-Singh, O.P. Meelu, and C.S. Khind. 2000. Long-term effects of organic manuring and crop residues on the productivity and sustainability of rice–wheat cropping system in Northwest India. p. 149–162. *In* I.P. Abrol et al. (ed.) Long-term soil fertility experiment in rice–wheat cropping systems. Rice–Wheat Consortium Paper Series No. 6. Rice–Wheat Consortium for the Indo-Gangetic Plains, New Delhi, India.

Yoshida, S., and F.T. Parao. 1976. Climatic influence on yield and yield components of lowland rice in the tropics. p. 471–494. *In* Climate and rice. IRRI, Manila, Philippines.

Zheng, J. 2000. Rice–wheat cropping system in China. p. 1–10. *In* P.R. Hobbs and R.K. Gupta (ed.) Soil and crop management practices for enhanced productivity of the rice–wheat cropping system in the Sichuan province of China. Rice–Wheat Consortium Paper Ser. 9. Rice–Wheat Consortium for the Indo-Gangetic Plains, New Delhi, India.

4 Sustainability of the Rice–Wheat System in Pakistan's Punjab: How Large is the Problem?

Derek Byerlee

World Bank
Washington, District of Columbia, USA

Mubarik Ali

Asian Vegetable Research and Development Center
Shanhua, Tainan, Taiwan

Akmal Siddiq

Asian Development Bank
Manila, Philippines

ABSTRACT

Evidence from studies carried out since 1984 on intensification, productivity, and resource degradation in the rice–wheat system of Pakistan's Punjab is reviewed in this chapter. Farm-level surveys in the 1980s showed that the major factors affecting wheat productivity were late planting after Basmati rice, continuous monocropping, weed infestation, inefficient use of fertilizer, and a shortage of irrigation water. For rice, the major factors identified were inappropriate land preparation, low plant density, and a shortage of irrigation water. In addition, a study of wheat showed that the widespread use of poor-quality tubewell water caused sodicity and reduced yields. Decomposition of yield trends since the advent of the Green Revolution also provided strong evidence of the negative effect of resource degradation on yields after accounting for the effects of the adoption of modern varieties and fertilizer. This was confirmed by analysis of soil data, where soil organic matter and soil P were declining, while salinity in soil and water samples was increasing. An analysis of total factor productivity showed a negative trend in the rice–wheat system from 1967 to 1994. Further decomposition of productivity through a production function analysis showed that the effects of resource degradation had cancelled the positive productivity effects of improved technology and investment in infrastructure and literacy. There is evidence in the Pakistan Punjab rice–wheat system that these trends may have been reversed to some extent in the 1990s with the use of earlier maturing varieties, improved weed control, mechanization, and diversification of cropping patterns.

 Improving the Productivity and Sustainability of Rice–Wheat Systems: Issues and Impacts. ASA Special Publication 65.

The introduction of Green Revolution technologies in wheat and rice in Asia in the mid 1960s produced impressive results in reversing the food crisis and stimulating agricultural and economic growth. However, beginning in the 1980s, scientists began to question the sustainability of this intensification strategy in light of concerns about slowing yield growth and degradation of the resource base (Hobbs & Morris, 1996; Cassman & Pingali, 1995). Much of this concern focused on areas where intensive rice–wheat monocropping was being continuously practiced.

To date, there has been little conclusive evidence on the sustainability of current intensification strategies in the rice–wheat system. Although the growth rate of yields has clearly slowed, there is no evidence of an absolute decline in yields in farmers' fields. Broader measures of productivity, such as total factor productivity (TFP), also generally show positive trends, although the evidence is sometimes conflicting (Ali & Byerlee, 2000). However, a positive trend in TFP does not necessarily imply a sustainable system in the long term, since rapid productivity gains from new technologies may mask the effects of serious resource degradation in the short to medium term. To interpret trends in productivity, it is necessary to also track indicators of agroecosystem health, especially measures of soil and water quality (Herdt & Lynam, 1992). Very few studies have provided data over time on trends in agroecosystem health in farmers' fields, and even fewer have attempted to relate productivity trends to such indicators (Lindert, 1999).

The main objective of this chapter is to review cumulative evidence on productivity changes in the rice–wheat system in the Pakistan Punjab, where the first intensive studies began in the mid 1980s. We provide evidence from different stages of intensification during the three decades since the advent of the Green Revolution as well as comparative information on productivity and sustainability in other major cropping systems in the Pakistan Punjab and the neighboring Indian Punjab where similar cropping systems are practiced. The issue is clearly critical to agricultural development strategies and food security in the Indo-Gangetic Plains, the major breadbasket of South Asia.

This review is not exhaustive but covers the main studies in which the authors have participated during the past 15 yr. We first summarize results from the initial studies that were based on farm survey data at one time, and then later studies that use secondary data on productivity trends, input use, and indicators of resource quality to analyze and decompose the components of long-term growth and quantify the effects, if any, of natural resource degradation. Finally, we discuss appropriate methods that can be used to track indicators of system sustainability in farmers' fields.

TRENDS IN THE RICE–WHEAT TRACT IN THE PUNJAB

More than 80% of the cropped area of Punjab Province is irrigated, and approximately 25% of the area is sown in a rice–wheat system. The Punjab was one of the earliest beneficiaries of the Green Revolution with the introduction of modern varieties of wheat and rice in the 1960s. Four cropping systems are commonly recognized in the Punjab, based on the dominant summer (kharif) crop: (i) rice–wheat, (ii) mixed summer crops (often maize or sugarcane)–wheat, (iii) cotton (*Gossypium hirsutum* L.)–wheat, and (iv) mungbean (or fallow)–wheat. The rice–wheat tract of the Punjab is concentrated on about one million hectares of cultivated land in the districts of Sheikhurpura, Gujranwala, and Sialkot. Although the rice–wheat system is often viewed as very intensive, the average cropping intensity is only 147, since land is often left fallow, mainly because of excess water or lack of water. Although the area is largely irrigated, annual rainfall, which varies from 425 to 800 mm, is an important supplement to irrigation water. Soil textures vary from well-drained loams to moderately well-drained and seasonally imperfectly drained clays and silty clays. Farm size averages 8 ha, with rice and wheat grown on 77 and 74% of the cropped area of the three districts in the summer and winter seasons, respectively. With the soils and temperatures, the area is ideally suited for producing high-quality Basmati rice, much of which is exported.

Trends in Input Use and Production at the System Level

Ali and Byerlee (2000) summarized quantitative trends in input and output use in different cropping systems of the Punjab and at different phases of Green Revolution technical change (Byerlee, 1992): the Green Revolution period, 1967 to 1974, when modern varieties were widely adopted with associated inputs; the input-intensification period, 1975 to 1984, when input use increased rapidly; and a post-Green Revolution period, 1985 to 1994, when input use leveled off.

Production growth in the Punjab has largely been through intensification. Land area has increased only 0.7% annually since 1967. Technological change in Pakistani Punjab's agriculture was triggered by the introduction of modern semidwarf varieties (MVs) of wheat in 1967, which covered almost all wheat area by 1983. Modern semidwarf varieties of rice were also adopted but were limited by farmers' specialization in older but high-quality Basmati rice varieties. However, an improved Basmati rice variety (Basmati 385) was released in 1985 and was rapidly adopted (Sharif et al., 1992).

The use of modern varieties stimulated rapid input intensification in the rice–wheat tract (Table 4–1). Fertilizer use jumped from an average of 12 kg of nutrient ha^{-1} of cropped area in the Green Revolution period 1967 to 1974 to an average of 65 kg ha^{-1} in the post-Green Revolution period 1985 to 1994. The total supply of water and its timely availability have also been greatly improved through investment (largely private) in tubewells, especially during the Green Revolution and intensification periods. The rice–wheat system is most dependent on tubewell water.

The Green Revolution technology increased the demand for labor in both rice and wheat so that average labor use for crops increased from 74 person-days ha^{-1} in the Green Revolution period to 89 d ha^{-1} in the intensification period (see Table 4–1). However, the additional labor demand together with off-farm demands, notably from the Middle East, pushed the wage rate up and induced the adoption of mechanical technology. This decreased labor use for crops to 69 d ha^{-1} in the post-Green Revolution period. Meanwhile, mechanical power (tractor, harvester, and thresher) increased from 1.6 h ha^{-1} in the first period to 15.7 h ha^{-1} in the most recent period, replacing draft power use, which decreased dramatically from 72.9 to 6.5 d ha^{-1} for the same period. Considerable variation across cropping systems in input intensification can be seen from Table 4–1, with intensification in the rice–wheat system generally lagging compared with other systems.

Table 4–1. Changes in input use per hectare in the Punjab, 1967 to 1994.†

System	Period‡	Fertilizer	Plant protection	Labor	Bullock	Machine	Water ($m^3 \times 10^{-2}$) Tubewell	Water ($m^3 \times 10^{-2}$) Canal	Irrigated area	Crop intensity
		kg	%	d	d	h			%	%
Rice–wheat	1	12.3	18.9	73.6	72.9	1.6	2.9	2.7	74.0	132
	2	44.8	28.3	88.7	38.1	7.1	5.1	2.6	85.2	139
	3	64.7	13.4	69.4	6.5	15.7	6.2	2.2	88.6	147
Mixed–wheat	1	14.1	18.2	89.5	74.4	1.4	1.1	5.0	81.0	113
	2	46.0	24.6	108.4	45.6	6.0	2.0	4.7	85.2	130
	3	76.0	23.9	84.6	10.8	14.7	2.6	4.3	84.7	132
Cotton–wheat	1	18.5	21.3	87.2	85.4	1.7	1.5	6.1	97.2	121
	2	62.3	35.6	100.7	49.6	6.4	2.5	5.6	98.1	128
	3	120.3	70.2	65.8	7.7	17.3	2.7	4.9	98.4	147
Mungbean–wheat	1	6.1	11.1	82.4	83.8	0.8	0.6	3.6	59.7	104
	2	26.0	16.6	86.4	45.8	2.6	1.4	3.3	61.5	109
	3	49.1	25.3	62.6	8.0	9.1	2.1	3.0	67.0	118

† Source: Ali and Byerlee (2000).
‡ 1 = Green Revolution 1967–1974, 2 = Intensification 1975–1984, 3 = Post-Green Revolution 1985–94.

Table 4–2. Wheat production practices and yields in the Punjab, 1977, 1988, and 2000.†

	1977	1988	2000
Number of survey observations	1159	293	240
Farm size, ha	7.2	8.6	8.1
Cropping intensity, %	127	144	180
Seed rate, kg ha^{-1}	89	100	124
Plant on time, %	68	43	56
Use MVs‡, %	94	na	100
Apply N, %	80	97	100
Apply P, %	39	79	95
Avg. dose, kg ha^{-1}			
N	48	93	124
P (P_2O_5)	19	50	64
Total	67	143	188
Use herbicide, %	Negligible	Negligible	67
Apply animal manure, %	46	36	28
Average yield (Mg ha^{-1})	1.55	1.80	3.45

† Source: Based on large-scale farm-level surveys reported in Byerlee and Siddiq (1994) and M.A. Khan, personal communication, 2000.
‡ MVs, modern semidwarf varieties.

From 1967 to 1994, the yield of all crops in the province increased at an average rate of 2.4% yr^{-1}, with the highest yield gain occurring in the Green Revolution period. However, the rate of gain in the rice–wheat system was only 0.9% over the whole period. The introduction of short-duration varieties of major crops, supported by increased water availability, also triggered double-crop cultivation on the same land. Overall, the cropping intensity rose about 30% during 1967 to 1994.

Trends at the Crop Level

Farm surveys at different times provide a more detailed picture of changes in management practices for the major crop, wheat. While such data are unavailable for the rice–wheat tract, summaries of results from large-scale surveys for all irrigated wheat in the Punjab from the late 1970s, 1980s, and 1990s probably reflect similar changes in the rice–wheat tract (Table 4–2). As a result of increased cropping intensity after the introduction of earlier-maturing varieties of cotton and rice, a large share of wheat area is planted late, even in 2000. The use of modern inputs has steadily increased, with the largest gain from 1977 to 1988 in P use, and in the 1990s in herbicide use. Intensification has been achieved in part through a reduction in the use of traditional inputs, especially farmyard manure, which may be important to sustainability. Finally, yields, which were stagnant from 1977 to 1988, jumped sharply from 1988 to 2000, in part reflecting increased inputs and partly the exceptional growing season in 2000.

Trends in Resource Quality

Ali and Byerlee (2000) also collected data on soil and water quality at the district level based on soil and water testing by the Punjab Soil Fertility Department.

Table 4–3. Average values of selected water and soil quality parameters, by period and region in Punjab, 1971 to 1995.†

System	Period‡	Soil				Water	
		Organic matter	Available P	Soil pH	Soluble salts	Residual carbonate	Electro-conductivity
		%	mg kg^{-1}		%	me L^{-1}	µmhos cm^{-1}
Rice–wheat	1	1.02	7.42	7.74	0.12	1.20	743
	2	0.72	6.23	8.26	0.33	1.72	800
	3	0.59	4.60	8.50	0.42	2.78	874
Mixed–wheat	1	0.85	5.94	8.01	0.28	1.35	848
	2	0.68	5.00	8.33	0.34	1.83	943
	3	0.57	5.01	8.39	0.45	2.66	1172
Cotton–wheat	1	0.86	5.94	8.30	0.20	2.16	1,010
	2	0.62	5.00	8.63	0.44	2.41	1099
	3	0.54	5.01	8.41	0.41	3.03	1214
Mungbean–wheat	1	0.90	7.53	8.08	0.27	1.18	939
	2	0.62	5.17	8.61	0.42	1.37	1006
	3	0.53	4.83	8.50	0.38	2.00	1141

† Source: Ali and Byerlee (2000).

‡ 1 = Green Revolution 1967–1974 (In this table, however, the mean is only for 1971–1974 as data for earlier years are not available), 2 = Intensification 1975–1984, 3 = Post-Green Revolution, 1985–1994.

These variables represent values, averaged by district and year, from thousands of soil tests (organic matter, P content, pH, and soluble salts) conducted by the department for farmers. Concern about secondary salinity and sodicity caused by the use of low-quality tubewell water was captured by a similar data set on tubewell water test values (residual carbonate and electroconductivity) by district and year.

There are strong indications from these data that soil and water quality in the province has deteriorated (Table 4–3). For example, average soil organic matter was already lower than 1% during the early 1970s, and this has further deteriorated in all production systems at an average of 2.3% annually, or a decline of more than 33% during the period. The rate of decrease was highest in the rice–wheat system. Available P has also decreased in all systems at about the same rate. Total soluble salts increased significantly in all the systems, while an increase in alkalinity (pH rising to about 8) occurred in two of the four systems, including the rice–wheat system.

Similarly, the data confirm the deterioration in tubewell water quality, reflected in a significant increase in residual carbonate and electroconductivity of tubewell water in all production systems. Residual carbonate more than doubled during the study period, reflecting the common observation that farmers are increasingly tapping poorer quality groundwater.

PRODUCTIVITY AND RESOURCE DEGRADATION

Evidence from Cross-Sectional Surveys

The first attempt to analyze productivity in the rice–wheat system was undertaken by a joint team of scientists and economists from the Pakistan Agricultural Research Council (PARC) and the International Maize and Wheat Improvement Center (CIMMYT) and is reported in Byerlee et al. (1984). A survey was undertaken of 152 wheat fields at harvest time in 1984 in the main rice–wheat districts of the Punjab. Yields were measured through "crop cuts" of four 1-m^2 samples. Visual observations were made on crop stand, weed density, and soil type, and farmers were interviewed on the production practices and crop rotation in that field.

Average wheat yields were a low 1.8 Mg ha^{-1}, with a high variability among fields as measured by a coefficient of variation of 45%. Subsequent analysis focused on identifying the factors influencing yield differences between fields through comparisons of high-and low-yielding fields and through regression analysis. Major findings of this survey were that

- More than one-half of the fields had been planted continuously to rice–wheat for five or more years. Continuous rice–wheat reduced yields by 23% relative to other rotations.
- Forty percent of wheat fields were planted late, especially after Basmati rice and in heavy soils, which further reduced wheat yields by an average of 1% d^{-1} (Aslam et al., 1989).
- Twenty-eight percent of fields were substantially or seriously infected with *Phalaris minor*, especially fields continuously planted to wheat.
- Many fields lacked sufficient irrigation water.

Subsequent surveys reviewed yield-limiting factors in Basmati rice, for which irrigation problems, tillage operations, and low plant density were the main yield-limiting factors (Amir & Aslam, 1992). One of the major problems in the system is the late planting of wheat after late-maturing Basmati rice. In 1985, a new earlier-maturing Basmati rice was introduced, which allowed rice to be harvested 14 d earlier. Farmers used this time savings by allocating one extra week for land preparation for wheat and by planting wheat 1 wk earlier (Sharif et al., 1992).

These surveys combined with the results of a subsequent extensive on-farm experimental program (Aslam et al., 1989) provided a good overview of the major yield-limiting factors in the rice–wheat system. However, the focus of these earlier studies was on the proximate factors affecting yields in the short term, with little emphasis on assessing the quality of land and water resources, their effects on yields, and trends over the long term.

In a subsequent in-depth survey of wheat fields in 1992, Siddiq (1994) collected information on both soil and water quality in 93 fields in the rice–wheat tract, with special attention to salinity and sodicity. Siddiq found that 75% of tubewells provided brackish water of varying quality that was unfit for irrigation. This was only partly mitigated by mixing with surface irrigation water, especially in tail-end locations where irrigation water was in short supply. Soil electroconductivity (mean of 1.62 dS m^{-1}) and Na adsorption ratio (mean of 7.99 mmol $L^{-0.5}$) were in turn highly correlated to their equivalent measures in tubewell water.

Siddiq (1994) identified the factors affecting productivity by including indices for soil quality in a wheat production function. Wheat yields were found to be negatively related to planting date and sodicity and positively related to the number of irrigations and the quantity of P and N applied. Average yield loss in the rice–wheat tract from sodicity derived from poor-quality tubewell water was estimated at 20%.

This survey confirmed earlier evidence from farm-level surveys on production factors affecting yields, such as planting date and fertilizer use, and provided strong evidence that soil and water quality were negatively affecting productivity, especially sodicity induced by the use of poor quality-tubewell water.

Evidence from Decomposition Analysis of Yield Trends

Cross-sectional studies have an advantage in data collection but provide limited evidence on productivity trends. One approach to understanding productivity trends is to decompose the change in yields over time into its various components, using available agronomic data from crop-response experiments, following the pioneering work of Cardwell (1982). Byerlee and Siddiq (1994) used this approach to analyze trends in irrigated yields in the Punjab from the advent of the Green Revolution in 1967 to the post-Green Revolution period up to 1986.

Average yields in irrigated areas, y^{i}, were first expressed as

$$y^{\mathrm{i}} = qy^{\mathrm{is}} + (1 - q)y^{\mathrm{iz}} \qquad [1]$$

where y^{is} and y^{iz} are yields of modern varieties (MVs), s, and tall varieties (TVs), z, in irrigated areas, respectively, and q is the proportion of irrigated area sown to MVs. The increase in yields of MVs in irrigated areas with time was then represented by

$$y_t^{\mathrm{is}} = y_0^{\mathrm{is}}\exp(g_t) + h(F_t - F_0) + K_t \qquad [2]$$

where $y_0{}^{\mathrm{is}}$ and $y_t{}^{\mathrm{is}}$ are yields of MVs in the base period and in year t, F_0 and F_t are fertilizer nutrients applied in the base period and in year t, g is the exponential rate of genetic gains in yields, h is the marginal grain-nutrient ratio for fertilizer application, and K_t is the combined effects of changes in other cultural practices and in the quality of the resource base. K_t may be positive (e.g., because of improvements in cultural practices) or negative (e.g., because of degradation of the resource base).

By substituting Eq. [2] into Eq. [1], the yield increase in irrigated areas was decomposed into four components as follows:

1. $(q_t - q_0)(y_0^{\mathrm{is}} - y_0^{\mathrm{iz}})$ is the component resulting from changing from tall varieties to MVs on $100(q_t - q_0)$ percent of the area, with an absolute yield increase of $y_0^{\mathrm{is}} - y_0^{\mathrm{iz}}$.
2. $y_0^{\mathrm{is}}q_{\mathrm{t}}[\exp(g_t) - 1]$ is the component resulting from genetic improvements in yields of newer MVs at a rate of $100g$ percent per year.
3. $q_t^{\mathrm{h}}(F_t - F_0)$ is the component resulting from increased fertilizer use on MVs and
4. $q_{t_t}^{K}$ is a residual effect of all other factors influencing yields of MVs.

Yield trends were decomposed for two periods; (i) from 1964–1966 to 1971–1973 when MVs were rapidly adopted and (ii) from 1971–1973 to 1984–1986 when input use intensified on MVs. Both survey and experimental data were reviewed to estimate values for key parameters, especially g and h, the rate of genetic gain in yields and the fertilizer response coefficient, respectively.

Table 4–4. Estimated contributions to yield increases, irrigated wheat, Punjab, 1971–1973 to 1984–1986.†

Component of yield increase	Estimated increase
	kg ha^{-1}
Further switching from tall varieties to MVs‡	141
Genetic gains in yield potential of new varieties	138
Increased fertilizer use on MVs	446
Total	725
Actual yield increase	375
Unexplained residual (K)	–350

† Source: Byerlee and Siddiq (1994).
‡ MVs, modern semidwarf varieties.

Byerlee and Siddiq (1994) found that the increase in irrigated wheat yields during the first period of 0.4 Mg ha^{-1} was explained almost entirely by the adoption of MVs (together with the initial dose of N). In the second period, 1971–1973 to 1984–1986, newer higher-yielding varieties became available in areas where farmers had already adopted modern varieties. The estimated contribution to the yield potential of these newer varieties, g, was about 0.75% yr^{-1}. Fertilizer applied to irrigated wheat in the second period increased by 73 kg nutrient ha^{-1}. On-farm fertilizer experiments suggest an average grain-nutrient ratio, h, of 8:1 associated with increasing fertilizer use from 40 kg nutrient ha^{-1} to 100 kg nutrient ha^{-1} on MVs (Aslam et al., 1989).

Substituting these various parameters into the yield decomposition model above provided the estimates of the various components of yield increases in irrigated wheat in the second period, summarized in Table 4–4. The results show that K_t is negative and large—about 20% of irrigated yields in 1984–1986.

Byerlee and Siddiq (1994) assembled evidence suggesting that several factors were contributing to the negative value of K, such as (i) increased cropping intensity and progressively later planting of wheat, (ii) increasing use of groundwater of marginal quality, (iii) low fertilizer use efficiency, (iv) declining use of organic manures, and (v) increased disease and weed losses, especially to *Phalaris minor.* While these results were computed for all irrigated wheat in the province, the negative effects on yields were probably larger in the rice–wheat tract, where yields were lower and the use of tubewell water and infestation with *Phalaris minor* was highest.

Evidence from Time-Series Studies of Total Factor Productivity

More comprehensive information on productivity trends is provided by measures of TFP. Following Lynam and Herdt (1989), a nonpositive trend in TFP has been widely accepted, at least by economists, as an indicator of unsustainable resource management (Cassman & Pingali, 1995; Ehui & Spencer, 1993). However, many practical problems are encountered in choosing an acceptable measure of TFP apart from the fact that positive TFP growth by itself does not necessarily indicate sustainable resource management.

Estimates of TFP as conventionally measured are the sum of (i) technological progress, (ii) changes in resource quality, and (iii) improvement in efficiency because of investments in human capital and infrastructure. Ali and Byerlee (2000) applied the chain-linked Tornqvist-Theil indexing procedure to measure TFP as follows:

$$\ln(Q_t/Q_{t-1}) = 1/2 \sum_{i=1}^{m} (s_{it} + s_{it-1})\ln(q_t/q_{t-1}) \quad [3a]$$

$$\ln(X_t/X_{t-1}) = 1/2 \sum_{z=1}^{m} (s_{zt} + s_{zt-1})\ln(x_{zt}/x_{zt-1}) \quad [3b]$$

and

$$\ln(\mathrm{TFP}_t/\mathrm{TFP}_{t-1}) = \ln(Q_t/Q_{t-1}) - \ln(X_t/X_{t-1}) \quad [3c]$$

where q_{it} is the quantity of the ith output ($i = 1,2,...,m$ number of product types) in the tth period ($t = 1,2,...,T$ number of years), x_{zt} is the quantity of the zth input ($z = 1,2,...,n$ number of inputs) in the tth period, $\ln(Q_t/Q_{t-1})$ and $\ln(X_t/X_{t-1})$ are the weighted rate of change in all outputs and inputs, respectively; and s_{zt} and s_{it} are the share of the zth input in total costs and of the ith output in total value of production, respectively.

Ali and Byerlee (2000) then decomposed productivity changes into various effects of technical change, resource degradation, and physical and human resources through the estimation of a cost function in translog form expressed as

$$\begin{aligned}\ln C = {} & \alpha + \sum_{i=1}^{m} \alpha_i \ln q_i + 1/2 \sum_{i=1}^{m}\sum_{j=1}^{m} \beta_{ij} \ln q_i \ln q_j + \sum_{z=1}^{n} \rho_z \ln w_z \\ & + 1/2 \sum_{z=1}^{n}\sum_{u=1}^{n} \chi_{zu} \ln w_z \ln w_u + 1/2 \sum_{i=1}^{m}\sum_{z=1}^{n} \delta_{iz} \ln q_i \ln w_z + \sum_{c=1}^{p} \mu_c \ln I_c \\ & + \sum_{o=1}^{r} \phi_o \ln S_o + \sum_{f=1}^{q} \gamma_f \ln H_f + \sum_{r=1}^{k} \lambda_r T_r + \sum_{d=1}^{l} v_d D_d \\ & + \sum_{o=1}^{r}\sum_{z=1}^{n} \Theta_{oz} \ln w_z \ln S_o + \sum_{i=1}^{m}\sum_{o=1}^{r} \Omega_{io} \ln q_i \ln S_o + \sum_{c=1}^{p}\sum_{o=1}^{r} \Psi_{co} \ln I_c \ln S_o \\ & + \sum_{z=1}^{n} \upsilon_z \ln w_z T + \sum_{j=1}^{m} \theta_j \ln q_j T \end{aligned} \quad [4]$$

where C is total cost, q_i is output, w_z is the price of the zth input, I_c is a variable related to technological innovations (such as proportion of MV area), S_o is a resource quality-related variable (such as organic matter content in soils), H_f is a human and infrastructure-related variable (such as literacy rate and distance from road), T_r is an annual index of time for production system r, and D_d is a dummy variable for district d. The effect of technological change on cost was controlled through the coefficient μ (expected to be negative), and the effect of changes in resource quality was measured through ϕ plus the coefficients of the interaction terms between resource quality and other variables evaluated at their mean value. The effect of improvements in human resource and infrastructure variables on productivity was estimated through γ (expected to be negative). The system-specific trends in technology, resource quality, or efficiency not captured by the above variables were estimated through λ_r, υ_z, and θ_z, with the last two evaluated at the mean price and output levels, respectively.

Table 4–5. Growth rate (%/yr) of partial factor productivity, input, output, and total factor productivity (TFP) indices by production system in the Punjab.†

	Period‡	Water productivity§	Land productivity§	Labor productivity§	Total output§	Total input§	TFP§
Rice–wheat	all	−3.03	0.89	1.03	1.79	2.30	−0.50
	1	−7.31	0.76	−2.11	3.44	5.88	−2.43
	2	−3.65	0.88	0.80	1.24	1.84	−0.60
	3	0.14	1.85	3.01	2.04	1.17	0.88
Mixed–wheat	all	−1.90	2.09	1.51	2.29	1.42	0.87
	1	−3.08	2.54	−5.09	3.35	4.46	−1.12ns
	2	−1.50	3.51	1.67	2.25	0.75	1.50
	3	−0.90	0.02	2.66	1.87	1.40	0.46ns
Cotton–wheat	all	−0.25	2.98	3.44	3.65	2.08	1.57
	1	−4.99	3.18	−1.81	3.66	3.96	−0.30
	2	0.10	2.32	3.48	3.55	1.77	1.79
	3	1.96	2.90	4.30	2.70	0.92	1.77
Mungbean–wheat	all	−3.49	1.89	3.08	3.68	2.36	1.32
	1	−7.85	4.36	0.95	6.79	4.55	2.24
	2	−2.81	−1.26	0.70	1.31ns	2.02	−0.70
	3	−0.48	3.68	7.70	4.80	1.56	3.24
All	all	−1.41	2.43	2.51	3.23	1.97	1.26
Punjab	1	−5.14	2.75	−2.85	4.00	4.49	−0.49ns
	2	−1.29	2.22	2.33	2.77	1.50	1.27
	3	0.61	1.96	4.14	2.85	1.25	1.60

† Source: Ali and Byerlee (2000).

‡ 1 = Green Revolution 1967–1974, 2 = Intensification 1975–1984, 3 = Post-Green Revolution, 1985–94; all = 1967–1994.

§ All the coefficients, except those bearing the ns sign, are statistically significant at least at the 0.10 level.

Ali and Byerlee (2000) estimated TFP for each major cropping system. They used district-level data from secondary sources for all irrigated districts of Punjab aggregated to the cropping-system level for the purpose of computing TFP. The study included 33 crops, 8 livestock products, 6 crop byproducts, and 17 input categories. Input categories for the crop sector included land, labor, water, machinery (separately for tractor, thresher, and harvester), draft animals, fertilizer (separately for N, P, and potash), and pesticide costs (separately for aerial and ground spray). For the livestock sector, categories were labor, fodder and feed, and interest and maintenance costs (shed, veterinary supplies, and other costs).

Trends in Productivity

The rate of growth of crop production in rice–wheat was highest in the Green Revolution period at 3.4% yr^{-1}, and then declined to <1.2% as yield growth in wheat and rice slowed sharply (Table 4–5). Production growth was lowest in the rice–wheat system and one-half of that in the cotton–wheat and mungbean–wheat systems (Table 4–5).

Table 4–5 summarizes various measures of productivity. Both land and labor productivity have increased about 2.5% annually for all irrigated Punjab, but only about 1% annually for the rice–wheat system. Land productivity has generally risen more slowly since the boost during the Green Revolution period, while labor productivity has generally increased in successive periods because of rapid mechanization. Again, rice–wheat is the exception, with land productivity improving in the post-Green Revolution period relative to earlier periods. An important result is the overall decline in water productivity in all systems, which reflects the inefficient use of irrigation water, in part because of underpricing.

The rice–wheat system experienced a significant negative growth in TFP of –0.5% yr^{-1} compared with positive and significant growth in other systems ranging from 0.9 to 1.6%. These results support widespread concerns that continuous double cropping of cereals, especially rice and wheat, which require very different soil and water management practices, has been an unsustainable cropping pattern (Hobbs & Morris, 1996). Deterioration in soil and water quality, discussed above, seems to be especially serious in this system. However, as in the other systems, performance has improved in the post-Green Revolution period as new higher-yielding varieties of wheat and rice were introduced, cropping patterns became somewhat more diverse, and labor productivity increased with mechanization (in part because of factors external to agriculture).

Decomposition of Productivity Growth

To estimate the cost function in Eq. [3], the effect of technological change was captured by two variables: (i) the proportion of area sown to modern wheat varieties and (ii) cropping intensity. Cropping intensity is a proxy for the adoption of MVs in summer crops (especially cotton and Basmati rice, for which varietal adoption data were not available). The effect of resource degradation was estimated through the soil and water quality variables described above. The adult literacy rate was used to measure the effect of changes in labor quality. The inverse of the distance of a village from the nearest paved road was used to quantify the effect of improvement in physical infrastructure. Region-specific time-trend dummies were included to capture the remaining unspecified effects of technological change or resource degradation not included in the function.

The estimated elasticities show a strong effect of soil and water quality. For example, organic matter and P have negative elasticities (i.e., reduced production costs and increased productivity) in all the systems, while total soluble salts in soils has a positive sign in the rice–wheat and mixed crops–wheat systems, confirming the findings from cross-sectional surveys of Siddiq (1994) described above. Of the two water quality variables, only electroconductivity produced a consistent and significant coefficient, indicating that soil salinity is in part caused by the application of poor-quality tubewell water.

Technological variables had the expected effects of lowering costs with the greatest effect for the cropping intensity variable, reflecting success in developing short-duration varieties of cotton and rice. Finally, the quality of human and physical resources was found to have an important positive effect on agricultural productivity, as expected in a dynamic agriculture like that of the irrigated Punjab.

Using coefficients of the cost function, productivity growth was decomposed into three effects: (i) technological change, (ii) degradation of soil and water quality, and (iii) improvements in human and physical infrastructure, all evaluated at the system-specific mean rate for each variable included in the cost function (Table 4–6). The combined effect of technological change, resource degradation, and im-

Table 4–6. Decomposition of the total change in crop sector productivity (%/yr) Punjab during 1971 to 1995.†

	Soil and water quality deterioration‡					Technological change‡		Public investment‡	
Region	Water electro-conductivity	Soil P	Soil organic matter	Total soil-soluble salts	Other factors§	Total	CI & MV¶	Roads and literacy	Net effects
Rice–wheat	–0.0109	–0.0868	–0.2123	–0.0583	–0.925	–1.2929	0.3394	0.3714	–0.5821
Mixed–wheat	–0.0241	–0.0268	–0.1547	–0.0567	–0.306	–0.5682	0.3499	0.4176	0.1992
Cotton–wheat	0.0033	–0.0642	–0.1090	0.0009	0.467	0.2982	0.7268	0.4984	1.5234
Mungbean–wheat	0.0013	–0.0043	–0.0655	0.0001	–0.557	–0.6257	0.5271	0.4643	0.3658
Overall Punjab	–0.0073	–0.0487	–0.1374	–0.0265	–0.314	–0.5343	0.4970	0.4434	0.4061

† Source: Ali and Byerlee (2000).
‡ The negative sign implies a negative effect on productivity. The rate of productivity change in each production system was evaluated by multiplying the negative of the coefficient in Table 4–6 with the system-level rate of change per year (in percentage) in each factor.
§ Based on the system-specific trend coefficients in Table 4–6 evaluated at regional mean values of the other variables used in the cost function.
¶ CI = cropping intensity, MV = percentage area in modern varieties of wheat.

provement in human and physical infrastructure on productivity was negative in the rice–wheat system, confirming the TFP analysis; that is, resource degradation more than cancelled the effects of technological change and improvements in education and infrastructure. The unspecified "other factors" captured by coefficients on the regional time-trend variable also reduced productivity quite strongly in this system, suggesting other degradation effects not captured in the soil and water quality variables. These environmental and management-related factors may include the development of pest populations (e.g., increased intensity of *Phalaris minor* in the rice–wheat system) because of the inappropriate use of pesticides and monocropping of cereals, soil health problems such as nematodes again because of monocropping, and the development of a hard plow pan caused by inappropriate mechanization (Byerlee, 1992: Hobbs & Morris, 1996). More research is needed to quantify the effects of these other factors.

Resource degradation had a negative effect in other systems, with the exception of the cotton–wheat system, but the effect was lower than in the rice–wheat system. For all Punjab, technological changes and improvements in human and physical infrastructure combined produced a growth of 0.94% yr^{-1}, which was partly offset by a resource degradation of 0.53% yr^{-1}. These results reinforce the conclusion that resource degradation is more serious in the rice–wheat system.

COMPARATIVE RESULTS FROM THE INDIAN PUNJAB

While the evidence for the Pakistan Punjab seems consistent, these results cannot be extrapolated to other rice–wheat areas. Even in the neighboring Indian Punjab, the situation is different, as shown by a comparable TFP study by Murgai (1999). Tables 4–7 and 4–8 compare the production record in the two Punjabs for wheat and rice during the same three periods. Although productivity trends exhibited similar patterns of change across periods in the two Punjabs, the gap between yields in the two states widened with time, even though the two states had similar cropping patterns and ecologies. Wheat yields in India during the post-Green Revolution period were nearly double those in Pakistan. Rice yields in India were also much higher during the post-Green Revolution period, although this was partly due to Pakistan's specialization in low-yielding, high-value Basmati rice.

The data indicate that the differences between the states in production performance are partly caused by differences in input use and cropping intensity (Table 4–7). Pakistan followed the same growth patterns with time, although cropping intensities and the use of fertilizer and machinery were considerably lower in all periods.

The Indian Punjab experienced a better overall growth rate for all systems combined than Pakistan, with the states having, respectively, 5 and 3.2% annual growth in output. However, the difference in growth of TFP was relatively small, 1.9 and 1.5%, respectively. In the Indian Punjab, the slowest growth in TFP was in the rice–wheat system, but compared with the negative performance in Pakistan, the rice–wheat system experienced a positive growth in TFP of 1.4% annually (Table 4–8). The better performance of this system in the Indian Punjab reflects, in part, the focus on early maturing coarse rice for local consumption as well as concerted

Table 4–7. Changes in cropping intensity, irrigated area, and input use by period, Indian and Pakistan Punjab.†

Variable and period‡	Indian Punjab§	Pakistan Punjab§
Cropping intensity, %	[1.0]	[0.8]
Green Revolution	138	117
Intensification	158	126
Post-Green Revolution	174	136
Irrigated area, %	[1.7]	[0.3]
Green Revolution	68.6	81.9
Intensification	81.9	85.1
Post-Green Revolution	90.8	86.3
Fertilizer, kg nutrient ha^{-1}	[13.1]	[12.6]
Green Revolution	33.0	14.1
Intensification	99.2	48.3
Post-Green Revolution	155.9	86.1
Machines, h ha^{-1}	[12.8]	[12.0]
Green Revolution	4.3	1.5
Intensification	15.3	5.7
Post-Green Revolution	41.0	14.8
Tubewells, no./1000 ha¶	[11.5]	[3.9]
Green Revolution	27.3	8.2
Intensification	80.6	16.2
Post-Green Revolution	104.4	26.0
Labor, d ha^{-1}	[−1.4]	[−0.9]
Green Revolution	84.2	85.0
Intensification	75.9	98.7
Post-Green Revolution	64.6	71.1

† Sources: Ali and Byerlee (2000) and Murgai (1999).
‡ Green Revolution = 1967–1974, Intensification = 1975–1984, Post-Green Revolution = 1985–1994.
§ The values in brackets indicate growth rate in the parameter value during overall period, 1967–1994.
¶ Tubewell numbers are not directly comparable. Tubewells in Pakistan are much larger.

Table 4–8. Growth rate of partial factor productivity and total factor productivity (TFP) indices in the rice–wheat systems of the Indian and Pakistan Punjabs, by period.†

	Total factor productivity		
	Output§	Input§	TFP§
		%	
Indian Punjab‡	[5.1]	[3.7]	[1.4]
Green Revolution	5.9	6.6	−0.6
Intensification	4.7	2.7	2.0
Post-Green Revolution	3.4	1.2	2.2
Pakistan Punjab	[1.8]	[2.3]	[−0.5]
Green Revolution	3.4	5.9	−2.4
Intensification	1.2	1.8	−0.6
Post-Green Revolution	2.0	1.2	0.9

† Sources: Ali and Byerlee (2000) and Murgai (1999).
‡ Green Revolution = 1967–1974, Intensification = 1975–1984, Post-Green Revolution = 1985–1994.
§ The values in brackets indicate values during overall study period.

efforts to arrest resource degradation (e.g., through the widespread use of gypsum to combat secondary salinity from tubewells).

IMPLICATIONS FOR ASSESSING SUSTAINABILITY

The results of these studies indicate the importance of selecting suitable indicators of sustainability and diagnostic approaches to the interpretation of trends. Economists have sought a single all-embracing measure of agricultural sustainability such as total social factor productivity (TSFP), which attempts to account for resource degradation and externalities in estimates of TFP (e.g., Herdt & Lynam, 1992). However, no single measure alone will be theoretically or empirically robust as an indicator of sustainability. In addition to theoretical problems in measuring TSFP (Byerlee & Murgai, 2001), productivity measures alone have limited value in diagnosing the nature of sustainability problems without relating changes in productivity to underlying changes in technology, human and physical infrastructure, and indicators of resource quality.

More attention needs to be given to defining key indicators of agroecosystem health, measuring changes with time, and relating these measures to trends in productivity. This analysis must be sufficiently disaggregated and for a long enough time period to allow for spatial and temporal variability inherent in agricultural production. Cross-sectional data, especially if they include agronomic, soil, and economic variables, can provide initial hypotheses on productivity-limiting factors and sustainability. However, sustainability is by definition a temporal concept and must be assessed from system trends.

Our findings strongly suggest a need for economists, agronomists, and soil scientists to collaborate in integrating approaches to provide more robust and informative measures of sustainability. There is an urgent need to invest in long-term data collection for key production systems that will allow full productivity accounting using more formal statistical procedures. Diagnosis and understanding of sustainability should be conducted within the framework of a production function that integrates economic, agronomic, and resource quality variables. Such a function would include conventional inputs (e.g., land, labor), nonconventional inputs (e.g., education, infrastructure), technology variables (e.g., the use of improved varieties), resource degradation variables (e.g., soil erosion, nutrient status), and weather variables (to reduce weather-induced variability and allow trends to be picked up with a shorter period of data). In a more advanced approach, resource quality would be modeled endogenously as a function of past productivity and management practices (Byerlee et al., 1991; Lindert, 1999).

The major challenge is to obtain time-series data that are long enough and complete enough to allow econometric estimation. District-level secondary data offer the best immediate opportunity to analyze long-term sustainability provided more emphasis is given to finding innovative means to track resource quality. For example, in Pakistan, soil and water test data had been collected continuously for more than 30 yr in response to demands from farmers and extensionists but had not been aggregated and made available. While not a random sample, these data explained important trends in the productivity of Punjab agriculture. Similar data for parts of

China have also provided new insights into long-term trends in soil quality and its effects on productivity (Lindert, 1999). Simulation models also offer much potential to infer trends in resource quality. For example, crop simulation models may help detect widening yield gaps between expected and actual yields, with the difference being attributed to unmeasured trends in resource quality.

More in-depth analysis will require panel data at the farm-plot level, including data from regular monitoring of resource quality at the plot level. We believe that a relatively modest investment in monitoring resource quality could provide high payoffs in identifying sustainability problems and allow more timely design of remedial actions by both farmers and policymakers. Collection of such data should become an integral part of budgets for natural resource management research programs at major research stations (Byerlee, 1991). The exact data to be collected will depend on hypotheses on the main factors likely to affect system sustainability and will necessarily be system-specific. In all situations, data would be collected on carefully selected soil physical, chemical, and biological parameters. In some cases, pest populations might be monitored through qualitative or quantitative scoring. Such data collection systems necessarily require considerable up-front discussion to define the key parameters, sample size, frequency of data collection, and benchmark sites. However, once established, the collection of data at regular intervals has been shown to be a relatively low-cost exercise, although institutionally difficult to sustain.

CONCLUSIONS

Although results achieved in Pakistan's rice–wheat system with Green Revolution technologies are impressive, this review raises important questions about the sustainability of those gains, especially in light of the evidence of degradation of land and water resources. On the surface, overall output growth in the sector over nearly three decades suggests significant technical change. However, the negative trend in TFP in the rice–wheat system almost certainly suggests problems of sustainability for much of the period analyzed.

This conclusion is reinforced by the considerable evidence of degradation in soil and water quality in the system. The analysis of productivity trends across districts and with time, including variables for resource quality and variables for technical change, suggests that resource degradation had important negative effects on productivity in all systems, and especially in the rice–wheat system where it more than cancelled gains from technical change. These trends underline growing concerns about degradation in Pakistan's most valuable asset, its irrigated land base.

Although the results from various studies are consistent in showing greater problems of resource degradation in the rice–wheat system relative to other systems, there is some room for optimism. In the 1990s, yields of Basmati rice and wheat increased at 3.7 and 2.2% annually, respectively, after a long period of relative stagnation. To some extent, this improved performance is evident in the TFP trends, in which growth was highest in the post-Green Revolution period 1985 to 1994. The improved performance in the 1990s reflects a combination of factors, including improved price incentives that raised fertilizer use, the adoption of chemical weed control, and the release of improved early maturing varieties. Cropping patterns may have become slightly more diversified with the introduction or reintroduction of oilseeds and pulses. Finally, mechanization of harvesting and the adoption of zero tillage on a significant area have reduced time conflicts between rice and wheat and, in the case of zero tillage, provided additional benefits in weed control, water management, and plant stand establishment (Hobbs & Gupta, 2003, Chapter 7 of this publication).

Nonetheless, resource degradation remains a serious threat to the system. Additional work is required to more precisely identify the causes of degradation and develop appropriate remedial efforts. Resource degradation in itself is not a reason for policy intervention if it is internalized into producer decision making, but there are several reasons to believe that this is not the case in the Punjab. First, much of the problem appears to arise from distorted policies that have caused a divergence in private and social costs. In particular, several of the modern inputs analyzed in this study have been subsidized for much of the period of analysis. Even now, electricity for tubewell operation is priced at a fixed annual rate, leading to the overuse of poor-quality tubewell water, which is a major contributor to soil salinity. Second, the information base on which farmers make decisions is inadequate with respect to internalizing rapid changes in soil and water quality variables by moving to more sustainable practices such as integrated nutrient and pest management, more diversified crop rotations, and the incorporation of legumes into the system. Third, public-sector research has, until recently, been biased toward the development of technologies based on packages of modern inputs, and has neglected research on integrated systems approaches to enhance diversification and sustainability, such as integrated pest management, conservation tillage, and improved on-farm water management. Thus, from a policy perspective, there is a case for public and private initiatives on several fronts to reverse degradation: increased investment in crop and resource management research and extension; research to develop diversified and more sustainable cropping patterns and rotations; removal of price distortions on key inputs, especially water; and special incentives to invest in inputs such as gypsum that can counteract the problem of poor-quality tubewell water.

REFERENCES

Ali, M., and D. Byerlee. 2000. Productivity growth and resource degradation in Pakistan's Punjab: A decomposition analysis. Policy Research Working Paper 2480. World Bank, Washington, DC.

Amir, P., and M. Aslam. 1992. The rice–wheat system of the Punjab: A conflict in crop management. p. 62–88. *In* D. Byerlee and T. Husain (ed.) Farming systems of Pakistan. Vanguard Publ., Lahore, Pakistan.

Aslam, M., A. Majid, P.R. Hobbs, N.I. Hashmi, and D. Byerlee. 1989. On-farm research on wheat in the rice–wheat systems of the Punjab: A synthesis of results from 1984 to 1988. PARC-CIMMYT Report No. 89-3. Pakistan Agricultural Research Council, Islamabad.

Byerlee, D. 1991. Monitoring farmers' practices and fields as a continuing activity in on-farm research for evolving cropping systems. J. Asian Farm. Syst. Res. 1:179–190.

Byerlee, D. 1992. Technical change, productivity, and sustainability in irrigated cropping systems of South Asia: Emerging Issues in the Post-Green Revolution Era. J. Int. Dev. 4:477–496.

Byerlee, D., and R. Murgai. 2001. Sense and sustainability revisited: The limits of total factor productivity measures of system sustainability. Agric. Econ. 26:227–236.

Byerlee, D., A.D. Sheikh, M. Aslam, and P. R. Hobbs. 1984. Wheat in the rice-based farming system of the Punjab: Implications for research and extension. National Agricultural Research Center, Islamabad, Pakistan.

Byerlee, D., and A.K. Siddiq. 1994. Has the Green Revolution been sustained?: The quantitative impacts of the seed-fertilizer technology in Pakistan revisited. World Dev. 22:1345–1361.

Byerlee, D., B. Triomphe, and M. Sebillotte. 1991. Integrating agronomic and economic perspectives into the diagnostic stage of on-farm research. Exp. Agric. 26:1–20.

Cardwell, B.V. 1982. Fifty years of Minnesota corn production: Sources of yield increase. Agron. J. 74:984–990.

Cassman, K.G., and P.L. Pingali. 1995. Extrapolating trends from long-term experiments to farmers' fields: The case of irrigated rice systems in Asia. p. 63–84. *In* V. Barnett et al. (ed.) Agricultural sustainability in economic, environmental, and statistical terms. John Wiley and Sons, Ltd., London.

Ehui, S.K., and D.S.C. Spencer. 1993. Measuring the sustainability and economic viability of tropical farming systems: A model from sub-Saharan Africa. Agric. Econ. 9:279–296.

Herdt, R.W., and J.K. Lynam. 1992. Sustainable development and the changing needs of international agricultural research. *In* D.R. Lee et al. (ed.) Assessing the importance of international agricultural research for sustainable development. Cornell University, Ithaca.

Hobbs, P.R., and R.K. Gupta. 2003. Resource-conserving technologies for wheat in the rice–wheat system. p. 149–172. *In* J.K. Ladha et al. (ed.) Improving the productivity and sustainability of rice–wheat systems: Issues and impacts. ASA Spec. Publ. 65. ASA, CSSA, and SSSA, Madison, WI.

Hobbs, P., and M.L. Morris. 1996. Meeting South Asia's future food requirements from rice–wheat cropping systems: Priority issues facing researchers in the post-Green Revolution era. Natural Resources Group Paper 96-01. CIMMYT, Mexico, D.F.

Lindert, P. 1999. Bad earth: China's soils and agricultural development since the 1930s. Econ. Dev. Cult. Change 47:701–736.

Lynam, J.K., and R.W. Herdt. 1989. Sense and sustainability: Sustainability as an objective in international agricultural research. Agric. Econ. 3:381–398.

Murgai, R. 1999. The Green Revolution and the productivity paradox: Evidence from the Indian Punjab. Policy Research Working Paper 2234. World Bank, Washington, DC.

Sharif, M., J. Longmire, M. Shafique, and Z. Ahmad. 1992. System implications of varietal adoption: Basmati-385 in the rice–wheat zone of the Punjab. p. 89–104. *In* D. Byerlee and T. Husain (ed.) Farming systems of Pakistan. Vanguard Publishing, Lahore, Pakistan.

Siddiq, A. 1994. Sustainability of the Indus basin: Impact of tertiary salinity/sodicity on wheat productivity, damage assessment, and future public policy. Ph.D. diss. Dep. of Agricultural Economics, University of Illinois, Urbana-Champaign, IL.

5 Soil Physical Effects of Puddling in Rice–Wheat Cropping Systems

Pradeep K. Sharma

Himachal Pradesh Agricultural University
Palampur, India

Jagdish K. Ladha

International Rice Research Institute (IRRI)
Manila, Philippines

Lav Bhushan

Rice Wheat Consortium
International Maize and Wheat Improvement Center
Modipuram, Meerut, India

ABSTRACT

Rice–wheat is the principal cropping system in Asian countries, covering about 24 million ha area. Puddling (wet tillage) is the most common technique of rice establishment in this system. However, controversies exist regarding the residual effects of puddling on wheat yield cultivated after rice. The reasons for these are that puddling effects on soil properties are texture related, and cultivation is done under variable conditions of climate and management. Post-rice wheat yields are generally low in those soils that undergo radical changes in their physical properties upon puddling. This chapter briefly reviews the effects of puddling on soil physical properties, and their influence on rice and wheat productivity. Wheat yields in post-rice soils may be constrained due to factors such as large turnaround time, poor soil tilth of seedbed, subsoil compaction, poor drainage, restricted aeration, and high mechanical impedance to roots. Adoption of a tillage system less intensive than puddling in rice may benefit the subsequent wheat crop. Organic residue management to enhance recovery of soil structure in post-rice soils, breaking hardpans by deep tillage, subsoiling or chiseling, or optimally regulating nutrients and water in the soil layer above the compact zone are other options to improve wheat yields in post-rice soils. In coarse-textured soils, as in the Indo-Gangetic plains, alternate tillage systems, including bed planting, zero or conservation tillage, and others, seem to have good potential and may be critically examined and adopted to reduce cost of cultivation and sustaining rice–wheat productivity.

The rice–wheat rotation is the principal agricultural production system in South Asian countries, which occupies around 13.5 million ha in the Indo-Gangetic

Table 5–1. Summary of advantages and disadvantages of puddling in rice–wheat cropping.

Rice

Advantages
- Land leveling for proper water control.
- Making soil soft for ease of transplanting.
- Enhanced water and nutrient use efficiency; reduction in water permeability leads to decrease in total water requirement and leaching losses of nutrients.
- Enhanced plant availability of soil nutrients; reduced conditions in puddled soils increase solubility of various macro- and micronutrients.
- Weed control.
- Takes advantage of monsoon rains in land preparation.

Disadvantages
- Labor-, capital-, and energy-intensive process.
- Leaching losses of residual soil NO_3–N during puddling.

Wheat

Advantage
- Moderate subsoil compaction due to puddling in highly permeable coarse-textured soils may improve nutrient and water use efficiency and wheat yield.

Disadvantages
- Destruction of soil structure leading to higher bulk density, higher soil penetration resistance, and enhanced surface cracking.
- Higher draft-power requirement.
- Clod formation, and difficulties in obtaining seedbed with fine tilth.
- Longer turnaround time, leading to delayed sowing.
- Development of subsurface hardpans, having high bulk density, low porosity, high soil penetration resistance, and low water permeability, which lead to problems of waterlogging, poor soil aeration, and restricted root growth.
- Reduced subsoil conditions decrease the availability of soil nutrients to wheat.

Plains (IGP) of Bangladesh, India, Nepal, and Pakistan, and about 10.5 million ha in China (Ladha et al., 2000). It provides food, income, and employment for millions of people. The annual system productivity in IGP is low (3–5 Mg ha^{-1}) compared with the climatic crop yield potential of the region (12.0–19.3 Mg ha^{-1}) (Ladha et al., 2003a; Pathak et al., 2003). Transplanting in puddled soils (wet tillage) is the most common method of crop establishment. Puddling, although a capital and energy-intensive process, has survived for centuries since it provides certain advantages to the rice crop. It helps in land leveling, makes soil soft for transplanting, establishes reduced soil conditions that improve availability of soil nutrients, provides weed control, and reduces percolation losses of water and nutrients. However, soil properties after rice are significantly different from soil properties after upland crops. Wheat, therefore, may respond differently in post-rice than in post-upland crop soils. Table 5–1 summarizes the relative advantages and disadvantages of puddling in a rice–wheat system.

A controversy exists regarding the effects of puddling in rice on the yield of a subsequent wheat crop. Some studies suggest a reduction in wheat yields in post-rice soils due to the destruction of soil structure and associated changes in soil physical properties caused by puddling. For example, investigations conducted at Punjab Agricultural University, Ludhiana (India) farm, reported 8 to 9% lower wheat yields after puddled rice than after maize (Meelu et al., 1979; Boparai et al., 1992; Aggarwal et al., 1995). Fujisaka et al. (1994), on the basis of diagnostic surveys con-

Table 5–2. Wheat yields in different crop rotations.†

Location	Crop rotation system	Wheat yield
		Mg ha^{-1}
Punjab, Pakistan	Rice–wheat	1.6
	Nonrice crops–wheat	2.1
Sindh, Pakistan	Rice–wheat	4.3
	Cotton–wheat	2.2
Uttar Pradesh, India	Rice–wheat	4.3
	Maize/soybean–wheat	3.9
China	Rice–wheat	2.7
	Nonrice crops–wheat	2.1
Sichuan, China	Rice–wheat	3.0
	Nonrice crops–wheat	3.0

† Source: Woodhead et al. (1994).

ducted in several rice–wheat areas in South Asia, also observed relatively low wheat yields in a rice–wheat system, mainly due to the deterioration in soil structure and development of subsurface hardpans. Studies conducted by Woodhead et al. (1994) in Pakistan, India, and China, on the other hand, did not explicitly support this conclusion, although comparisons have been made for a wide range of soil, climatic, and input factors. Of five locations, wheat yields (average of several trials at each location) in rice–wheat rotation were higher than in nonrice crop–wheat rotation at three locations, the same at one location, and lower at one location (Table 5–2). Similarly in a recently published report from India (Yadav et al., 2001), the average yield of wheat in rice–wheat cropping areas was 4.27 Mg ha^{-1} compared with 4.53 Mg ha^{-1} in maize–wheat cropping areas (Table 5–3). However, if wheat yields in two cropping sequences at the same locations were compared, wheat yield in the rice–wheat sequence was either higher (Mandi district of Himachal Pradesh) or was

Table 5–3. Wheat yields in rice and maize-based systems in the Indo-Gangetic Plains of India.†

		Avg. wheat grain yield		Fertilizer applied to wheat	
District	State	R–W‡	M–W§	After rice	After maize
		— Mg ha^{-1} —		— kg ha^{-1} —	
Mandi	HP¶	4.40	3.49	120-60-30	80-40-40
Kullu	HP	--	4.61	--	80-40-40
Jalandhar/Kapurthala	Panjab	4.76	4.89	120-60-30	120-60-60
Hoshiarpur	Panjab	4.70	--	120-60-30	--
Jind	Haryana	4.14	--	120-60-60 + 25††	--
Badaun	UP#	3.66	--	120-60-40	--
Ghazipur	UP	4.09	--	120-60-40	--
Gonda	UP	4.16	--	120-60-40	--
Ghaziabad	UP	--	5.19	-	120-60-40
Aligarh	UP	--	4.47	-	120-60-60
Avg.		4.27	4.53		

† Source: compiled from Yadav et al. (2001).
‡ R–W, Rice–wheat.
§ M–W, Maize–wheat.
¶ HP, Himachal Pradesh.
# UP, Uttar Pradesh.
†† $ZnSO_4$.

almost the same (Jalandhar/Kapurthala districts of Panjab) as in the maize–wheat sequence (Yadav et al., 2001). Nevertheless, these data should be used with caution as fertilizer-input levels were different in two cropping sequences (Table 5–3).

On-farm yield gap analyses in rice–wheat rotation conducted by Ladha et al. (2003b, Chapter 3 of this publication) for several locations in IGP and adjoining areas give some clue to the effects of puddling on wheat yields. The yield gap for wheat ranged from 35 to 66%. They ascribed these yield gaps to nutritional disorders, and climatic and biotic stresses. But puddling may also be partly responsible for low on-farm wheat yields. In a recent yield trend analysis of 33 long-term experiments in Asia, Ladha et al. (2003a) found only two experiments where wheat yield trends were significantly lower. Conversely, more data sets of rice had significantly negative yield trends. Critical analysis of the yield data further indicate that the negative yield trends in wheat were mostly observed in medium- to fine-textured soils, which undergo radical changes in their physical properties upon puddling. In coarse-textured soils of Punjab and Haryana, wheat-yield trends were positive (Ladha et al., 2003a), indicating that puddling in rice may favor the following wheat crop in coarse-textured soils. Nevertheless, these data do indicate that the effects of puddling on wheat are site specific, and hence must be examined in relation to soil conditions, such as soil texture and clay mineralogy.

Yield comparisons of wheat in rice and maize growing areas may sometimes be risky because of ecological variations in these regions. Rice dominates in areas where water availability is adequate, while maize dominates in areas with low water supply. Nutrient management is different in rice–wheat and upland crop–wheat systems (Table 5–3). Variations in soil properties may further complicate the yield comparisons. Even the effects of puddling on soil properties vary with soil type, and rice is cultivated in a wide range of texturally different soils, with different clay mineralogy. Further, the speed of regeneration of puddle soil structure after rice harvest is linked with soil texture, clay mineralogy, organic matter, environmental conditions, and the tillage system followed for seedbed preparation for wheat. Hence, even in post-rice soils wheat performance would show a large variation.

Sharma and De Datta (1986) reviewed the physical properties and processes that occur in typical rice soils. This chapter reviews the physical effects of puddling on rice and wheat yields in relation to soil texture, and compiles information on strategies to overcome constraints on rice–wheat productivity in puddled soils.

PHYSICAL CHARACTERISTICS OF PUDDLED RICE SOILS

Puddling refers to a tillage system in which soil is repeatedly plowed and harrowed under submerged conditions to make the soil soft for transplanting and less permeable to water. It is an intensive tillage system. Puddling breaks the soil aggregates and peds into fine plastic mud, thereby practically eliminating the water transmission (macro) pores (Bodman & Rubin, 1948; Sharma & De Datta, 1986; Acharya & Sood, 1992). Dispersed clay particles settle in horizontal orientation in the puddled layer. Fine clay particles move downwards with percolation water, and clog the remaining macropores (Adachi, 1990). This results in drastic reduction in percolation losses of water and nutrients (Sharma & De Datta, 1985; Sharma et al.,

1988). Depth of puddling plays a significant role in reducing percolation losses. Sharma and Bhagat (1993a) reported exponential reduction in water flux through soils with increase in puddling depth.

Puddled soils shrink on drying, become compact and hard, and produce surface fissures varying in size, depth, and pattern (Bolton & De Datta, 1979; Sur et al., 1981; Bhagat & Acharya, 1987; Prihar et al., 1990). Sharma et al. (1995b) observed the development of wide and deep cracks in a hexagonal pattern upon drying of a puddled silty clay loam soil after rice harvest. The total length of the cracks of width <5, 5 to 10, 10 to 15, and 15 to 20 mm was 0.9, 2.5, 2.6, and 2.1 m m^{-2} area, respectively, with a corresponding volume of 0.2×10^{-4}, 7.0×10^{-4}, 1.8×10^{-4} and 2.5×10^{-4} m^3 m^{-2}, respectively. The cracking width was positively and linearly correlated with cracking depth. The cracking pattern of the soil surface affects the hydrodynamic properties of soil. Cracking extends the soil–air interface into the soil profile, thereby increasing the moisture loss through evaporation. It also influences the extent and size distribution of clod formation during land preparation and, thus, the tilth of seed bed for wheat.

Reduction in water transmission (macro) pores results in a corresponding increase in water retention and residual pores. Water retention in puddled soils always exceeds that in nonpuddled soils, usually below –0.01 MPa water potential. Evaporation losses from puddled soils are relatively low, and it may take weeks to months before a puddled soil reaches the optimum moisture content for tillage (Sharma & De Datta, 1986).

The draft-power requirement of dried post-rice soils is very high. Bhagat et al. (1994) reported 5.4 GJ ha^{-1} energy requirements for plowing a silty clay loam soil after rice harvest. Further, puddling, decreases the soil moisture range over which an optimum soil tilth can be obtained for sowing the following upland crop. According to Utomo and Dexter (1981), for some soils, a gravimetric moisture content around 0.9 times the plastic limit provided the maximum friability. At this moisture content, soil tillage maximizes the proportion of small aggregates (Ojeniyi & Dexter, 1979).

The long-term effect of puddling is the formation of hardpan in subsoil below the puddled layer at 10- to 45-cm depths (Sur et al., 1981; Sharma & De Datta, 1986; Sawhney & Sehgal, 1989; Chenkual & Acharya, 1990; Rahman, 1991; Boparai et al., 1992). Chenkual and Acharya (1990) observed higher bulk density throughout the 0- to 30-cm profile under rice–wheat than under maize–wheat in silt loam, silty clay loam, and silty clay soils. Titkov and Gausev (1989) recorded an increase in bulk density from 1.27–1.29 to 1.45–1.54 Mg m^{-3}, and a decrease in porosity from 48–53 to 42–46% at the 10- to 36-cm depth in soil under rice monoculture system. According to a survey of 35 rice-growing locations in South and South-East Asia, the soil horizon below the 10- to 15-cm depth had mechanical impedance >2.8 MPa, high enough to inhibit root elongation of most crop plants (Hasegawa et al., 1985).

Subsurface hardpans may develop due to physical compaction caused by the use of heavy machinery; animal and human processes during tillage, transplanting, weeding, and other processes; and chemical precipitation of Fe, Mn, and Si in the subsoil layers. It may take as little as 3 yr, as in fine sandy loam, and as many as 200 yr, as in polder lands, to develop a hardpan (Sharma & De Datta, 1986).

Hardpans are characterized by high bulk density and penetration resistance, and low porosity and water permeability (Hasegawa et al., 1985; Sharma & De Datta, 1985; Sharma et al., 1988). In one study (Sur et al., 1981), the saturated hydraulic conductivity of a 15- to 20-cm soil layer under rice–wheat rotation was less than one-half (1.14 cm h^{-1}) that under maize–wheat rotation (2.58 cm h^{-1}). Sawhney and Sehgal (1989) observed lower cumulative water intake under rice–wheat than under maize–wheat rotation due to the presence of a compact layer at the 12- to 45-cm depth. Meelu et al. (1979) observed 1.86 cm h^{-1} infiltrability of soil under rice–wheat–mungbean rotation compared with 2.76 cm h^{-1} in soil under maize–wheat–mungbean rotation. The low infiltrability in the former case was attributed to the subsoil compaction. Yun-Sheng (1983) observed a decline in percolation rate from 0.360–4.212 to 0.072–0.432 cm h^{-1} due to a shift in cropping system from rice–wheat to rice–rice–wheat. The saturated hydraulic conductivity (K_s) of the plow layer was as high as 43.2 cm h^{-1}, compared with 0.07 cm h^{-1} in the plow pan; it was 0.13 cm h^{-1} in soil below the plow pan. Thus, the K_s value of the plow pan was about 0.15% of that in the plow layer. Chenkual and Acharya (1990) also observed lower infiltrability due to high bulk density under rice–wheat than under maize–wheat cropping system.

The effects of puddling on soil properties, however, vary with soil texture. Fine-textured soils, dominated by sesquioxides and nonexpanding clay minerals, generally have well-developed soil structure, with a high proportion of water-stable aggregates. If not puddled, these soils maintain a high proportion of soil aggregation even under submergence. Intensive puddling is needed to destroy soil aggregates in these soils. Destruction of aggregates significantly affects physical and chemical (e.g., redox potential, pH, solubility of plant nutrients) properties of these soils. Fine-textured soils, which dominate in expanding types of clay minerals, easily disperse under submergence even without wet tillage. These are self-puddling soils and generally have low water permeability. Puddling per se causes less changes in their physical properties. Coarse-textured soils generally have single-grain loose structure, without much soil aggregation, and are highly permeable. Puddling does little to change the permeability of these soils. For puddling to be effective in decreasing water permeability of soils, the sand content of soil should be <70%, and the finer fraction should be dominated by clay (13–20%) (Sharma & Bhagat, 1993a). In one study (Aggarwal et al., 1995), percolation rates in a sandy loam soil were 14 mm d^{-1} with low puddling (1 disking + 1 planking) and 10 mm d^{-1} with high puddling (4 disking + 1 planking). Whatever reduction was achieved in water permeability was due to the settling of clay particles suspended in floodwater during puddling, which sealed the macropores between sand particles. The migration of clay particles into lower layers, coupled with physical soil compaction, however, increases the vulnerability of coarse-textured soils to develop subsurface hardpans. Coarse-textured soils, therefore, may develop subsurface hardpans faster than fine-textured soils. Sur et al. (1981) and Aggarwal et al. (1995) observed the development of a subsurface hardpan within 6 yr of puddling in a sandy loam soil. In that situation, coarse-textured soils will show reduced percolation losses, and benefit the rice crop. Sharma et al. (1995a) artificially created compaction below the 20-cm depth in a loamy sand soil using a road roller for harvesting rainwater in situ for growing a rainfed lowland rice crop. It significantly increased rice yield by 59 to 88% over the puddled control.

Lal (1985) developed a tentative system for assessing soil tillage requirements for a rice-based system. The system is based on five parameters including clay content, cation exchange capacity, soil permeability, endurance to traffic for wet tillage, and days available for seedbed preparation. The soils are rated from 1 to 5. Soils with a rating of 1 (coarse-textured soils, having <10% clay, with lower contents of high-activity clay minerals, and >50 $\mu m\ s^{-1}$ water permeability) may not respond to puddling (but to compaction). Soils with 2 to 3 rating (medium- to fine-textured soils, having 10 to 40% clay, higher proportion of water-stable aggregates, and 17 to 50 $\mu m\ s^{-1}$ water permeability) may require varying intensities of puddling, and soils with 4 to 5 rating (fine-textured soils, having >40% clay with higher content of high-activity clays, and low water permeability of <17 $\mu m\ s^{-1}$) may not require puddling.

RESPONSE OF RICE TO PUDDLING

In most Asian countries, puddling is such a common feature in rice cultivation that it has become almost synonymous with rice culture. Puddling favors the rice crop in many ways, as was summarized in Table 5–1. The important point, however, is that rice may not respond to puddling in all soil types. Rice response to puddling is related to the effects of puddling on soil properties. Puddling benefits rice only in soils that undergo radical changes in physical properties due to wet tillage; in others, rice yields remain the same with or without puddling. Data in Table 5–4 do not show the significant effect of puddling on rice yield. If these soils are examined based on tillage criteria as suggested by Lal (1985), soils at No. 5 to 8 probably had a rating of 1, soils at No. 1 to 4 would fall into the 4 or 5 rating group, and both of these soil groups would not respond to puddling. Hence, puddling did not significantly affect rice yield in these soils.

It is important, therefore, for rice soils to be critically examined before adopting puddling; otherwise it will be a waste of time, energy, and capital, in addition

Table 5–4. Effect of puddling on rice yield in relation to soil texture.

Site No.	Rice yield†		Soil texture	Location	Source
	Puddled	Nonpuddled			
	— Mg ha^{-1} —				
1	5.6	5.6	Clay loam	IRRI, Philippines	Mabbayad & Buencosa (1967)
2	5.5	5.5	Clay	IRRI, Philippines	Sharma et al. (1988)
3	5.4	5.1	Vertisol	Kenya	Croon (1978)
4	3.9	3.8	Clay	Surinam	Scheltema (1974)
5	2.5	2.5	Sandy loam	Senegal	FAO (1976)
6	5.9	5.5	Sandy loam	Nigeria	FAO (1976)
7	5.0	4.8	Sandy loam	Nigeria	Lal (1985)
8	5.9	5.7	Sandy loam	Panjab, India	Aggarwal et al. (1995)

† Rice grain yields in puddled and nonpuddled soils do not differ significantly ($P = 0.05$).

to creating soil conditions unfavorable for the succeeding wheat crop. The majority of soils in the IGP, where more than one-half of the area is devoted to rice–wheat rotation, are medium textured (sandy loam, silt loam, silty clay loam, and clay loam), and have low-activity clay minerals and low aggregation. The soil texture becomes finer and rainfall increases as one moves from west to east in the IGP. In western parts of the IGP (Panjab and Haryana), rice is also cultivated in coarse-textured soils. Most of the soils exhibit subsurface compaction at depths below 15 to 20 cm. About 2 million ha in the IGP is salt affected. Soils are highly permeable to water, except those that have subsurface compact layers, and/or are sodic in nature. Majority of these soils may fall under tillage rating 1, and some in 4 to 5 (Lal, 1985), and may not require puddling for rice cultivation. It may be possible to replace puddling with a less intensive tillage system in these soils. The tillage requirements for rice in the IGP, therefore, should be seen with a new perspective.

WHEAT YIELDS IN POST-RICE SOILS

Wheat, in general, requires a well-drained soil with good aeration for producing potential yield. Poor soil aeration is the single most important soil constraint responsible for low wheat yields after wetland rice, especially in medium- to fine-textured soils (Bhushan & Sharma, 1999). The significance of soil physical properties on wheat yield, however, may be seen probably at relatively higher production levels. In a long-term field experiment, improvement in soil physical properties contributed only <20% toward increased wheat yields at yield levels ≥3 Mg ha^{-1}; the rest of the yield increase came from improved nutrient status of the soil (Sharma & Verma, 2001; Sharad & Verma, 2001; Bhushan & Sharma, unpublished data). Other factors, which are the indirect consequence of puddling, are responsible for reduced wheat yields in post-rice soils. Different factors that adversely affect wheat yields in post-rice soils are described below.

Large Turnaround Time

Puddled soils have relatively high water retention because of increased microporosity (Taylor, 1972; Sharma & De Datta, 1986). Relatively more energy is required to evaporate the same quantity of water in puddled than in unpuddled soils, while the atmospheric evaporativity after rice harvest is relatively low. In addition, higher unsaturated hydraulic conductivity of puddled soils may also help keep the surface soil wet longer by transporting more water upward to the soil surface from the lower profile (Sharma & De Datta, 1986). As a result, puddled soils remain more wet than unpuddled soils for a longer time (Sharma et al., 1988). The situation is aggravated further if a shallow water table exists that is capable of keeping the soil surface wet through capillarity. In many low-lying areas, therefore, it takes several weeks before poorly drained puddled soils reach a state with workable moisture content.

Longer turnaround time delays wheat planting, which results in low yields. Delayed planting has been identified as one of the serious causes of low wheat productivity in rice–wheat areas of South Asia (Fujisaka et al., 1994). According to an estimate, each day of delay in planting after 18 November lowered wheat yield by 0.04 Mg ha^{-1} (Regmi et al., 2002). Hobbs et al. (1996) reported a yield decline of 1% (equivalent to 0.035 Mg ha^{-1} d^{-1}) for each day of delay in wheat planting beyond the optimum sowing date (15–20 Nov.). According to Aggarwal et al. (2000), potential wheat yields in the IGP (except West Bengal) under late-sown conditions were lower than in optimal-sown conditions by almost 0.6 to 1.4 Mg ha^{-1}. In West Bengal, no reduction in potential wheat yield was apparent. These data emphasize that timely sowing of wheat is crucial for maximizing wheat yields in post-rice soils. Data from various farmers' participatory trials in the IGP have also shown that wheat yields may be increased in rice soils by enhancing the date of planting of wheat.

Poor Soil Tilth

Medium- (dominating in clay fraction) to fine-textured soils are difficult to till after rice harvest because of their massive structure, high bulk density, and high draft-power requirement. Sometimes the draft-power requirement is so high that it becomes difficult to till them with the animal-drawn countryside plow or even small tractors. Upon tillage, they break into medium to large clods, having a high breaking strength (Sharma & Bhagat, 1993b). To crush these clods into a fine tilth is a time and energy intensive process. In spite of the energy and time spent on crushing, the mean weight diameter of soil aggregates of the seed bed, most of the time, remains large, resulting in poor seed–soil contact. Larger clods also permit greater water losses from the surface layer and may cause seed-zone moisture to fall below the level needed for seed germination (Prihar et al., 1985). The large clods cause poor germination and seedling emergence, resulting in unsatisfactory crop stand of wheat (Prihar et al., 1985; Sawhney & Sehgal, 1989; Chenkual & Acharya, 1990; Hobbs, 1994). Therefore, cloddy soil tilth is unfavorable for the wheat crop.

Subsoil Compaction

Subsurface compaction is beneficial for rice, as it lowers percolation losses and increases water and nutrient use efficiency. However, its effect on wheat yield is determined by the penetration resistance and water permeability of the compact layer. Some degree of subsurface compaction in coarse-textured soils (below the critical value that restricts root growth) may favor wheat yields by decreasing leaching losses of water and nutrients, especially N. If subsurface compaction is so high that it restricts the root growth, then to achieve the same wheat yield precise regulation of water and nutrients is a must in soil above the compact layer (Bhushan & Sharma, 1997). Otherwise, a crop with restricted root growth may experience nutritional stress and produce lower yield. Occasional or permanent waterlogging in soils with subsurface compaction may cause oxygen stress, decrease soil redox potential, and favor accumulation of phytotoxins in the root zone. It lowers wheat yields. Under such circumstances, precise regulation of soil moisture through improvement in soil drainage properties is needed to improve wheat yields. Soil penetration resistance in a given soil is influenced by the bulk density, pore-size distribution, rigidity of pore system, and soil moisture content. Sharma et al. (1971)

observed a decline in root mass of wheat in a clay soil above a bulk density of 1.42 Mg m^{-3} in the surface layer and 1.55 Mg m^{-3} in the subsurface layer. Root mass in the subsurface layer was the same at bulk densities from 1.10 to 1.42 Mg m^{-3}. They concluded that 1.2 to 1.30 Mg m^{-3} bulk density is ideal for root growth of wheat in clay soils. Tomar et al. (1981) observed 80 to 85% of the total root mass of wheat within the top 10-cm soil layer, with a bulk density of 1.25 Mg m^{-3}. Root growth declined when the bulk density of soil below 10-cm depth increased from 1.45 to 1.55 Mg m^{-3}. Barraclough and Weir (1988) reported a reduction in vertical extension of winter wheat root system in a sandy loam soil layer at the 35-cm depth compacted to a bulk density of 1.8 Mg m^{-3}. Roots were largely confined above the pan during the early growth period of wheat. Around anthesis, roots reached a maximum depth of 100 cm, compared with 140 cm in pan-free soil. However, the total root length above the pan was the same in soil with and without a hardpan. Early shoot growth was reduced due to pan.

Sur et al. (1981) observed reduced root growth of wheat below a 15- to 20-cm depth after rice than after a maize crop in a sandy loam soil. Total root mass within 0- to 180-cm profile was 48% higher after maize than after rice. Restricted root growth of wheat after rice than after maize was also observed by other workers (Prihar et al., 1975; Chenkual & Acharya, 1990; Boparai et al., 1992). Restricted root growth may result into restricted nutrient uptake and low crop yields. Barraclough and Weir (1988) obtained a 5% lower grain yield in winter wheat in a soil with a pan at the 35-cm depth than in the pan-free soil. Rao and Kathavate (1972) observed a decline in wheat yield when the 7.5- to 30-cm soil layer was compacted to 1.5 Mg m^{-3} bulk density.

Waterlogging adversely affects growth and yield of wheat due to depletion of dissolved oxygen, accumulation of toxins, and/or changes in concentrations of nutrient ions in soil solution (Trought & Drew, 1980). Bhushan and Sharma (1999) concluded that poor soil aeration was the most limiting factor for wheat yield in post-rice soils. Wheat is sensitive to waterlogging, particularly during early growth periods (Hobbs, 1994); it decreases with age and duration of waterlogging (Belford, 1981; Belford et al., 1984; Cannell et al., 1984; Meyer & Barrs, 1988). In one study, very few wheat plants survived more than 6 d of waterlogging during germination and emergence stage. After emergence, waterlogging reduced tillering (Belford et al., 1984). Wheat plants under excess-water conditions turn yellow, and show reduced tillering (Hobbs, 1994), shoot biomass (Guyot & Prioul, 1985), leaf number on the main stem (Cai & Cao, 1987), and apparent photosynthesis (Meyer & Barrs, 1988).

It is generally agreed that the aeration porosity of soil for the successful cultivation of upland crops must be more than 10%. Proper drainage in waterlogged soils, therefore, must be provided to keep a proper balance of air and water in the rhizosphere.

IMPROVING POST-RICE WHEAT YIELDS

Decline in wheat yields due to puddling in rice is not universal. In many soils, wheat yields in rice–wheat systems are either the same or even higher than in nonrice–wheat cropping systems (Table 5–2). The constraints to wheat production in rice soils, if any, are site specific and necessitate specific technique(s) that need to be followed to mitigate them. This section reviews different strategies adopted by different researchers to improve wheat production in post-rice soils.

Reduced Tillage in Rice

Substitution of puddling with a less intensive tillage system or at least decreasing the frequency of puddling is one way of arresting yield decline in wheat grown in sequence with rice, since it is possible to avoid puddling in many soils without sacrificing rice yields (Lal, 1985). Coarse-textured soils with loose single-grain structure, fine-textured soils dominant in smectite clays, or soils with shallow water tables do not require puddling for growing rice (Table 5–4). Ghildyal (1969) suggested soil compaction as an alternative to puddling, especially in coarse-textured soils. Compaction is more effective than puddling in decreasing soil permeability if sand content in soil is >70% (Sharma & Bhagat, 1993a). Direct seeding of rice in unpuddled flat or on permanent raised beds may be an alternate option in fine-textured soils (Balasubramanian et al., 2003, Chapter 6 of this publication; Connor et al., 2003, Chapter 9 of this publication). A lot of work on direct seeding of rice on flat and raised beds with reduced tillage is being undertaken in North India. Cultivating rice on raised beds aims at saving water, time, fuel and energy on tillage; and maintaining soil structure for the subsequent wheat crop (Balasubramanian et al., 2003, Chapter 6 of this publication; Gupta et al., 2003, Chapter 1 of this publication). However, on-station and on-farm research is needed to refine and standardize the bed planting technique. There are issues such as salinization of beds due to the net upward movement of water and dissolved solutes, weed control (mechanically or with herbicides), crop residue management, micronutrient management (Fe deficiency), varietal selection for bed planting, and others that need to be studied.

Organic Residue Management

In some soils puddling is unavoidable to achieve higher rice yields. In that case, techniques should be adopted to speed up the regeneration of soil structure. This can be achieved by adopting suitable tillage practices for land preparation for wheat at appropriate moisture contents, and through proper organic matter management. Tillage operations should be performed when soil moisture is in a friable range. In puddled soils, friable moisture range is relatively narrow (Sharma & De Datta, 1986).

Soil incorporation of organic residues, such as rice straw, compost, waste organic residues such as *Lantana* spp., and others have been found to improve puddlability (Sharma & Bhushan, 2002) and aggregation in puddled soils (Chaudhary & Ghildyal, 1969; Sahoo et al., 1970; Bhagat & Verma, 1991; Sharma et al., 1995b; Sharma & Bhushan, 2002). Boparai et al., (1992) observed an increase in water-stable aggregates between 0.1 and 0.5 mm size by 62% and reduction in soil bulk density with the addition of green manure (*Sesbania cannabina* Linn & Merill) before wetland rice. Sharma et al. (1995b) observed a significant improvement in the structural status of a silty clay loam soil, under rice–wheat cropping, with six annual additions of fresh lantana residues at 10 to 30 Mg ha^{-1}. The surface crack density decreased from 5×10^{-4} to 1.2×10^{-4} m^3 m^{-2}, clod breaking strength from 420 to 216 kPa (Sharma et al., 1995b), and draft-power requirement to prepare seedbeds for wheat from 5.3 to 3.5 GJ ha^{-1} with lantana additions (Bhagat et al., 1994). Rice and wheat yields in lantana-treated soil were 16 and 35% higher than in the control, respectively. Recently, Bhushan and Sharma (2002, 2003) published data on the improvement of structural, mechanical, water retention, and water transmission properties of a silty clay loam soil under rice–wheat cropping due to addition of lantana residues for 10 yr. Ten annual additions of lantana significantly improved soil-physical productivity and wheat yield (Sharma & Bhushan, 2001).

The other alternative for improving the structure of puddled rice soils is including legumes, such as sesbania in rice–wheat system (Boparai et al., 1992) or cultivating chickpea, in rotation with wetland rice (Padma Raju & Deb, 1969). According to Prasad et al. (1980), broadcasting sprouted seeds on puddled soils resulted in less soil cracking and better aggregation than transplanting onto a puddled soil or direct seeding in a moist soil. This probably resulted from the positioning of plants in the soil, as has been reported for other crops, such as wheat in Vertisols (Johnson, 1962; Fox, 1964). Rice soils with better soil structure are relatively easy to till for wheat.

Reduced Tillage in Wheat

Post-rice tillage and wheat establishment have been studied by various workers with diverse conclusions. Many findings are necessarily soil and site specific, but some generalizations seem possible. Wheat yields with zero or reduced tillage comparable with conventional tillage are possible with proper organic residue and water management (Jongdee, 1994). Recently, scientists under the Rice–Wheat Consortium program experimented with different resource conservation tillage techniques, such as surface seeding, zero tillage, reduced tillage, conservation tillage, and bed planting, in the IGP and China (Hobbs & Gupta, 2003, Chapter 7 of this publication). The objective is to develop tillage systems that reduce turnaround time and enhance timely sowing of wheat, in addition to being economic and energy efficient. With these techniques, wheat is planted in soils at rather high moisture content. Direct drilling of wheat into post-rice soil without tillage has been found feasible in some areas in northwestern India, Bangladesh, Nepal, and Pakistan (Aslam et al., 1993; Mehla et al., 2000). Zero-till planting produced yields that were equal or better than those obtained under normal tillage because of timely planting of wheat, efficient fertilizer use, and better weed control. In addition, zero-till planting was cost effective due to reduced use of tractor and fuel and overcame the problem of obtaining a seed-bed with fine tilth. These benefits may, however, get offset if the fields are weedy, as after direct-seeded rice, and weeds have to be controlled by using weedicides prior to wheat seeding with zero tillage. Further, precise water management is very essential in zero-till plots to maintain lower soil penetration resistance to favor a deeper root system. Jongdee (1994) suggested that zero tillage may not be an appropriate technology for wheat cultivation in very compact soils.

Management of Subsoil Compaction

Technology that aims at improving post-rice wheat yields should focus on avoiding the formation of hardpans, breaking the hardpans if they are already formed, or managing soil above the hardpan in such a way that the wheat crop does not experience nutrient, water, and oxygen stress. Subsoil compaction may be avoided or delayed by resorting to a less intensive tillage system than puddling in rice and/or use of organic amendments (Sharma & De Datta, 1986). With proper nutrient and water management, a good crop of wheat is possible even in a soil with a hardpan at the 15-cm depth (Bhushan & Sharma, 1997). Under field conditions, however, it may be difficult to precisely regulate soil moisture regime because of irrigation or rainfall constraints; hence, breaking the hardpan may be a more practical alternative.

El-Sharkawy and Sgaier (1975) suggested deep tillage for improving wheat yields in compacted soils. In their studies, compared with no tillage (control), 15- and 30-cm-deep disking, and 50- to 70-cm-deep subsoiling increased rooting depth by 24, 48, 100, and 132%, respectively, over the control. Increases in grain yields for the four treatments were 28, 35, 39, and 54%, respectively. Increases in wheat yields through breaking of hardpans in rice soils with deep tillage, subsoiling, or chiseling have also been reported by various other workers (McEvan & Johnston, 1979; Mickelson, 1983; Van Ouwerkerk & Raats, 1986; Barraclough & Weir, 1988; Ahmed & Maurya, 1989; Ahmed, 1990). Increases in wheat yields were attributed mainly to reduced soil penetration resistance, improved soil drainage properties, and total and air-filled porosity of soils. These techniques, however, require heavy machinery to break the hardpans, which may be unavailable at the site or not feasible to operate. They are also expensive. The effects of subsoiling, chiseling, or deep tillage are short lived, and they may have to be repeated every 2 to 3 yr, depending on soil texture.

CONCLUSIONS

Although puddling is the most common technique of land preparation for rice in South Asia, its effects on soil properties, water and nutrient economy, and rice yield are not universal. Similarly, the residual effect of puddling on wheat yield in rice–wheat cropping system is not always detrimental, as is believed by many rice workers, although their conclusions are based on wheat yields averaged over different input levels, soil types, environmental conditions, water availability, crop cultivars, and other reasons. Such controversies have arisen because puddling effects on soil properties and thus on crop yields are texture related. In many soils, such as coarse- to medium-textured soils, with a dominant sand fraction, puddling does little to affect soil structural properties. In such cases, as in major parts of the IGP, alternate tillage systems that are less intensive than puddling, may be evaluated and adopted. However, in areas where puddling is absolutely necessary for getting higher water productivity and rice yields, it is essential to adopt soil management practices that hasten the recovery of soil structure after puddled rice. Different strategies are available and discussed in this chapter to make appropriate choices for cultivating a rice–wheat sequence in different soil types.

REFERENCES

Acharya, C.L., and M.C. Sood. 1992. Effect of tillage methods on soil physical properties and water expense of rice on an acidic alfisol. J. Indian Soc. Soil Sci. 40:409–414.

Adachi, T. 1990. Effect of rice-soil puddling on water percolation. p. 146–151. Trans. 14th Int. Congr. Soil Science. Kyoto, Japan.

Aggarwal, G.C., A.S. Sidhu, N.K. Sekhon, K.S. Sandhu, and H.S. Sur. 1995. Puddling and N management effects on crop response in a rice–wheat cropping system. Soil Tillage Res. 36:129–139.

Aggarwal, P.K., K.K. Talukdar, and R.K. Mall. 2000. Potential yields of rice–wheat system in the Indo-Gangetic Plains of India. Rice–Wheat Consortium Paper Series 10. Rice–Wheat Consortium for Indo-Gangetic Plains. New Delhi, India.

Ahmed, A. 1990. Effect of plowsol on soil infiltration rate and crop yield in irrigated soil. Agric. Mech. Asia Africa Latin Am. 21:24–26.

Ahmed, A., and P.R. Maurya. 1989. The effect of chiseling, subsoiling and irrigation frequency on wheat production at Kadaway Nigeria. Samaru J. Agric. Res. 6:15–21.

Aslam, M., A. Majid, N.I. Hashmi, and P.R. Hobbs. 1993. Improving wheat yield in the rice–wheat cropping system of the Punjab through zero tillage. Pakistan J. Agric. Res. 14:8–11.

Balasubramanian, V., J.K. Ladha, R.K. Gupta, R.K. Naresh, R.S. Mehla, Bijay-Singh, and Yadvinder-Singh. 2003. Technology options for rice in the rice–wheat system in South Asia. p. 115–148. *In* J.K. Ladha et al. (ed.) Improving the productivity and sustainability of rice–wheat systems: Issues and impacts. ASA Spec. Publ. 65. ASA, CSSA, and SSSA, Madison, WI.

Barraclough, P.B., and A.H. Weir. 1988. Effect of a compacted subsoil layer on root and shoot growth, water use efficiency and nutrient uptake of winter wheat. J. Agric. Sci. 110:207–216.

Belford, R.K., D.G. Christian, R.Q. Cannell, and M.J. Goss. 1984. Drainage, cultivation and growth of winter cereals in clay soils. p. 167–174. *In* Fonctionnement hybrique et Comportement des sols. Association Franciase pour I'Etu du sol (AFES), Plaisir, France.

Belford, R.K. 1981. Response of winter wheat to prolonged waterlogging under outdoor conditions. J. Agric. Sci. Camb. 97:557–568.

Bhagat, R.M., and C.L. Acharya. 1987. Effect of soil management on wheat in northern India. I. Hydrothermal regime and root growth. Soil Tillage Res. 9:65–77.

Bhagat, R.M., P.K. Sharma, and T.S. Verma. 1994. Tillage and residue management effects on soil physical properties and rice yields in northwestern Himalayan soils. Soil Tillage Res. 29:327–334.

Bhagat, R.M., and T.S. Verma. 1991. Impact of rice straw management on soil physical properties and wheat yield. Soil Sci. 152:108–115.

Bhushan, L., and P.K. Sharma. 1997. Effect of soil moisture regime and nutrient supplies on root growth and yield of wheat in a subsurface compacted soil. J. Indian Soc. Soil Sci. 45:213–219.

Bhushan, L., and P.K. Sharma. 1999. Effect of depth, bulk density and aeration status of root zone on productivity of wheat. J. Indian Soc. Soil Sci. 47:29–34.

Bhushan, L., and P.K. Sharma. 2002. Long-term effects of lantana (*Lantana* sp. L.) residue additions on soil physical properties under rice–wheat cropping. I. Soil consistency, surface cracking and clod formation. Soil Tillage Res. 61:157–167.

Bhushan, L., and P.K. Sharma. 2003. Long-term effects of lantana (*Lantana* sp. L.) residue additions on soil physical properties under rice–wheat cropping. II. Water retention and transmission properties. Soil Tillage Res. (in press).

Bodman, G.B., and J. Rubin. 1948. Soil puddling. Soil Sci. Soc. Am. Proc. 13:27–36.

Bolton, F.R., and S.K. De Datta. 1979. Dry soil mulching in tropical rice. Soil Sci. Plant Nutr. 25:173–181.

Boparai, B.S., Y. Singh, and B.D. Sharma. 1992. Effect of green manuring with *Sesbania aculeata* on physical properties of soil and on growth of wheat in rice–wheat and maize–wheat cropping systems in a semiarid region of India. Arid Soil Res. Rehab. 6:135–143.

Cai, S.B., and Y. Cao. 1987. Effect of waterlogging and high temperature on senescence in wheat leaves during booting stage. Jiangsu Academy of Agricultural Sciences. J. Agric. Sci. 3:45–47.

Cannell, R.Q., R.K. Belford, K. Gales, R.J. Thomson, and O.P. Webster. 1984. Effect of waterlogging and drought on winter wheat and winter barley grown on a clay and sandy loam soil. I. Crop growth and yield. Plant Soil 80:53–66.

Chaudhary, T.N., and B.P. Ghildyal. 1969. Aggregate stability of puddled soils during rice growth. J. Indian Soc. Soil Sci. 17:261–265.

Chenkual, V., and C.L. Acharya. 1990. Effect of rice–wheat and maize–wheat rotations on soil physical properties including soil-water behavior in an acidic Alfisol. J. Indian Soc. Soil Sci. 38:574–582.

Connor, D.J., J. Timsina, and E. Humphreys. 2003. Prospects for permanent beds for the rice–wheat system. p. 197–210. *In* J.K. Ladha et al. (ed.) Improving the productivity and sustainability of rice–wheat systems: Issues and impacts. ASA Spec. Publ. 65. ASA, CSSA, and SSSA, Madison, WI.

Croon, F.W. 1978. Zero-tillage for rice on Vertisols. World Crops 30:12–16.

El-Sharkawy, M.A., and K. Sgair. 1975. Effect of tillage on root penetration and grain yield in Mexican wheat (*Triticum aestivum* L.) grown in the Libyan desert. Libyan J. Agric. 4:25–28.

Food and Agriculture Organization. 1976. Mechanization of rice production: An international Co-ordinated Research Project. FAO, Rome, Italy.

Fox, W.E. 1964. Cracking characteristics and field capacity on a swelling soil. Soil Sci. 98:413.

Fujisaka, S., L. Harrington, and P.R. Hobbs. 1994. Rice–wheat in South Asia: Systems and long-term priorities established through diagnostic research. Agric. Syst. 46:169–187.

Ghildyal, B.P. 1969. Soil compaction—A new system of rice cultivation. Indian Farm. 19:19–22.

Gupta, R.K., R.K. Naresh, P.R. Hobbs, Z. Jiaguo, and J.K. Ladha. 2003. Sustainability of post-Green Revolution agriculture: The rice–wheat cropping system of the Indo-Gangetic Plains and China. p. 1–26. *In* J.K. Ladha et al. (ed.) Improving the productivity and sustainability of rice–wheat systems: Issues and impacts. ASA Spec. Publ.65. ASA, CSSA, and SSSA, Madison, WI.

Guyot, C., and J.L. Prioul. 1985. Alleviation of waterlogging effects in winter wheat by mineral fertilizer application. I. Experiment on soil. Agron. 5:743–750.

Hasegawa, S., M. Thangaraj, and J.C. O'Tool. 1985. Root behavior: Field and laboratory studies for rice and non-rice crops. p. 383–395. *In* Soil physics and rice. IRRI, Los Banõs, Philippines.

Hobbs, P.R. 1994. Rice–wheat system in South Asia. p. 61–76. *In* Proc. Symp. Sustainability of rice–wheat cropping in India. Haryana Agricultural University, Hissar, India.

Hobbs, P.R., and R.K. Gupta. 2003. Resource-conserving technologies for wheat in the rice–wheat system. p. 149–172. *In* J.K. Ladha et al. (ed.) Improving the productivity and sustainability of rice–wheat systems: Issues and impacts. ASA Spec. Publ. 65. ASA, CSSA, and SSSA, Madison, WI.

Hobbs, P.R., L.W. Harrington, C. Adhikari, G.S. Giri, S.R. Upadhya, and B. Adhikari. 1996. Wheat and rice in the Nepal Tarai: Farm resources and production practices in Rupandehi District. Nepal Agriculture Research Council and CIMMYT, Mexico.

Johnson, W.C. 1962. Controlled soil cracking as a possible means of moisture conservation on wheat lands of the Southwestern Great Plains. Agron. J. 54:323–325.

Jongdee, B. 1994. Tillage methods for wheat after rice in paddy soils in Thailand. p. 272–275. *In* D.A. Saunders and G.P. Hettel (ed.) Wheat in heat-stressed environments: Irrigated, dry areas and rice–wheat farming systems. CIMMYT, Mexico.

Ladha, J.K, D. Dawe, H. Pathak, A.T. Padre, R.L. Yadav, Bijay-Singh, Yadvinder-Singh, Y. Singh, P. Singh, A.L. Kundu, R. Sakal, N. Ram, A.P. Regmi, S.K. Gami, A.L. Bhandari, R. Amin, C.R. Yadav, E.M. Bhattarai, S. Das, H.P. Aggarwal, R.K. Gupta, and P.R. Hobbs. 2003a. How extensive are yield declines in long-term rice–wheat experiments in Asia? Field Crops Res. (in press).

Ladha, J.K., K.S. Fischer, M. Hossain, P.R. Hobbs, and B. Hardy (ed.) 2000. Improving the productivity and sustainability of rice–wheat systems of the Indo-Gangetic Plains: A synthesis of NARS-IRRI partnership research. Discussion Paper 40. IRRI, Los Baños, Philippines.

Ladha, J.K., H. Pathak, A.T. Padre, D. Dawe, and R.K. Gupta. 2003b. Productivity trends in intensive rice–wheat cropping systems in Asia. p. 45–76. *In* J.K. Ladha et al. (ed.) Improving the productivity and sustainability of rice–wheat systems: Issues and impacts. ASA Spec. Publ. 65. ASA, CSSA, and SSSA, Madison, WI.

Lal, R. 1985. A soil suitability guide for different tillage systems in the tropics. Soil Tillage Res. 5:179–196.

Mabbayad, B.B., and I.A. Buencosa. 1967. Tests on minimum tillage of transplanted rice. Philipp. Agric. 5:541–551.

McEvan, J., and A.E. Johnston. 1979. The effect of subsoiling and deep incorporation of P and K fertilizers on the yield and nutrient uptake of barley, potato, wheat and sugarbeet grown in rotation. J. Agric. Sci. Camb. 92:695–705.

Meelu, O.P., V. Beri, K.N. Sharma, S.K. Jalota, and B.S. Sandhu. 1979. Influence of paddy and corn in different rotations on wheat yield, nutrient removal and soil properties. Plant Soil 51:51–57.

Mehla, R.S., J.K. Verma, R.K. Gupta, and P.R. Hobbs 2000. Stagnation in productivity of wheat in the Indo-Gangetic Plains: Zero-till-seed-cum-fertilizer drill as an integrated solution. Rice–Wheat Consortium Paper Series 8. Rice–Wheat Consortium for the Indo-Gangetic Plains, New Delhi, India.

Meyer, W.S., and H.D. Barrs. 1988. Response of wheat to single short-term waterlogging during and after stem elongation. Austr. J. Agric. Res. 39:11–20.

Mickelson, R.H. 1983. Deep chiseling for improving soil water storage and grain yield. Paper 83–1523. Am. Soc. Agric. Eng., St. Joseph, MI.

Ojeniyi, S.O., and A.R. Dexter. 1979. Soil factors affecting the micro-structure produced by tillage. Trans. ASAE 22:339–343.

Padma Raju, A., and A.R. Deb. 1969. Influence of crop rotations on the structure of paddy soils. Indian J. Agric. Sci. 39:81–87.

Pathak, H., J.K. Ladha, P.K. Aggarwal, S. Peng, S. Das, Yadvinder-Singh, Bijay-Singh, S.K. Kamra, B. Mishra, A.S.R.A.S. Sastri, H.P. Aggarwal, D.K. Das, and R.K. Gupta. 2003. Climatic potential and on-farm yield trends of rice and wheat in the Indo-Gangetic plains. Field Crops Res. (in press).

Prasad, M., R. Prasad, and T.N. Chaudhary. 1980. Effect of planting methods on some physical properties of rice soils. Indian J. Agron. 25:213–218.

Prihar, S.S., B.P. Ghildyal, D.K. Painuli, and H.S. Sur. 1985. Physical properties of mineral soils affecting rice-based cropping systems. p. 57–70. *In* Soil physics and rice. IRRI, Los Banõs, Philippines.

Prihar, S.S., S.D. Khapra, R. Singh, S.S. Grewal, and S.K. Sodhi. 1990. Water resources of Punjab—A critical concern for the future of its agriculture. Punjab Agricultural University, Ludhiana, India.

Prihar, S.S., K.K. Khera, and Gajri RR. 1975. Effect of puddling with different implements on the water and yield of paddy. J. Res. 13:249–254.

Rahman, S.M. 1991. Tillage effects on some physical properties. Annals Agric. Res. 12:196–199.

Rao, M.S.R.M., and Y.V. Kathavate. 1972. Effect of soil compaction on the yield of wheat and maize. India J. Agron. 17:199–205.

Regmi, A.P., J.K. Ladha, H. Pathak, E. Pasuquin, C. Bueno, D. Dawe, P.R. Hobbs, D. Joshy, S.L. Maskey, and S.P. Pandey. 2002. Yield and soil fertility trends in a 20-year rice–rice–wheat experiment in Nepal. Soil Sci. Soc. Am. J. 66:857–867.

Sahoo, R., A.K. Bandhopadhya, and B.B. Nanda. 1970. Effect of organic manures with and without drainage on rice yield, nutrient uptake and soil aggregation. J. Indian Soc. Soil Sci. 18:51–55.

Sawhney, J.S., and J.L. Sehgal. 1989. Effect of rice–wheat and maize–wheat crop rotations on aggregation, bulk density and infiltration characteristics of some alluvium derived soils. J. Indian Soc. Soil Sci. 37:235–239.

Scheltema, W. 1974. Puddling against dry ploughing for lowland rice culture in Surinam. Cent. Agric. Publ. Doc., Wageningen. Agric. Res. Rep. 828.

Sharad, N., and T.S. Verma. 2001. Improvement in soil productivity with long-term lantana (*Lantana camara* L.) additions in rice–wheat cropping. I. Effects on some soil properties and zinc fractions. J. Indian Soc. Soil Sci. 49:462–466.

Sharma, P.K., and R.M. Bhagat. 1993a. Puddling and compaction effects on water permeability of texturally different soils. J. Indian Soc. Soil Sci. 41:1–6.

Sharma, P.K., and R.M. Bhagat. 1993b. A simple apparatus for measuring clod breaking strength. J. Indian Soc. Soil Sci. 41:422–425.

Sharma, P.K., and L. Bhushan. 2002. Puddlability and structural regeneration of a rice soil amended with waste organic residues. J. Indian Soc. Soil Sci. (in press).

Sharma, P.K. and L. Bhushan. 2001. Physical characterization of a soil amended with organic residues in a rice–wheat cropping system using a single value soil physical index. Soil and Tillage Res. 60:143–152.

Sharma, P.K., and S.K. De Datta. 1985. Puddling influence on soil, rice development and yield. Soil Sci. Soc Am. J. 49:1451–1457.

Sharma, P.K., and S.K. De Datta. 1986. Physical properties and processes of puddled rice soils. Adv. Soil Sci. 5:139–178.

Sharma, P.K., S.K. De Datta, and C.A. Redulla. 1988. Tillage effects on soil physical properties and wetland rice yield. Agron. J. 80:34–39.

Sharma, P.K., K.T. Ingram, and D. Harnpichitvitaya. 1995a. Subsoil compaction to improve water use efficiency and yields of rainfed lowland rice in coarse-textured soils. Soil Tillage Res. 36:33–44.

Sharma, P.K., T.S. Verma., and R.M. Bhagat. 1995b. Soil structural improvements with the addition of *Lantana camara* biomass in rice-wheat cropping. Soil Use Manage. 11:199–203.

Sharma, R.B., G.P. Verma, and M.P. Russell. 1971. Effect of varying bulk densities of surface and subsurface soil on wheat growth under field conditions. Proc. Int. Symp. Soil Fert. Evaluation, New Delhi. Indian Soc. Soil Sci. 1:519–527.

Sharma, R.P., and T.S. Verma. 2001. Dynamics of nitrogen fractions with long-term additions of *Lantana camara* biomass in rice–wheat cropping sequence. J. Indian Soc. Soil Sci. 49:407–412.

Sur, H.S., S.S. Prihar, and S.K. Jalota. 1981. Effect of rice–wheat and maize–wheat rotations on water transmission and wheat root development in a sandy loam soil of Punjab, India. Soil Tillage Res. 1:361–371.

Taylor, H.M. 1972. Effect of drying on water retention of a puddle soil. Soil Sci. Soc. Am. Proc. 36:972–973.

Titkov, A.A., and P.G. Gusev. 1989. Composition and properties of soils of rice rotation in Sivash region of Crimea. Soviet Soil Sci. 21:110–116.

Tomar, S.S., M.B. Russell, and A.S. Tomar. 1981. Effect of subsurface compaction on root distribution and growth of wheat. Zeitschriff fur Acker-und Pflan Zenban. 150:62–67.

Trought, M.C.T., and M.C. Drew. 1980. The development of waterlogging damage in wheat seedlings (*Triticum aestivum* L.) I. Shoot and root growth in relation to change in concentration of dissolved gases and solute in the soil solution. Plant Soil. 54:77–94.

Utomo, W.H., and A.R. Dexter. 1981. Soil friability. J. Soil Sci. 32:203–213.

Van Ouwerkerk, C., and P.A.C. Raats. 1986. Experiences with deep tillage in the Netherlands. Soil Tillage Res. 7:273–283.

Woodhead, T., R. Huke, and E. Huke. 1994. Areas, locations, and ongoing collaborative research for the rice–wheat systems in Asia. p. 68–96. *In* R.S. Paroda et al. (ed.) Sustainability of rice–wheat production systems in Asia. Regional Office for Asia and the Pacific. Food and Agriculture Organization of the United Nations, Bangkok.

Yadav, R.L., K. Prasad, and D.V. Singh. 2001. On-farm Integrated Nutrient Management in Cropping Systems. p. 72. PDCSR Bulletin No. 2001–3, Project Directorate for Cropping Systems Research, Modipuram, Meerut, India.

Yun-Sheng, C. 1983. Drainage of paddy soils in Taihu Lake region and its effects. Soil Res. Rep. No. 8. Inst. Soil Sci., Academia Sinica, Nanjing, China.

6 Technology Options for Rice in the Rice–Wheat System in South Asia

Vethaiya Balasubramanian and Jagdish K. Ladha

International Rice Research Institute
Manila, Philippines

Raj K. Gupta

Rice–Wheat Consortium for Indo-Gangetic Plains
New Delhi, India

Ram K. Naresh

Sardar Vallabh Bhai Patel Univeristy of Agriculture and Technology, Meerut, India

Randhir S. Mehla

Department of Agriculture, Government of Haryana, India

Bijay-Singh and Yadvinder-Singh

Punjab Agricultural University
Ludhiana, India

ABSTRACT

The rice–wheat rotation is a dominant system that contributes immensely to the food security in South Asia. Intensification of the rice–wheat system requires innovative knowledge-based technologies such as precision land leveling; mechanical transplanting, seedling broadcasting, and wet seeding on puddled soil; transplanting on beds and dry drill seeding on flat land or on beds with conventional or minimum/zero tillage for nonpuddled soil; integrated weed control strategies; site-specific nutrient management; and improved machinery for crop establishment and fertilizer placement. Transplanting or direct drill seeding of rice on beds is more efficient in irrigation water use than traditional transplanting of rice on puddled soil. Laser leveling can further improve water use efficiency. The low-till or zero-till rice planting system seems to be promising. Higher yield and the efficient use of inputs in new systems mean lower cultivation costs and higher profit to farmers, lesser adverse effects of crop intensification on the environment, and better resource conservation and system sustainability. Research is ongoing to improve planting machines, to standardize soil moisture for dry seeding in different soil types, to tackle the problem of bird damage of seeds sown in open slits, to prevent micronutrient deficiency and termite attack, and to manage crop residues in low-till systems. Long-term monitoring of grain yield and profit, the evolution of weeds and other pests, soil physical and chemical conditions, and water productivity at the field and watershed level are critical to providing solutions to the sustainability problems of the rice–wheat rotation in nonpuddled soil.

Rice is the staple food for about one-half of the world's population that resides in Asia. More than 90% of the world's rice (596 million megagrams in 1999) is produced and consumed in this region. Hossain and Pingali (1998) estimated that about 250 million smallholders in Asia are engaged in rice farming, and most of them use modern rice varieties and production technologies. Thus, rice is an integral part of Asian culture.

Two to three rice crops are grown per year if water is available for such intensive, continuous rice systems. Rice is also rotated with other crops. The rice–wheat rotation is one of the most important rice-based systems in South Asia, covering an area of about 13.5 million ha. Both rice and wheat contribute greatly to food security in this region.

In South Asia, the rate of rice production kept pace with or even exceeded population growth from the mid 1960s to the early 1990s (Hossain & Fischer, 1995). Wheat experienced a similar trend in production during the same period. However, if we become complacent about food production because of recent food surpluses experienced in some Asian countries, a food crisis could be inevitable in the near future. The South Asian population still grows around 2% annually, land and water resources available for rice and wheat cultivation are declining rapidly, farmers are losing interest in food crop farming because of decreasing profit, and communities are increasingly concerned about farming-related pollution of the environment. Thus, we urgently need new strategies and management techniques to ensure adequate food security for the growing population, to promote efficient use of inputs and other resources, to reduce costs and increase profits to farmers, and to minimize the adverse effects of farming on the environment. This means that increasing the yield per unit of all inputs, particularly land, water, fertilizer, and labor, is a must to produce the additional food more efficiently. This effort, if successful, will help lower the cost of food production and limit the adverse effects of crop intensification on the environment.

A systems approach is necessary to increase the productivity of both rice and wheat crops. The drastically different seedbed requirements for rice and wheat create problems in tillage, the timeliness of wheat sowing, the maintenance of soil structure, and the management of irrigation water, weeds and other pests, fertilizers, and crop residues (Ladha et al., 2000). Reduced wheat yield because of the late planting of wheat is a serious problem in the rice–wheat system (Ortiz-Monasterio et al., 1994). The response of late-planted wheat to N application is also poor (Saunders, 1990). Planting medium-duration (140 d) basmati rice or a third crop between rice and wheat and/or the long turnaround time required for elaborate tillage in traditional wheat crop establishment after puddled rice are the root causes of delayed wheat planting (Hobbs & Gupta, 2003, Chapter 7 of this publication). Post-rice soil conditions must be favorable for succeeding nonrice crops, including wheat, and tillage methods have to be modified to reduce the turnaround time between rice and wheat if we are to improve the total system productivity. Technologies proposed for either crop must solve these problems, in addition to increasing the input use efficiency and reducing the cultivation cost of individual crops (Hobbs & Gupta, 2003, Chapter 7 of this publication). This chapter reviews the technical advances in rice production in the context of improving rice–wheat system productivity.

TECHNOLOGY OPTIONS FOR RICE PRODUCTION

Rice scientists of national agricultural research and extension systems (NARES), the International Rice Research Institute (IRRI), and advanced research laboratories continue to develop improved and innovative rice production technologies that can help increase the productivity of rice lands worldwide. Selected technologies and strategies in land preparation and leveling, crop establishment, weed control, site-specific nutrient management, and improved planting equipment are discussed below. Chapter 7 of this publication (Hobbs & Gupta, 2003) presents the resource-conserving technologies for wheat. Fitting of these technologies for rice and wheat crops is also considered to increase the total productivity of the system.

LAND PREPARATION

Puddling is the traditional method of land preparation for establishing rice. It produces soft mud for easy transplanting of rice seedlings, reduces water percolation losses, provides anaerobic conditions that promote rice growth (Sanchez, 1973), lessens weed competition, and increases the availability of certain nutrients such as P and Zn. Repeated puddling leads to formation of a hardpan at shallow depth that helps retain water in flooded rice fields. However, puddling destroys soil aggregates, breaks capillary pores, disperses clay particles, and lowers soil strength in the puddled layer (Hobbs & Gupta, 2000; Sharma & De Datta, 1985). The hardpan created by repeated puddling may restrict root growth and performance of crops rotated with rice. Nonrice crops such as wheat require contrasting soil conditions such as aerobic soil, good soil structure, and a deep soil layer for root growth. In traditional cultivation systems, extensive tillage is used to establish nonrice crops after puddled rice.

Alternative tillage and crop establishment methods are needed to facilitate the timely establishment and growth of both rice and wheat crops. Nonpuddled flat-surface or raised-bed and/or zero-till systems are undergoing experiments for direct seeding and/or transplanting of rice.

Conventional Wet Tillage and Land Leveling

Farmers use animal-drawn plows, power tillers, and/or four-wheel tractors for wet tillage in rice fields. Water leveling is the most common method in Asia, and typical leveling devices include wooden or metal harrows, wooden planks, sand-filled plastic piles (Fig. 6–1a), bamboo poles, and even banana stems (Bell et al., 1998). The method is highly laborious, and leveling precision strongly depends on operator skill and field size. The smaller the fields, the better will be the manual leveling. Farmers achieve fairly uniform seeding, good water control, and relatively higher yields in leveled than in unleveled rice fields.

Laser-Assisted Precision Leveling

Land leveling has become faster, easier, more precise, and less labor-intensive with the advent of the laser technology. The equipment for the laser-leveling system is described below.

Fig. 6–1. Manual (Andhra Pradesh, India) and laser leveling on puddled soil (IRRI, Philippines).

Dry Laser Leveling

The laser-leveling system has been well adapted to the commonly practiced dry leveling. Work rates depend on tractor capacity and the amount of soil to be moved. Tractor leveling takes about 8 h ha^{-1} with a back-blade and 4 h ha^{-1} with a drag bucket. In northeast Thailand, the variability in soil elevation was reduced from 21 ± 10 cm before leveling to 6.9 ± 3.1 cm after leveling. Releveling is not necessary for 8 to 10 yr if appropriate plowing techniques are followed. Farmers need to plow from the center of the field outward rather than the conventional practice of plowing from the periphery inward to keep the land level for many years (Bell et al., 1998). The cost of leveling is proportional to the amount of soil moved plus the fixed cost of machinery. The higher the variability in within-field elevation or surface level, the higher will be the cost of leveling. Estimated leveling cost was $80 to $90 ha^{-1}, while commercial contractors charged $160 to $180 ha^{-1} for land leveling in Cambodia (Rickman et al., 1998).

Wet Laser Leveling

Rice farmers prefer wet leveling in puddled fields. Therefore, a wet laser-leveling system (Fig. 6–1b) has been successfully developed at the Philippines-based IRRI in collaboration with Spectra-Precision of the United States (Bell et al., 1998). Selection of the appropriate tractor with proper tire configuration is critical for wet leveling. In addition, the texture of the soil and the timing of the introduction of water into the field for land preparation and wet leveling are critical for tractor mobility. Shallow tillage immediately after the rice harvest (when soil moisture is favorable) or dry shallow tillage before the soaking of cracked soil has been shown to save 30 to 50% water (≈100 mm) for wetland preparation (IRRI, 1998). Such tillage practices will also facilitate tractor mobility during wet leveling.

Benefits of Laser Leveling

Laser-assisted precision leveling has been shown to improve water management, crop stand, and productivity in direct-seeded rice in temperate countries (Hill et al., 1991) as well as in the tropics (Bell et al., 1998; Rickman et al., 1998). Hill et al. (1991) rated the development and use of laser technology for precision leveling as second only to the breeding of high-yielding semidwarf rice varieties for increased yields in California during the 1980s. With laser leveling, significant savings in irrigation water, efficient use of inputs, and higher yields are reported in temperate countries (Barnes, 1996; Luzes, 1991). In rainfed lowlands of Cambodia, land leveling facilitated faster crop establishment by direct seeding, reduced the labor for weeding from 21 to 5 person d ha^{-1}, improved water use efficiency, and increased rice yields by more than 30% (Rickman et al., 1998). In irrigated rice fields of Pakistan, leveling has been shown to reduce water use by up to 50%, facilitate uniform germination and crop establishment, and increase yields by as much as 25% [Rice–Wheat (Pakistan–Punjab) Interim Report, 1999]. Kahlown et al. (2000) reported that dry laser leveling improved crop performance in nonpuddled soil with zero-till surface seeding or seeding on permanent beds. Water productivity was also better in laser-leveled, zero-till, and bed-planted wheat compared with the conventional check.

Nonpuddled Wet Soil System for Transplanted Rice

Certain public and private sector organizations are experimenting with less tillage and/or no puddling and using nonselective herbicides such as glyphosate [*N*-(phosphonomethyl)glycine] to kill weeds before crop establishment. Preliminary results show (i) increased land and water productivity, (ii) higher rice yields, and (iii) better soil quality (IRRI, 1998). The prolonged absence of puddling in flooded soil may lead to high percolation losses of water and leaching of nutrients beyond the root zone, profuse algal growth, and difficulty in manual transplanting (possible finger injury because of hard soil). Water use and weed control appear to be the key problems that will confront the nonpuddled flooded rice system. There is much less soil cracking in the nonpuddled wet rice system when soils dry after the rice harvest, and this will have implications for water requirement and water use efficiency. We need to quantify the total water use and water use efficiency of transplanted, flooded rice in puddled and nonpuddled soils. Weeds are a serious problem in the nonpuddled wet soil system. Research is ongoing to develop and evaluate a suitable combination of tillage, flooding, and row planting coupled with the use of the rotary or conical weeder, and/or crop rotations with a minimum supplementary use of herbicides to control weeds in the nonpuddled wet rice system.

CROP ESTABLISHMENT

Transplanting on Puddled Soil

In certain soils (Vertisols), puddling followed by transplanting of seedlings is essential to establish rice crops. In South Asia, rice seedlings are grown in wet or dry nursery beds and seedlings are pulled out and transplanted in the main field 21 to 40 d (average 30 d) after seeding. Rice seedlings can be transplanted either manually or by machine.

Manual Transplanting

The traditional method of manual transplanting is one of the most time-consuming and laborious operations in rice cultivation. It consumes more than 30 person d ha^{-1}. Grain yield is limited by low plant population as a result of contract manual transplanting on an area basis, as practiced in western India (Gupta et al., 2000b) and Pakistan [Rice–Wheat (Pakistan–Punjab) Interim Report, 1999]. It is difficult to maintain optimum plant–plant spacing and crop density under random transplanting practiced in many Asian countries.

Mechanical Transplanting

Mechanical rice transplanters are being tried in parts of India and Pakistan to tackle the increasing labor scarcity and to maintain plant spacing and crop density. The most cost-effective machine tried so far is the Chinese rice transplanter, which costs about $2750, compared with $15 000 for an imported Japanese model (Gupta et al., 2000b). Rice seedlings are raised as mats on plastic sheets (mat nursery). Seedling mats are placed in the transplanter that is used to plant the seedlings in the field at predetermined spacing. The key to success is the raising of seedlings in the mat nursery. If mat nurseries are raised commercially, they may provide an income-generating opportunity for landless labor in some countries. Good land preparation and leveling are essential for successful transplanting by machine. The settling time after the last puddling (1–2 d depending on the soil type) and the water level (<10 mm) in the field are critical for machine transplanting.

Field evaluation of the Chinese transplanter in Karnal, India showed faster recovery of seedlings from transplanting shock, higher plant population and tillering, and uniform maturity—all resulting in higher grain yields (R.K. Malik, 2000, personal communication). Another experience in Pakistan indicates that machine transplanting saves labor, reduces stress and drudgery, and ensures timely planting and optimum spacing and plant density [Rice–Wheat (Pakistan–Punjab) Interim Report, 1999]. Machine transplanting is highly successful, and crop establishment is uniform in laser-leveled fields. Missing hills are common with machine transplanting in poorly prepared and leveled fields. Operators have to be trained on the proper use of the machine, and the transplanter has to be modified for local conditions to minimize missing hills and other problems.

Seedling Broadcasting

The system of seedling broadcasting has been tested and used in Sichuan Province of China. Seedlings are raised on plastic bubble sheets placed on raised beds. The resulting seedlings are easily uprooted and removed from individual depressions as soil plugs (i.e., each seedling is attached with a clod of soil) and broadcast onto the puddled fields. A mechanical blower is also available to help with seedling broadcasting. After broadcasting, some gap filling and turning of some seedlings upright are necessary to ensure a more uniform distribution of seedlings. Because the roots are encased in soil, the period of recovery is reduced. However, flotation and uneven distribution of seedlings, high labor cost for raising seedlings, disposal of the plastic sheet after use for two seasons, and difficulty in weed control are some of the problems encountered in this method. In some parts of China, plastic bubbled trays are used to prepare seedlings for broadcasting in the main field, and this method saves on seed and labor cost for uprooting seedlings. It is especially suitable for hybrid rice where the seed cost is very high. Gupta et al. (2000a) provided a detailed description of the seedling broadcasting method and its suitability for South Asian farmers, particularly in Bangladesh and Nepal.

Wet Seeding of Rice on Puddled Soil and the Drum Seeder

In this system, the fields are plowed and puddled and then sprouted seeds are sown on the wet soil either by broadcasting or by using a drum seeder (Fig. 6–2). Many farmers adopt wet seeding to save on labor cost, to reduce drudgery, to establish crops in a timely manner, and to improve crop stand (De Datta, 1986; De

Datta & Flinn, 1986). Good land preparation and leveling, the use of good-quality seed, precision water management for the first 7 to 10 d, and effective weed control are critical for success in wet-seeded rice (WSR) (Balasubramanian & Hill, 2002). It is becoming increasingly popular in Asia because of the availability of suitable rice varieties for wet seeding and effective herbicides for weed control (Pandey & Velasco, 1999). Wet seeding could become an ecofriendly method when sprouted seeds are sown in rows and rotary or conical weeders are used to eradicate weed, instead of applying herbicides for weed control. Periodic interrow cultivation followed by field drying for 2 to 6 d may also aerate the root zone of rice plants. This system, however, extends the main field time by 12 to 15 d for growing rice (Yoong & Hui, 1982), an important consideration where triple cropping (e.g.,

Fig. 6–2. Broadcasting and drum seeding in rows of pregerminated seeds on puddled soil (Andhra Pradesh, India).

rice–rice–wheat as in eastern India) is practiced or the climate is cooler. There are two types of wet seeding: surface and subsurface wet seeding.

Surface Wet Seeding

In this case, sprouted seeds are broadcast or sown in rows by a drum seeder on the surface of the puddled soil. The person sowing the seed must walk backward without making too many depressions in the field that will collect water and rot the seeds. The seed rate varies from 100 to 150 kg ha^{-1} for broadcasting and one-half as much for drum seeding. Surface-sown seeds lay half buried in the mud and half exposed to the air. Thus they are vulnerable to damage by birds, rats, snails, ants, torrential rain, or dry weather.

Subsurface Wet Seeding

In this system, pregerminated seeds are broadcast immediately after puddling and the suspended mud is allowed to settle down and form a thin protective cover over the seeds sown. Thus, the seeds are protected against damage by weather and/or pests. Alternately, pregerminated seeds can be sown in rows 1 to 2 d after puddling by an anaerobic seeder fitted with furrow openers and closers. The seed rates are similar to that of surface seeding (Balasubramanian & Hill, 2002). Rice varieties tolerant of low oxygen conditions during germination are needed for successful crop establishment by subsurface wet seeding (Yamauchi et al., 1993).

Dry Seeding of Rice on Nonpuddled Soil

In this system, rice is grown as any other upland crop by sowing the dry seed on well-prepared dry or moist, but not puddled, soil. The Chinese hand tractor with tiller–seeder–planking attachment or the Pant Nagar seed-cum-fertilizer drill could be used for drilling dry rice seeds in rows. The dry-seeded crop can be raised as aerobic rice with no flooding throughout the crop cycle or as semidry rice with flooding after the crop has reached a certain stage. *Aerobic rice* is defined as high-yielding rice that grows in nonpuddled soil with no standing water in bunded fields that can be supplementary irrigated, if necessary, and/or flooded by rain occasionally. However, this system will extend the main field time by about 15 d for the crop to mature. In addition, weeds are a major problem in dry-seeded rice (DSR), especially under no flooding after crop establishment. If weeds can be controlled by suitable methods, this system would have many benefits. It would do away with the labor-intensive seedling nursery and transplanting systems. Postrice upland crops would have better soil conditions for establishment and growth and therefore higher yields (Hobbs et al., 2002). Water used for puddling can be saved, but deep percolation losses may increase in nonpuddled soil. Thus, the total water balance needs to be evaluated on different soil types. Nitrate accumulated in the soil during the dry fallow period will not be denitrified because of the absence of early flooding in nonpuddled soil. However, fertilizer chemistry would be different in moist aerobic soil, and a strategy to reduce percolation losses of nitrate would be needed. Micronutrients such as Zn and Fe may become less available under aerobic conditions, especially in the high-pH saline soils that are widespread in the Indo-Gangetic Plains.

Soil insects such as termites may become serious. Long-term trials of DSR should evaluate the sustainability of this system by looking at yield trends, the dynamics of weeds and other pests, water retention and water use efficiency, nitrate leaching and nutrient use efficiency, and changes in soil quality with time.

Zero-Till Dry-Seeded Rice

One possible subsystem of DSR could be zero-till rice. In this system, fields are flush irrigated and weeds are allowed to germinate and are then controlled by applying nonselective herbicides such as glyphosate. A zero-till drill is then used to sow the rice seed. Since the soil is not disturbed, fewer weeds may germinate and methane emissions will also decrease because of the residues not being incorporated into the soil as under puddled conditions. Mulching of residues on the soil surface may also suppress weeds in the zero-till system.

Another version of zero-till rice is practiced in China. After the wheat harvest, the fields are flooded and seedlings are transplanted or broadcast into the softened soil after a few days. Continuous use of this system for 10 yr in China gave good yields and weeds became fewer with time, possibly because of the depletion of the weed seed bank in the soil (Hobbs et al., 2000). This system would reduce the cost of land preparation. It would be good for areas with clay soils (Vertisols) that cannot be tilled easily.

Planting of Rice on Raised Beds

This is similar to bed planting of wheat, which has attracted the attention of farmers in India (Haryana and Punjab) and Pakistan (Punjab). A bed-planting machine is used to prepare raised beds and furrows (Fig. 6–3). A 27-cm furrow separates each 47-cm-wide bed. Two to three rows of wheat are planted on the bed. Several advantages cited for bed-planted wheat are reduced seed rate, water use, and

Fig. 6–3. Bed formation and dry drill seeding on raised beds, Parwanipur, Nepal.

crop lodging; more efficient weed control and fertilizer placement; conservation of rainwater and avoidance of flooding; and facilitation of a last irrigation at grain filling (Gupta et al., 2000b).

Bed planting was tried for rice for the first time in Haryana, India, during the 2000 wet season (kharif) and in Parwanipur, Nepal, during the 2001 wet season. The preliminary results are highly encouraging. The beds can be made fresh or the beds from the previous wheat crop can be used after reforming. Farmers who tried the bed planting of rice in 2000 reported good crop performance, high yields, and considerable savings in irrigation water in the raised-bed system compared with transplanted flooded rice. Maybe the controlled traffic of the tractor in the furrow bottoms is providing sufficient compaction to reduce water percolation in furrows. This needs further study to see if it exists over years. The advantages of fertilizer placement and ease of access for spraying and weeding are obvious for bed-planted rice. Hybrid rice appears to perform well in raised beds (R.K. Gupta, Rice–Wheat Consortium, New Delhi, India, personal communication, 2001). There are two ways to establish rice on beds: transplanting of seedlings and dry seeding.

Transplanted Rice on Beds

To establish transplanted rice on beds (TPR-B), seedlings are transplanted on moist beds and furrows are kept flooded in early stages to suppress weeds. If necessary, postemergence herbicides are applied in later stages to control weeds. Seedlings transplanted on beds recovered fast and tillered profusely.

Dry-Seeded Rice on Beds

With the dry-seeded rice on beds (DSR-B) system, when dry rice seeds are drilled on beds, seed germination and initial crop establishment appear to be problematic, especially in the absence of complete saturation of beds for the first 20 to 25 d. The bed planter could be improved to ensure seeding at the right depth with adequate moisture on beds. Weeds are a serious problem in DSR-B. Thus there is an urgent need to develop an integrated weed management system for DSR-B as discussed below in "Weed Control in Puddled and Nonpuddled Rice Systems". Mechanical weeding in the furrows is possible only if the fields are dry enough to allow the entry of a tractor. Suitable herbicides have to be identified for DSR-B. There were problems of Zn and Fe chlorosis and termite attack in a few fields with DSR-B. Experiments are ongoing to solve the various problems of DSR-B and to evaluate the effects of DSR-B on the rice–wheat system. These trials will also evaluate the continuous use of permanent beds for rice–wheat planting and its effects on system sustainability.

Effects of Crop Establishment Methods on Rice Yields

In general, crop performance was better for transplanted than for DSR-B. One farmer, Mr. Mann, in Karnal District of Haryana, India, reported a grain yield of more than 8 Mg ha^{-1} for TPR-B. He told us that the yield of TPR-B was slightly higher than that of TPR on puddled soil. During the 2001 wet season, 19 farmers evaluated TPR-B and 34 evaluated TPR on puddled soil; the mean yield was sig-

Table 6–1. Rice grain yield and yield components under different crop establishment methods, Ghaziabad, India, 2000 and 2001 wet (kharif) season.

Method of crop establishment	Farmer collab-orators	Grain yield	No. of plants at 30 DAS†	Productive tillers per plant	Panicle length	Grains per panicle
	no.	Mg ha^{-1}	m^{-2}	no.	cm	no.
			2000 wet (kharif) season			
TPR-puddled soil‡	8	4.64	--	11	19.4	179
DSR-B§	8	4.45	28	15	20.4	186
LSD¶	--	0.08	--	1.5	0.96	NS#
			2001 wet (kharif) season			
TPR-puddled soil	34	5.29	26	12	21.8	163
TPR-B††	19	5.62	34	17	23.3	172
DSR-B	22	5.02	35	15	22.9	165
DSR-low till‡‡	7	5.19	32	13	22.8	168
DSR-zero till§§	10	5.69	56	13	21.9	164
LSD	--	0.25	1.2	1.2	0.50	4

† DAS, days after seeding.
‡ TPR, transplanted rice on puddled soil (traditional method).
§ DSR-B, dry drill-seeded rice on beds.
¶ LSD, least significant difference at $P = 0.05$.
# NS, not significant at $P = 0.05$.
†† TPR-B, transplanted rice on beds.
‡‡ DSR-low till, dry drill-seeded rice on flat land with reduced tillage.
§§ DSR-zero till, dry drill-seeded rice on flat land with zero tillage.

nificantly higher for TPR-B (5.62 Mg ha^{-1}) than for the traditional TPR on puddled soil (5.29 Mg ha^{-1}) as shown in Table 6–1.

The performance of DSR-B was poorer than that of TPR on puddled soil. During the 2000 and 2001 wet seasons, grain yield was significantly lower in DSR-B in contrast to TPR on beds or on puddled soil (Table 6–1). Dry-seeded rice on beds produced significantly more productive tillers per plant and longer panicles than TPR on puddled soil, but did not result in higher grain yields. The number of grains per panicle was not different between the two methods of crop establishment. It is difficult to explain why grain yield was low in DSR-B. Maybe grain filling was poor in DSR-B in contrast to TPR on puddled soil. We need to examine the extent of grain filling (percentage filled grains) and grain weight to better explain the data. If grain filling is a problem, we may have to improve the seed rate and/or crop nutrition to improve grain filling and increase grain weight. Another system that holds high promise is DSR on flat land. Dry-seeded rice on flat land with reduced or zero tillage produced rice yields similar to or higher than that of TPR on puddled soil (Table 6–1).

Irrigation Water Use under Different Crop Establishment Methods

Traditional transplanted rice (TPR) on puddled soil required more than 24 irrigations per season. With a mean time of 10 h per irrigation ha^{-1}, the total pumping hours per hectare per season worked out to be 244 h (Table 6–2). Zero-till TPR on flat land reduced the total pumping hours per hectare per season by as much as

Table 6–2. Details of irrigation and water savings under different crop establishment methods, Ghaziabad, India, 2001 wet (kharif) season.

Method of crop establishment	Farmer participants	Mean time per irrigation	Mean number of irrigations	Total pumping hours	Savings in pumping time over traditional method
	no.	h ha^{-1}	—— ha^{-1}	——	%
TPR-puddled soil† (traditional method)	33	10.0	24.4	243.7	--
TPR-zero till‡	3	8.5	24.3	208.4	15
TPR-B§	18	7.5	19.7	147.5	40
DSR-B¶	19	7.4	21.1	154.4	37
LSD#	--	0.5	1.3	15.7	--

† TPR, transplanted rice on puddled soil (traditional method).
‡ TPR-zero till, transplanted rice on flat land with zero tillage.
§ TPR-B, transplanted rice on beds.
¶ DSR-B, dry drill-seeded rice on beds.
# LSD, least significant difference at $P = 0.05$.

15%. However, the bed-planting methods were significantly more efficient than the traditional transplanting of rice on puddled soil. The number of irrigations per season, the time taken per irrigation per hectare, and the total pumping hours per hectare per season were significantly lower for TPR-B than for TPR on puddled soil (Table 6–2). Overall, there is a 40% savings in pumping hours when rice is transplanted on beds instead of on flat land with puddled soil. Similarly, DSR-B reduced the total pumping hours per hectare per season by 37%.

Equipment Needs for Rice Crop Establishment

Mechanical Transplanter for Puddled Soil

It is difficult to develop cost-effective equipment for transplanting rice seedlings on puddled soil. Various transplanters imported from China, Korea, and Japan are being tried in India and Pakistan. The cost varies from $600 for the IRRI-designed manual transplanter to $2750, $3500 and more than $15 000 for the Chinese, Korean, and Japanese machines, respectively. Their capacity ranges from 0.15 to 1 ha d^{-1} depending on the type of machine used. Local manufacturers have started modifying the Chinese transplanter to improve its performance under local conditions. The transplanters require young (about 15-d-old) seedlings raised in a mat-type nursery. Shallow puddling is beneficial for mechanical transplanting.

Drum Seeders for Row Wet Seeding on Puddled Soil

The drum seeder is a simple machine designed to sow pregerminated seeds in rows on puddled soil. The distance between rows can be adjusted easily. Generally, the drum seeder opens shallow furrows and sows the seeds in a continuous row; the speed of pulling must be maintained at a constant rate to drop the seeds uniformly into the rows. If necessary, the seeder can be fitted with furrow closers to cover the seeds sown in furrows. The number of holes in the drums can be altered by closing certain holes with a tape to control the distance between hills in a row. The drum seeder can sow 0.5 to 0.6 ha d^{-1}. It can be manufactured in small metal workshops at a cost of $30 to $50 per unit.

Seed-cum-Fertilizer Drills for Dry Drill Seeding on Flat Land or on Raised Beds

The university at Pantnagar, India has developed the seed-cum-fertilizer drill for conventional-till and zero-till planting of wheat. A Chinese hand tractor with a tiller–seeder–planking attachment is available for tilling, sowing, and planking at the same time. Both machines can be modified for dry drill seeding of rice on flat land. The seed tube has been redesigned to facilitate the free flow of rough rice (paddy) seeds. The seed drill attachments cost $400 to $500 per unit.

Machines have been developed for making raised beds, drilling seeds, and firming the soil after sowing of wheat. The wheat bed-planter has been modified for rice seeding on beds by redesigning the seed tube. The bed-planter has been fitted with fertilizer drills to place fertilizer between rows or within seed rows to complete the bed formation, seeding, and fertilizer placement operations in one pass. The planter-cum-fertilizer drill can also place urea briquettes at optimum soil depths. The bed-planter-cum-fertilizer drills cost about $600 per unit.

Laser-Leveling System

The laser-leveling system consists of a laser transmitter, a laser receiver and mast, a control box for tilt control, and a hydraulic interface. The tractor-leveling system consists of (i) mounted-type equipment such as a back-blade or drag bucket or box suitable for small tractors (30–80 HP), (ii) an easy-to-maneuver dragged scraper for tractors of 50 HP or higher, and (iii) a carry-all scraper for heavy tractors of 110 HP or higher (Bell et al., 1998). The laser-leveling system has been well adapted to dry leveling. Experience from Cambodia indicates that tractor leveling takes about 8 h ha^{-1} with a back-blade and 4 h ha^{-1} with a drag bucket (Rickman et al., 1998). The laser system costs $7500 to $15 000 per unit.

Impact of Rice Crop Establishment Methods on System Productivity

Various rice crop establishment methods, their characteristics, and their effects on the wheat crop are summarized in Table 6–3. With the traditional transplanting of rice on puddled soil, wheat planting is delayed because of the late rice harvest, the need for elaborate tillage in traditional systems, and the long turnaround time between the rice harvest and wheat planting (Fujisaka et al., 1994). Late planting of wheat after the end of November has been shown to decrease wheat yield by 1 to 1.5% d^{-1} of delay (Ortiz-Monasterio et al., 1994). Poor soil physical conditions after puddled rice lead to poor wheat germination and crop stand, while the hardpan formed by repeated puddling often restricts root growth in wheat, especially in fine-textured soils. Immediately after the first irrigation, yellowing of wheat is commonly observed in poorly drained and/or waterlogged soils with hardpans. This is not a problem in well-drained light-textured soils. Hobbs and Gupta (2000) reported that puddling had an adverse effect on the following wheat crop on fine-textured silty loam or silty clay loam soils, but not on coarse-textured sandy loam soils of the Indo-Gangetic Plains of India and Nepal.

Alternative rice crop establishment methods on puddled soil, such as machine transplanting, seedling broadcasting, and direct wet seeding (Table 6–3), are being evaluated to enable the timely planting of rice and wheat, which can enhance the total productivity of both crops. A critical issue in WSR is that the crop will occupy the main field for about 15 d longer than TPR, and it may delay wheat seeding in areas where two rice crops or one rice crop plus a nonrice crop precede a wheat crop, as in eastern India. Preparation of a stale seedbed (allowing initial weeds to germinate with a flush irrigation and killing them with a nonspecific herbicide such as glyphosate) for rice and zero tillage for wheat are also being evaluated for the timely planting of both crops. Weeds could become a serious problem in WSR and low-till systems. Row seeding with a drum seeder and the use of a rotary or conical weeder will help control weeds in WSR. Supplemental use of herbicides may be needed for WSR and reduced-tillage planting systems on puddled soil.

Nonpuddling of soil for rice crop establishment is another way to ensure the timely planting of both crops and to maintain good soil physical conditions for nonrice crops after rice. Nonpuddling will also prevent the formation of a root growth–restricting hardpan at shallow depth. Data from India and Nepal show that wheat yield is higher after nonpuddled than puddled soil cultivation for rice, whereas rice yields are equal under both conditions if weeds are effectively controlled (Hobbs et al., 2002). Thus, nonpuddled cultivation is likely to increase the total productivity of the rice–wheat system. Salient features and potential effects on wheat of various rice crop establishment methods on nonpuddled soil are summarized in Table 6–3. Dry-seeded flooded rice on flat land seems to be attractive to farmers if weeds could be controlled effectively. On the other hand, dry-seeded aerobic rice (no standing water throughout the crop growth period) suffers from poor germination, inadequate crop stand, serious weed problems, deficiency of micronutrients (Zn, Fe, Cu), and termite attack. Planting of rice and wheat on permanent beds is another alternative to ensure the timely planting of both crops and to maintain good soil structure for nonrice crops. The raised-bed system reduces seed rate, saves on irrigation water by up to 50%, conserves rainwater and avoids flooding, facilitates more efficient weed control and fertilizer placement, and minimizes lodging in the wheat crop (Gupta et al., 2000b). Transplanting of rice seedlings on beds has been successful, whereas dry seeding of rice on beds suffers from problems similar to those of dry-seeded aerobic rice on flat land. Our experience shows that we need to keep the furrows flooded and the beds saturated with water for the first 25 to 30 d to ensure uniform germination and crop stand in DSR-B. The effectiveness of applied herbicides will be higher when soils are saturated during the first 3 to 7 d after seeding on beds.

Short-term problems that confront the new crop establishment methods include the standardization of soil moisture for dry seeding in different soil types, micronutrient and termite problems on dry-seeded aerobic rice, and bird damage of seeds sown in open slits or on the soil surface. Appropriate seed-cum-fertilizer drills have been developed for proper seeding and fertilizer placement on flat land and on raised beds. Precision leveling is a prerequisite for uniform crop establishment, weed suppression, and efficient water management in all the systems.

The long-term effects of nonpuddling and reduced irrigation on crop yield, overall profit to farmers, percolation loss of water, input-use efficiency, weed ecol-

Table 6–3. Rice crop establishment methods, their characteristics, and their effect on wheat yield and total productivity in the rice–wheat system in South Asia.

Rice crop establishment (CE) methods	Characteristics	Research and development issues	Total system productivity
		Puddled soil	
Manual transplanting	High labor need, high drudgery, time-consuming, variable spacing and variable & often low plant density, high water use	Adaptation of machine transplanting, development of alternative CE methods	Less total productivity because of reduced rice & wheat yields, poor soil structure for wheat, high water use, high fuel cost for tillage, less profit, poor system sustainability.
Seedling broadcasting	Fast & easy CE, less drudgery, uneven distribution of seedlings & need for gap filling, high labor need for nursery, difficulty in weeding	Evaluation of zero-till seedling broadcasting as practiced in China, weed control research in zero-till wet soil system	Less total productivity because of reduced rice yield, poor soil structure for wheat, high water use, high fuel cost for tillage, less profit, poor system sustainability
Machine transplanting	Timely rice planting, optimum plant spacing and density, less labor and drudgery	Local development and commercialization of suitable transplanters, training of farmers in mat-type nursery	Fairly high total yield, poor soil structure, high water use, high fuel cost for tillage, less profit, poor sustainability of the system
Direct wet seeding (DWS)	Same as above, but more weed problems, initial water control should be good, longer main field occupation	Evaluation & promotion of row seeding by drum seeder, determination of total water use, long-term shift in weeds	Same or slightly less total yield, poor soil structure, high water use, high fuel cost for tillage, less profit, poor sustainability of the system

Nonpuddled soil			
Zero-till transplanted or direct-seeded rice in wet soil	Fast CE, less drudgery, less energy use, less methane emissions, but less uniform plant stand, more weed problems, possible only in heavy clay soils	Weed control methods, total water use and productivity, monitoring of long-term effect on soil quality & system sustainability	Fairly high total yield, more profit, good soil structure, less fuel cost for tillage, medium to high water use, long-term effect not known
Dry drill-seeded (DDS) flooded rice on flat land with conventional or minimum tillage	Fast & timely seeding, less labor need and less drudgery, fairly good crop stand, weed problems at CE stage, longer main field occupation, high water percolation loss	Weed control methods, especially for low-till systems, percolation losses, total water use, & N leaching, nutrient management, system sustainability	Fairly high total productivity, good soil structure, possibly higher water percolation loss, shift in weed ecology, system sustainability not known
Dry drill-seeded (DDS) aerobic rice on flat land with conventional or minimum tillage	Same as above, plus: poor or uneven crop stand, severe weed problems, micronutrient deficiency, termite attack	Same as above, plus: how to ensure good crop stand, management of micronutrient deficiencies & termite problems	Same as above plus: slightly less total productivity because of less rice yield, shift in weed ecology & possible pest buildup, especially in low-till systems
Transplanted rice (TPR) on beds	Timely CE, but requires more labor, optimum plant density & good tillering, less water use, easy for weeding and fertilizer–pesticide application, less flood damage & lodging, high yield	Long-term yield trends, efficiency of inputs, total water use, shifts in weed species & pests, integrated weed management methods, nutrient management options	High total yield, more profit, less water use, good soil structure, possibly higher water percolation loss, less machinery time and fuel cost, system sustainability of permanent beds not known
Dry drill-seeded (DDS) aerobic rice on beds	Timely & easy CE, but poor germination & crop stand, easy for weeding and fertilizer–pesticide application, less flood damage, less water use, serious weed & termite problems, micronutrient deficiency, low yield	Same as above, plus: how to manage micronutrient deficiencies & termite problems, machinery for drill seeding & fertilizer placement on beds	Same as above, but severe weed problems and shift in weed ecology

ogy and management, pest buildup, micronutrient needs, soil physical and chemical properties, environmental impact, and finally system sustainability have to be assessed.

WEED CONTROL IN PUDDLED AND NONPUDDLED RICE SYSTEMS

Weeds become a serious problem in rice when puddling and/or flooding is no longer practiced. Effective weed control strategies are needed to manage weeds in alternative crop establishment systems.

Strategies for Cultural Control of Weeds in Rice Systems

The three major factors that determine the severity of weed infestation in rice systems are (i) flooding time, flooding duration, and water depth; (ii) type of tillage and puddling; and (iii) land leveling. Generally, puddling and early flooding are used to suppress weeds in TPR. Weed infestation is more severe in shallow (5 cm) water depth than in deep (10 cm) flooding in both transplanted and WSR on puddled soil (Janiya et al., 1999). Weed infestation is high in WSR because of the initial nonflooding for 10 to 15 d to ensure good germination and seedling establishment. Farmers in the Philippines and Vietnam use high seed rates (150–250 kg ha^{-1}) to suppress weeds in early stages, thin the rice plants later, and use the excess seedlings to fill gaps in patchy areas (Balasubramanian & Hill, 2002). Inadequate land preparation and leveling result in poor conditions for seeding and water control that in turn aggravate weed problems in direct-seeded rice. The relative dominance of weed flora changes dramatically with field level and water depth (Hill et al., 2001). Nonpuddled aerobic conditions generally lead to the dominance of grassy weeds such as *Echinochloa* spp. and *Leptochloa sinensis* (L.) Nees. Another factor that determines the severity of weed infestation is the buildup of a weed seed bank in soils under continuous direct seeding (Ismael et al., 1994). The type and frequency of tillage have been found to determine the size and distribution of the weed seed bank in soil and influence the breakdown of seed dormancy of certain weed species (Azmi & Mortimer, 1999). Thus, the severity of weed infestation in rice system depends on the interaction among tillage and/or puddling, land leveling, and flooding and water management. In general, weed infestation increases from puddled transplanted to puddled direct-seeded to nonpuddled direct-seeded rice systems. A good knowledge of this interaction is critical for cultural control of weeds in rice systems. This can be supplemented by the use of submergence-tolerant rice varieties that could germinate and grow well under flooded conditions that suppress grassy weeds and sedges (Hill et al., 2001).

In stale seedbed preparation, weeds are induced to germinate by flush irrigation and then killed by herbicide spraying or by plowing the weeds into the soil and allowing them to decompose under flooded conditions before sowing pregerminated seeds or transplanting rice seedlings. In some areas of South Asia and China with poorly drained heavy clay soils, rice seedlings are transplanted into nonpuddled wet soils and allowed to grow under flooded conditions (Hobbs et al., 2002).

Continuous use of this system for more than 10 yr in China gave good yields, and weeds became fewer with time (Hobbs et al., 2000). In nonflooded soil, minimum or zero tillage coupled with mulching of crop residues could minimize weed infestation. Effective weed control strategies have to be developed based on evolving weed dynamics in ongoing field trials with zero-till DSR on flat land or on raised beds. In addition, mechanical weeding is possible in row-seeded rice. A rotary or conical weeder can be used to control weeds in row-seeded rice on puddled soil, while tractor-attached scrapers can be used to remove weeds in row-seeded rice in nonpuddled soil. Laser-assisted leveling improves water control and reduces weed problems in both puddled and nonpuddled soils (Hill et al., 1991; Kahlown et al., 2000; Rickman et al., 1998). Manipulation of the crop rotation is another way to reduce weed problems in cropping systems. Farmers in South Asia and China observe fewer weed problems in rice after potato or sweet potato or vegetables.

Herbicides and Integrated Weed Control Strategies for Rice Systems

Finally, the above nonchemical weed control methods may have to be supplemented with herbicides, wherever necessary. Suitable preemergence herbicides have been developed and used widely for controlling weeds in transplanted and WSR in some Asian countries. Foliar-active herbicides, such as bentazon [3-isopropyl-1*H*-2,1,3-benzothiadiazin-4(3*H*)-one 2,2-dioxide], glyphosate, and 2,4-D [(2,4-dichlorophenoxy)acetic acid], require draining of the field and exposure of weeds before application for effective weed control, whereas herbicides such as molinate [*S*-ethyl azepane-1-carbothioate] and sulfonylureas are most effective applied into the floodwater. Herbicides such as fenoxaprop {(*RS*)-2-[4-(6-chloro-1,3-benzoxazol-2-yloxy)phenoxy]propionic acid} that are effective on *Echinochloa* and *Leptochloa* species may cause injury to rice plants if submerged under water immediately after application (Hill et al., 2001). Thus, a working knowledge of the conditions under which various herbicides will be most effective on weeds and least injurious to rice plants is essential for developing successful chemical weed control recommendations. A suitable combination of herbicides and water depth has to be worked out by experimentation to control weeds effectively in flooded rice systems. For nonflooded rice systems, especially for aerobic rice, suitable herbicides have yet to be identified and/or developed.

Repeated use of selected herbicides has resulted in the development of weed species resistant to some chemical groups of herbicides (Fischer et al., 2000; Gressel & Baltazar, 1996). Therefore, the use of herbicides must be integrated with other nonchemical methods, especially water management, to reduce the cost of herbicides, minimize and/or delay the development of herbicide-resistant weeds, and reduce the environmental risk from off-site herbicide runoff (Hill et al., 2001).

NUTRIENT MANAGEMENT FOR RICE IN THE RICE–WHEAT SYSTEM

Both rice and wheat crops require nutrients such as N, P, and K in large amounts to produce high and stable yields. Nutrients such as S, Ca, Zn, Fe, Cu, B, and others are not adequate for crop production only in certain locations (deficient sites) and they can be taken care of by specific or local recommendations based on the best management practices developed for deficient areas.

Soil and irrigation water provide some amounts of N, P, and K to crops. Soil nutrient supply varies greatly from field to field and sometimes even within a field (Doberman et al., 1996; Dobermann & White, 1999; Olk et al., 1999). In addition, plant growth and the nutrient requirement of crops vary with season, location, and crop growing conditions. In the rice–wheat system in northwestern India, farmers preferentially apply high doses of P to wheat and other winter crops, and this should be considered while determining the rates of P application for rice. All these factors make general or blanket fertilizer recommendations highly ineffective for most situations. The solution has to be site-specific, meaning that each farmer should know the nutrient status of his or her field in order to decide which nutrients and how much to apply for each field. We therefore need simple and effective strategies to enable farmers to determine the nutrient status of their own fields and design variable fertilizer rates for their rice and other crops.

Site-Specific Nutrient Management

Site-specific nutrient management (SSNM) is a component of precision farming that takes into account site, season, and crop growth variability in making soil and crop management decisions (Dobermann et al., 1996; Dobermann & White, 1999; Olk et al., 1999). Dobermann and White (1999) defined SSNM as the dynamic site-specific management of nutrients in a particular cropping season to optimize the supply and demand of nutrients according to their differences in cycling through soil–plant systems. It provides an approach to optimize the use of nutrients from all sources—soil, irrigation water, biological fixation, organic matter, and fertilizers. The SSNM approach enables farmers to apply the right amount of nutrients at the right time when the crops need them the most. Nutrient applications are thus matched properly with crop demand, thereby minimizing the risk of overapplication of fertilizer. This also ensures balanced application of all nutrients to maintain productivity and soil quality over time. IRRI has developed some strategies and tools for practicing SSNM on the small rice farms of Asia. The major components of the small-farm-based SSNM approach are:

- Balanced application of NPK and micronutrients
- Real-time N management using the leaf color chart (LCC) or chlorophyll meter
- Soil- and plant-based P and K management using the nutrient omission plot method
- Application of K in two splits
- Soil-test or local recommendation-based rates of nutrients such as S, Zn, and Fe at deficient sites

Soil- and plant-based tests are used to make fertilizer recommendations for rice and other crops. They have variable rates of success in terms of matching fertilizer application with crop demand as discussed below.

Soil Testing

Soil tests have been commonly used to make fertilizer recommendations at enormous costs to national governments. Some countries (e.g., India) use mobile soil-testing laboratories for analysis of soil samples and to make fertilizer recommendations on-site at the request of farmer groups. They provide fertilizer recommendations based on the classification of soils into three categories—low, medium, and high—in relation to the availability of major nutrients (N, P, and K). For nutrients such as S, Zn, Fe, and others, only one rate is recommended for sites deficient in such nutrients. The success of soil testing is highly variable, depending on farmers' confidence in soil test-based fertilizer recommendations. A widespread use of soil testing is not practical because of the lack of analytical facilities in many countries and the high cost of providing such services to millions of farmers on a regular basis.

Plant-Based Tools for Real-Time Nitrogen Management in Rice

Matching N supply with crop demand optimizes N use efficiency and reduces N losses to the environment. In many field situations, up to 50% of applied N is lost in part because of the lack of synchrony of plant N demand with N supply. This lack of synchrony reduces the physiological use of the N available to the crop. A growing body of evidence indicates large site-to-site variation in N supply capacity of soils to meet crop demand (Dobermann et al., 1996; Dobermann and White, 1999; Olk et al., 1999). In developed-world agriculture, the concept of precision farming is rapidly expanding. A problem in Asia is the extent to which this principle, that is, crop demand-driven site-specific nutrient application, can add to farmers' productivity and profits. At IRRI, we are trying to identify crop demand indicators for major nutrients. Two decision tools, the chlorophyll meter and leaf color chart (LCC), have been tried to regulate the timing of N application in rice. Using such tools reduces N fertilizer needs of crops, increases N use efficiency (Balasubramanian et al., 1999, 2000a,b; Peng et al., 1996), and improves plant health, thus reducing the crop's susceptibility to pest attack.

Chlorophyll Meter Method

We use the chlorophyll meter or soil plant analysis development (SPAD) meter method (Fig. 6–4), a reliable and nondestructive method, to determine the right time of N topdressing for rice (Balasubramanian et al., 1999, 2000a; Peng et al., 1996; Turner & Jund, 1994). The concept is based on results that show a close link between leaf chlorophyll content and leaf N content. Thus, the chlorophyll meter can provide a quick and accurate measurement of the leaf N status of rice plants *in situ* in the field.

The SPAD threshold or critical value indicates the leaf area-based critical N concentration in rice leaves. Whenever SPAD readings fall below the critical SPAD value, the crop will suffer from N deficiency and yields will decline if N fertilizer is not applied immediately. For example, a SPAD threshold value of 35 is equal to 1.4 to 1.5 g N m^{-2} of leaf area in semidwarf indica varieties (Peng et al., 1996). Dif-

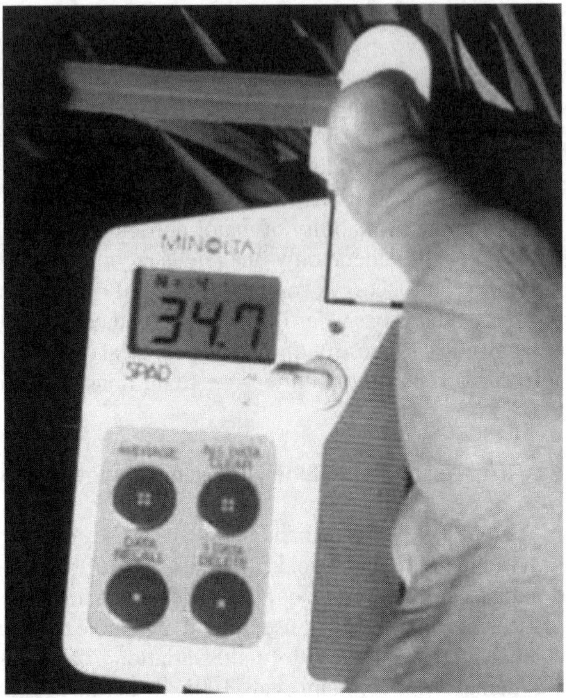

Fig. 6–4. Chlorophyll meter for determining the right time of N fertilizer application to rice crops and integrating variable soil N supply and crop nutrient needs for real-time N management in rice.

ferent SPAD threshold values are needed to optimize rice yields in different conditions. Suggested critical values are summarized in Table 6–4 (Balasubramanian et al., 2000a). These values can be refined by one to two seasons of testing for locally important rice varieties and the crop's environmental, nutritional, and cultural conditions.

The SPAD meter is highly useful as a research and training tool for researchers, extension specialists, and crop consultants. For example, agronomists can study the N requirement of new rice varieties and develop precise N recommendations for well-defined domains. It is a quick and accurate method for breeders and biotechnologists to choose N-efficient lines and identify genes responsible for high N use efficiency. Extension agents and crop consultants can provide precise advice to their farmer clients on need-based N fertilization for rice crops after testing the N status of standing rice crops in the fields. They can also verify the validity of existing N recommendations by monitoring crop N status for one to two seasons and then refining them as necessary.

Many experiments conducted either at research stations or in farmers' fields in Asia have demonstrated the utility of the SPAD technique in increasing rice yield per unit of N applied (Balasubramanian et al., 2000a). These trials compared the SPAD technique with other N management practices using two efficiency indicators: (i) agronomic efficiency of applied N (AEN) calculated as the additional grain yield per kilogram of N applied beyond the control and (ii) partial factor pro-

Table 6–4. Suggested critical SPAD values for different seasons, cropping conditions, and rice varieties.†

Crop establishment	Varietal group	Panicle density per m^2	Wet season	Dry season
Transplanted rice	Traditional, improved local aromatic rice	300–400	30‡–32	32
	Semidwarf indica varieties	400–500	32‡–35	35¶–37
	Hybrid rice	400–500	32‡–35	35–37
Wet direct-seeded rice				
Broadcast-sown	All varieties	High: ≈800	029–30¶	30¶
		Medium: ≈400–500	32‡	35
Row-seeded (drum-seeded)	All varieties	High: ≈600–650	32¶	32¶
		Medium: ≈400–500	32–35	32–35

† Source: Balasubramanian et al. (2000a).
‡ Under Philippine conditions where radiation is low in wet season because of continuous cloud cover for most part of the growing season (mean wet season yield is <60% of mean dry season yield).
¶ Validated in the Philippines.

ductivity of applied N (PFP-N) calculated as the grain yield divided by the amount of N applied. Two typical cases are presented below:

An Increase in Grain Yield, but with a Higher Fertilizer Nitrogen Use. This observation was recorded in the case of 12 on-farm trials conducted in Nueva Ecija Province of the Philippines during the 1996 and 1997 dry seasons (Table 6–5). The efficiency values were similar for the farmers' practice and the SPAD method, indicating the efficient fertilizer use by farmers. However, the SPAD method revealed that farmers in this case were not applying adequate N to reach high yields with the same high efficiency in N use.

A Savings in Nitrogen Fertilizer Use Without Reducing Grain Yield. Here the grain yields were similar for both farmers' and SPAD methods, but the amount of N used was much lower in SPAD plots compared with the farmers' practice. Therefore, the efficiency values were higher for the SPAD method than for the farmers' practice or local recommendation as observed in India and Vietnam (Table 6–5). In this case, the farmers' N fertilization practice has to be improved.

Thus, the on-farm trials have demonstrated the advantage of using the SPAD technique to synchronize N application with crop demand in rice. The increase in N use efficiency values (AEN and PFP-N) was due to the same or higher grain yield with lesser N application in the SPAD-N plots. The average savings in N fertilizer ranged from 32 to 65 kg N ha^{-1} for various locations in India and Vietnam, while there was no savings in N fertilizer use for the Nueva Ecija site in the Philippines (Table 6–5).

Leaf Color Chart Method

The high cost (>$1000 per unit) of the chlorophyll meter restricts its widespread use by farmers. Because the chlorophyll meter measures leaf color as a proxy for leaf N, a simplification of this is to use the green color as an indicator of plant N status. Farmers generally use leaf color as a visual and subjective indicator of the

Table 6–5. Comparison of chlorophyll meter (SPAD) method with farmers' practice or local recommendation for N management in rice at selected sites in India, Philippines, and Vietnam.†

Treatments	N used	Grain yield	AEN‡	PFP-N§	N saved
	kg ha^{-1}	Mg ha^{-1}			kg ha^{-1}
Philippines—Nueva Ecija, 1996 DS¶, 12 farms					
Control	0	3.75c#	--	--	--
Farmers' practice	126	6.05b	18a	48a	--
SPAD-35††	150	6.70a	20a	45a	−24
Philippines—Nueva Ecija, 1997 DS, 12 farms					
Control	0	5.03b	--	--	--
Farmers' practice	139	7.49a	18b	54b	--
SPAD-35	138	7.20a	16b	52b	1
UT/DP‡‡	87	7.24a	25a	83a	52
India—Old Cauvery Delta, Padugai soil series, mean of 1997 & 1998 DS (Kuruvai)					
Control	0	4.94b	--	--	--
Local recommend	125	7.26a	19c	58c	--
SPAD-35	65	7.62a	41a	117a	60
Basal N + SPAD-35	85	7.35a	28b	87b	40
India—New Cauvery Delta, 1996 DS (kuruvai), 4 farms					
Control	0	5.34b	--	--	--
Local recommend	125	6.45a	9b	52b	--
SPAD-35	60	7.11a	51a	118a	65
Vietnam—Cai Lay District, 1996 WS§§, 6 replications					
Control	0	2.78b	--	--	--
Local recommend	120	3.96a	10b	33b	--
SPAD-32¶¶	70	4.03a	18a	58a	50

† Source: Balasubramanian et al. (2000a).
‡ AEN, agronomic efficiency of applied N = kg additional grain yield over control per kg N applied.
§ PFP-N, partial factor productivity for applied N = grain yield divided by applied N.
¶ DS, dry season.
# Values followed by the same letters within a column and a set are not significantly different at $P >$ 0.05 by Duncan's Multiple Range Test.
†† SPAD-35, chlorophyll meter or SPAD critical value set at 35.
‡‡ UT/DP = urea tablet deep placement.
§§ WS, wet season.
¶¶ SPAD-32, chlorophyll meter or SPAD critical value set at 32.

rice crop's N status. Hence, we have adapted and simplified the concept. A leaf color chart (LCC; Fig. 6–5) was developed from a Japanese prototype to help farmers measure leaf color intensity and relate it to plant N status (Balasubramanian et al., 2000b). A simple instruction sheet in the local language accompanies the chart and explains to farmers how to use the LCC to determine the right time of N application to their rice crops. The color chart is an ideal tool to optimize N use in rice cropping, irrespective of the source of N applied: inorganic, organic, or biofertilizers. It is thus an ecofriendly tool in the hands of small farmers. It costs about $1 per piece.

The LCC contains six green strips with the color ranging from yellowish green (No. 1) to dark green (No. 6). The critical leaf color reading for N topdressing in TPR ranges from 3 for varieties with light green foliage (e.g., aromatic rice varieties) to 4 for semidwarf indica varieties and hybrids. Similarly, the critical LCC value is 3 for high-density (600–700 panicles m^{-2}) direct WSR under Philippine conditions. Crops showing a leaf color below the critical values suffer from N de-

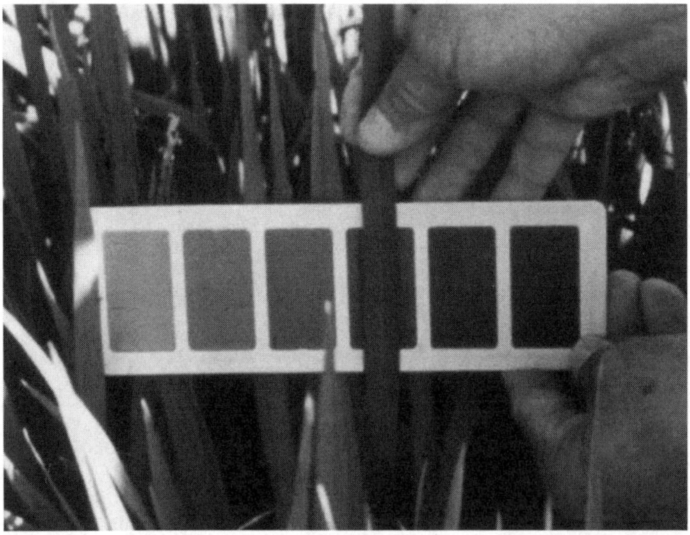

Fig. 6–5. Leaf color chart (LCC), a simple, easy-to-use, and inexpensive tool for determining the right time of N application to rice crops and integrating variable soil N supply and crop needs for real-time N management in rice.

ficiency and require immediate N fertilizer application to prevent yield loss. Users can refine the critical LCC values after one or two seasons of testing for locally important varieties and crop establishment methods.

Three factors affect leaf color measurement by the LCC: (i) a subjective decision on color comparison by the field person, (ii) the perceptible difference in color between panels for facilitating leaf color measurement by the field person, and (iii) the quality of the LCC in terms of the maintenance of color shades for different panels.

Subjective Measurement. We recommend that the same person take the LCC readings at the same time of the day to minimize variations in readings (subjective decisions) and minimize reading errors. We also recommend that the national research and extension staff fix the critical LCC value for locally important rice varieties and growing conditions by testing for one to two seasons.

Perceptible Differences in Color Between LCC Panels. In the IRRI standard LCC, the critical color panels 3, 4, and 5 produce spectral signatures at about 540 nm wavelength. Panels 3, 4, and 5 are also easily distinguishable by farmers when they measure leaf color in the field to determine the timing of N application for rice crops.

Quality of Color Shades of Panels. We try to maintain quality among different batches of LCC production by standardizing many factors (e.g., quality of plastic, quality and composition of colorant, temperature of panel molding) that affect the panel color.

In a set of 17 on-farm trials conducted in the Philippines during the 1996 wet season, the leaf color chart method produced grain yield similar to that of the more precise chlorophyll meter method, but with less N (Table 6–6). In another set of 20

Table 6–6. Comparison of chlorophyll meter (SPAD) and leaf color chart methods in on-farm trials in the Philippines (1996) and India (1998).

Treatment	N used	Grain yield	AEN†	PFP-N‡	N saved
	kg ha^{-1}	Mg ha^{-1}			kg ha^{-1}
Philippines—Maligaya: TPR§ 1996 WS¶, 17 farms					
Control	0	3.16c#	--	--	--
Farmers' practice	101	4.16a	10c	41	--
SPAD-32 N††	73	4.17a	14b	57	28
LCC-4‡‡	48	4.18a	21a	87	53
India—New Cauvery Delta, 1998 DS§§ (Kuruvai), 20 farms					
Control	0	3.57b	--	--	--
Soil test recommendation	142	5.03a	10b	35	--
SPAD-35¶¶	110	4.99a	13a	45	32
LCC-4	108	4.93a	13a	46	34

† AEN, agronomic efficiency of applied N = kg additional grain yield over control per kg N applied.
‡ PFP-N, partial factor productivity for applied N = grain yield divided by applied N.
§ TPR, transplanted rice.
¶ WS, wet season.
# Values followed by the same letters within a column and a set are not significantly different at $P >$ 0.05 by Duncan's Multiple Range Test.
†† SPAD-32, chlorophyll meter or SPAD critical value set at 32.
‡‡ LCC-4, leaf color chart critical value set at 4.
§§ DS, dry season.
¶¶ SPAD-35, chlorophyll meter or SPAD critical value set at 35.

on-farm evaluation trials conducted in the new Cauvery Delta Zone of Tamil Nadu, India, during the 1998 dry season, the LCC compared very well with the chlorophyll meter method in terms of grain yield and N use efficiency and savings in N fertilizer use (Table 6–6). Thus, the simple LCC can serve as an effective tool for farmers to determine the right time of N fertilizer application to their rice crops. A recent study by Buresh et al. (2001) compared various N management techniques for rice and found that the LCC method produced the highest total grain yield and N fertilizer use efficiency for two seasons during 2000 and 2001.

Thousands of rice farmers in Asia have been using this chart to make correct decisions to apply N fertilizer at the right time, with substantial savings in N fertilizer compared with their traditional practice. Balasubramanian et al. (2000b) demonstrated three possibilities of using the LCC in evaluation trials as shown below (Table 6–7).

An Increase in Grain Yield but with Higher Fertilizer Nitrogen Use. The efficiency values were similar for both farmers' and LCC methods (Table 6–7). This case confirms the local farmers' efficient use of N fertilizers, but at a level lower than the amount needed to reach maximum yields. The LCC method shows that yields can be further increased without compromising on N use efficiency.

Same Grain Yield, but with Less Nitrogen Fertilizer Use in the LCC Method. Here, grain yields were similar for both farmers' and LCC methods. However, the amount of N used was much lower and N use efficiency much higher for the LCC technique (Table 6–7). In this case, the farmers' N management practice has to be improved.

Table 6–7. Comparison of leaf color chart (LCC) method with farmers' practice or local recommendation for N management in rice in Asian countries.

Treatment	N used	Grain yield	AEN‡	PFP-N§	N saved
	kg ha^{-1}	Mg ha^{-1}			kg ha^{-1}
An increase in grain yield but with higher fertilizer N use in the LCC method					
Philippines—Maligaya: TPR¶, 1998 DS#, 14 farms					
Control	0	3.84b††	--	--	--
Farmers' practice	116	5.75a	19a	49	--
LCC-4‡‡	130	6.05a	19a	46	−14
Philippines—Maligaya: TPR, 1999 DS, 9 farms					
Control	0	3.33b	--	--	--
Farmers' practice	121	5.10a	15a	42	--
LCC-4	135	5.30a	16a	39	−14
Same grain yield, but with less N fertilizer use in the LCC method					
Philippines—Maligaya: TPR, 1998 WS#, 11 farms					
Control	0	3.43b	--	--	--
Farmers' practice	78	3.97a	9b	51	--
LCC-4	33	3.87a	20a	117	45
Philippines—Maligaya: TPR, 1999 WS, 11farms					
Control	0	3.71b	--	--	--
Farmers' practice	74	4.49a	12b	61	--
LCC-4	46	4.68a	19a	102	28
Vietnam—Cai Lay District: B-WSR¶, 1998 DS, 28 farms					
Farmers' practice	120	5.24a	-	44	--
LCC-3‡‡	82	5.26a	--	64	38
Vietnam—Cai Lay District: B-WSR, 1999 DS, 7 farms					
Farmers' practice	99	6.34a	--	64	--
LCC-3	70	6.31a	--	90	29
India—Haryana State: Random TPR, 2001 WS, 165 farms					
Farmers' practice	149	6.36a	--	43	--
LCC-4	124	6.37a	--	51	25
Increase in grain yield and a reduction in N fertilizer use in the LCC method					
Philippines—Maligaya: B-WSR, 1998 DS, 6 farms					
Control	0	3.62c	--	--	--
Farmers' practice	151	4.53b	6b	30	--
LCC-3	125	5.15a	14a	41	26
Vietnam—Huyen District: B-WSR, 1999 DS, 18 farms					
Farmers' practice	98	4.63b	--	47	--
LCC-3	80	4.92a	--	62	18

† Source: Balasubramanian et al. (2000b).

‡ AEN, agronomic efficiency of applied N = kg additional grain yield over control per kg N applied.

§ PFP-N, partial factor productivity for applied N = grain yield divided by applied N.

¶ TPR, transplanted rice; B-WSR, Broadcast wet-seeded rice.

DS, dry season; WS, wet season.

†† Values followed by the same letters within a column and a set are not significantly different at $P < 0.05$ by Duncan's Multiple Range Test.

‡‡ LCC-4, leaf color chart critical value set at 4; LCC-3, Leaf color chart critical value set at 3.

Fig. 6–6. Omission plot technique to determine the nutrient-supplying capacity of soils.

An Increase in Grain Yield and a Reduction in Nitrogen Fertilizer Use in the LCC Method. Nitrogen use efficiency was much higher for the LCC method than for the farmers' practice (Table 6–7). In this case, much improvement in the farmers' N fertilizer use practice is needed.

Site-Specific Phosphorus and Potassium Management

Nutrient Omission Plot Technique

The design of field-specific recommendations for P and K is more difficult because of the lack of simple plant diagnostic tools such as the chlorophyll meter or LCC. One possibility is to use the nutrient omission plot (OP) technique (Dobermann & Fairhurst, 2000) to determine the soil-supplying capacity of and crop requirement for P and K in individual rice fields (Fig. 6–6). Some farmers may be reluctant to keep OPs because of potential yield loss, although yield losses in P and K omission plots are too low to affect the total yield of the field or farm. The estimation of P and K will be accurate only if OPs are managed adequately and the weather does not affect the crop adversely.

The OP technique should be used during the higher-yielding season [e.g., dry or winter (boro) season in most cases or wet (kharif) season in rice–wheat areas]. The size of the OP should be 5 by 5 m (25 m^2) minimum. It should be adequately fertilized with all nutrients except the "omission nutrient" (P or K) and managed adequately to attain maximum yields. Farmers must be trained to measure the grain yields in OPs as accurately as possible. This technique can be used for one to two seasons to obtain a fair value of soil P and K status. The process should be repeated once every 4 to 5 yr to assess the soil P and K status periodically and to adjust the application rates of P and K fertilizers to rice crops. The OP technique is being evaluated in the rice–wheat area of India.

Table 6–8. Phosphorus and K fertilizer calculation chart for rice based on nutrient omission plot yield.†

Yield in P or K omission plot	Soil P or K supply status	Fertilizer rate									
		P_2O_5					K_2O				
		Yield goal (Mg ha^{-1})									
		4–5	5–6	6–7	7–8	8–9	4–5	5–6	6–7	7–8	8–9
Mg ha^{-1}		kg ha^{-1}									
3–4	Very low	15–20	30–40	45–60	--	--	30	60	90	--	--
4–5	Low	--	15–20	30–40	45–60	--	--	30	60	90	--
5–6	Medium	--		15–20	30–40	45–60	--		30	60	90
6–7	High		--		15–20	30–40		--		30	60
7–8	Very high			--		15–20			--		30

† Assumptions: A maximum additional yield of 3 Mg ha^{-1} over P or K omission plot yields is considered possible. Aiming for yield goals of more than 3 Mg ha^{-1} additional yield over omission plot yield is highly unrealistic. Such a soil is likely to have many other fertility constraints that cannot be overcome by applying heavy doses of P or K or PK alone. A nutrient requirement of 15 to 20 kg P_2O_5 and 30 kg K_2O for every Mg of increase in grain yield over that of P and K omission plots, respectively.

Developing Phosphorus and Potassium Recommendations for Individual Farmers

Generally, two separate OPs, one each for P and K, are used to measure independently the soil supply status of P and K. Adequate rates of N and K are applied to the P omission plot and N and P to the K omission plot. Adequate crop management practices are followed in the OPs to achieve maximum yields. If P is not judged as a problem in the target area, a sole K omission plot can be used to measure the soil K status.

At the end of the season, farmers can measure crop yield in the OPs and use the chart given in Table 6–8 to estimate the P and K status of their rice fields. They can then decide on the application rates of P and K for required yield targets of their rice crops.

Developing Phosphorus and Potassium Recommendations for a Domain

The P and K omission plots can be installed in 15 to 20 representative sites or fields of a target area to estimate the mean soil-supplying capacity of P and K in the target area. Then, the mean soil P and K supply is used to make the P and K recommendation for the domain, using the chart given in Table 6–8. This can be refined once every 4 to 5 yr.

Nutrient Addition Plot Technique

If farmers are reluctant to use the OP technique for fear of losing yields, they can use the nutrient addition plot (AP) technique to verify whether their fields and crops respond to additional P or K application. Generally, 50 to 100% more fertilizer P or K is applied in APs (5 by 5 m size); all other nutrients are applied at an adequate level. This is a simple technique that farmers themselves can use to adjust the P and K rates for their fields to adequately supplement the soil nutrient supply. The AP technique will be useful only in areas where farmers apply inadequate amounts of P and K fertilizers to their rice crops, but not in areas with overapplication of P and K fertilizers.

Adjustment of Phosphorus and Potassium Rates in the Rice–Wheat System

In rice–wheat areas of northern India, farmers preferentially apply high doses of P to the wheat crop, which has resulted in an accumulation of P in cultivated soils. Sixty percent of the 105 soil samples taken from eight villages of Rampur District of Uttar Pradesh, India showed soil P levels of 80 to 300 kg ha^{-1} of P_2O_5 (S.K. Srivastava, personal communication, 2001). Some farmers of Rampur District apply no P to their rice crops and still get high rice yields because of medium to high levels of residual P in the soil from wheat fertilization. Thus, soil P availability changes with P fertilization patterns of different crops in rotation. Although organic P plays an important role in soil P supply in soils with high organic matter level (Haygarth & Jarvis, 1999), it is not important in the light-textured soils of the Indo-Gangetic Plains with low organic matter content. The P-omission plot technique described above is expected to accurately assess the availability of soil P in dynamic soil–plant systems and provide good guidance for need-based P fertilization of crops planted in sequence.

Crop residues provide substantial amounts of K to the soil. Potassium fertilization should take into account the amount of K added through crop residues in the rice–wheat system. Farmers require improved techniques to fast decompose the straw in situ in the field. If efficient organic matter decomposition methods are made available to farmers, they will reduce the burning of straw and allow it to decompose in the field. If farmers use crop residues as mulch in the zero-till system of planting of rice and/or wheat, this will greatly enhance the levels of soil organic matter and K in the soil. Again, the K-omission plot technique is expected to take into account the contribution of K from crop residues and provide accurate estimations of soil K supply to rice and wheat crops in the system.

Split Application of Potassium

It is recommended to apply K fertilizer in two splits, 50% at planting and 50% at panicle initiation (PI) for flooded rice systems. Placement of K along with N and P in seed furrows or between rows is being evaluated for direct drill-seeded rice in nonpuddled soil. Potassium topdressing at PI has been found to decrease diseases and lodging in direct-seeded rice in Vietnam.

CONCLUSIONS

Farmers face problems of increasing cultivation cost, stagnant yields, and low profit in rice–wheat areas covering about 13.5 m ha in South Asia. Alternative crop establishment methods are needed to maximize total productivity and input use efficiency in the rice–wheat system and to conserve resources such as energy and water. Establishment of rice on nonpuddled flat land or raised beds is promising. Bed-planting methods reduce the total water pumping hours per hectare per season by 37 to 40% and produce high yields similar to those of TPR on puddled soil. However, weeds are a serious problem and micronutrient deficiency is conspicuous when rice is direct-seeded in nonpuddled soil. At some sites, termites also attack the rice crops on beds. In general, weed infestation increases from puddled transplanted to puddled direct-seeded to nonpuddled direct-seeded rice systems. Farmers require a good knowledge of the interaction among tillage and/or puddling, land leveling, and water management to effectively control weeds in various rice systems. Similarly, a working knowledge of the conditions under which various herbicides will be most effective on weeds and least injurious to rice plants is essential to develop successful chemical weed control strategies as a supplement to cultural weed control methods. Further research is needed to develop improved integrated weed control strategies for direct-seeded rice in nonpuddled soil. Promising site-specific nutrient management techniques such as real-time N management with the leaf color chart and nutrient omission or addition plot techniques for P and K are being evaluated for rice planted in nonpuddled soil. Methods are being developed to address micronutrient deficiency and termite attack on bed-planted rice. Suitable technology is being developed for in-field decomposition of crop residues without resorting to hazardous burning. Long-term monitoring of grain yield and profit, water use and productivity at the field and watershed level, the dynamics of weeds and other pests, and soil physical and chemical conditions is critical to solve the sustainability problems of the rice–wheat rotation in nonpuddled soil.

REFERENCES

Azmi, M., and A.M. Mortimer. 1999. Effect of tillage practices, seeding rates, and herbicides on weed infestations in direct seeded rice. p. 199–204. *In* Weeds and environmental impact. Proc. I (A) 17th Asian-Pacific Weed Sci. Soc. Conf. 2. Bangkok, Thailand. 22–27 Nov. 2001.

Balasubramanian, V., and J.E. Hill. 2002. Direct seeding of rice in Asia: Emerging issues and strategic research needs for the 21st century. p. 15–39. *In* S. Pandey et al. (ed.) Direct seeding: Research strategies and opportunities. IRRI, Manila, Philippines.

Balasubramanian, V., A.C. Morales, R.T. Cruz, and S. Abdulrachman. 1999. On-farm adaptation of knowledge-intensive nitrogen management technologies for rice systems. Nutr. Cycl. Agroecosyst. 53:59–69.

Balasubramanian, V., A.C. Morales, R.T. Cruz, T.M. Thiyagarajan, R. Nagarajan, M. Babu, S. Abdulrachman, and L.H. Hai. 2000a. Adaptation of chlorophyll meter (SPAD) technology for real-time nitrogen management in rice: A review. Int. Rice Res. Notes 25(1):4–8.

Balasubramanian, V., A.C. Morales, R.T. Cruz, N.N. De, P.S. Tan, and Z. Zaini. 2000b. Leaf color chart as a simple decision tool for N management in lowland rice. p. 278. *In* 2000 Annual Meetings Abstr. ASA–CSSA–SSSA, Madison, WI.

Barnes, H. 1996. Irrigation strategies. Rice Farm. 30:8–12.

Bell, M.A., J.F. Rickman, E.C. Castro, Jr., L.B. Aclan, and J. McNamara. 1998. Precision land leveling for rice production in Asia. p. 257–264. *In* V.M. Salokhe and Z. Jianxia (ed.) International Agricultural Engineering Conferencce. Proc. Int. Agric. Eng. Conf., Asian Institute of Technology (AIT). Bangkok, Thailand. 7–10 Dec. 1998. AIT, Bangkok, Thailand.

Buresh, R.J., S. Peng, C. Witt, V. Balasubramanian, and E. Laureles. 2001. Improving nitrogen fertilizer-use efficiency for rice: How do management options compare? Poster paper presented at the 2001 ASA–CSSA–SSSA Meeting. Charlotte, NC. 21–25 Oct. 2001.

De Datta, S.K. 1986. Technology development and the spread of direct-seeded flooded rice in Southeast Asia. Exp. Agric. 22:417–426.

De Datta, S.K., and J.C. Flinn. 1986. Technology and economics of weed control in broadcast-seeded flooded tropical rice. *In* K. Noda and B.L. Mercado (ed.) Weeds and the environment in the tropics. Proc. 10th Conf. Asian-Pacific Weed Sci. Soc. Chiang Mai, Thailand.

Dobermann, A., K.G. Cassman, S. Peng, P.S. Tan, C.V. Phung, P.C. Sta. Cruz, J.B. Bajita, M.A.A. Adviento, and D.C. Olk. 1996. Precision nutrient management in intensive irrigated rice systems. p. 133–154. *In* Maximizing sustainable rice yields through improved soil and environmental management. Int. Symp. Khon Kaen, Thailand. 11–17 Nov. 1986. Dep. Agric., Soil and Fert. Soc. Thailand, Dept. Land Dev., and Int. Soc. Soil Sci., Bangkok, Thailand.

Dobermann, A., and T. Fairhurst. 2000. Rice: Nutrient disorders and nutrient management. Handbook Series. IRRI, Manila, Philippines.

Dobermann, A., and P.F. White. 1999. Strategies for nutrient management in irrigated and rainfed lowland rice systems. Nutr. Cycl. Agroecosyst. 53:1–18.

Fischer, A.J., C.M. Ateh, and J.E. Hill. 2000. Herbicide-resistant *Echinochloa oryzoides* and *E. phyllopogon* in California *Oryza sativa* fields. Weed Sci. 48:225–230.

Fujisaka, S., L.W. Harrington, and P.R. Hobbs. 1994. Rice–wheat in South Asia: System and long-term priorities established through diagnostic research. Agric. Syst. 46:69–187.

Gressel, J., and A.M. Baltazar. 1996. Herbicide resistance in rice: Status, causes and prevention. p. 195–238. *In* Weed management in rice. FAO Plant Production and Protection Paper 139. FAO, Rome, Italy.

Gupta, R.K., P.R. Hobbs, M. Salim, N.H. Chowdhary, and S.I. Bhuiyan. 2000a. Study of research and extension issues in the Sichuan Province of China for farm-level impact on the productivity of rice–wheat system. Rice–Wheat Consortium Traveling Seminar Rep. Ser. 2. Rice–Wheat Facilitation Unit, IARI, New Delhi, India, and CIMMYT, Mexico.

Gupta, R.K., P.R. Hobbs, M. Salim, R.K. Malik, M.R. Varma, T.P. Pokharel, T.C. Thakur, and J. Tripathy. 2000b. Research and extension issues for farm-level impact on the productivity of the rice–wheat systems in the Indo-Gangetic Plains of India and Pakistan. Rice–Wheat Consortium Traveling Seminar Rep. Ser. 1. Rice–Wheat Facilitation Unit, IARI, New Delhi, India, and CIMMYT, Mexico.

Haygarth, P.M., and S.C. Jarvis. 1999. Transfer of phosphorus from agricultural soils. Adv. Agron. 66:195–249.

Hill, J.E., D.E. Bayer, S. Bocchi, and W.S. Clampett. 1991. Direct-seeded rice in the temperate climates of Australia, Italy, and the United States. p. 91–102. *In* Direct-seeded flooded rice in the tropics. IRRI, Manila, Philippines.

Hill, J.E., A.M. Mortimer, O.S. Namuco, and J.D. Janiya. 2001. Water and weed management in direct-seeded rice: Are we headed in the right direction? p. 491–510. *In* S. Peng and B. Hardy (ed.) Rice research for food security and poverty alleviation. IRRI, Manila, Philippines.

Hobbs, P.R., and R.K. Gupta. 2000. Sustainable resource management in intensively cultivated irrigated rice–wheat cropping systems of the Indo-Gangetic Plains of South Asia: Strategies and options. Paper presented at the Int. Conf. on "Managing natural resources for sustainable production in the 21st century". New Delhi, India. 14–18 Feb. 2000. Rice–Wheat Facilitation Unit, IARI, New Delhi, India, and CIMMYT, Mexico.

Hobbs, P.R., and R.K. Gupta. 2003. Resource-conserving technologies for wheat in the rice–wheat system. p. 149–172. *In* J.K. Ladha et al. (ed.) Improving the productivity and sustainability of rice–wheat systems: Issues and impacts. ASA Spec. Publ. 65. ASA, CSSA, SSSA, Madison, WI.

Hobbs, P.R., R.K. Gupta, J.K. Ladha, and V. Balasubramanian. 2000. Crop establishment and management: New opportunities for enhancing rice–wheat system productivity. Paper presented at the International workshop on "Developing an Action Program for Farm-level Impact in Rice–Wheat Systems of the Indo-Gangetic Plains". 25–27 Sept. 2000. New Delhi, India. Rice–Wheat Facilitation Unit, IARI, New Delhi, India, and CIMMYT, Mexico.

Hobbs, P.R., Y. Singh, G.S. Giri, J. Lauren, and J. Duxbury. 2002. Direct seeding and reduced tillage options in the rice–wheat systems of the Indo-Gangetic plains of South Asia. p. 201–215. *In* S. Pandey et al. (ed.) Direct seeding in Asian rice systems: Strategic research issues and opportunities. IRRI, Los Baños, Philippines.

Hossain, M., and K.S. Fischer. 1995. Rice research for food security and sustainable agricultural development in Asia. GeoJournal 35:286–298.

Hossain, M., and P.L. Pingali. 1998. Rice research, technology progress, and impact on productivity and poverty: An overview. p. 1–25. *In* P.L. Pingali and M. Hossain (ed.) Impact of rice research. Thailand Dev. Res. Inst., Bangkok, Thailand, and IRRI, Los Baños, Philippines.

IRRI. 1998. Program Rep. to the Board Program Comm. 1–2 Apr. 1998. IRRI, Los Baños, Philippines.

Ismael, S., Z. Noor Faezah, and N.K. Ho. 1994. Weed populations and their buried seeds in the rice fields of the Muda area, Malaysia. p. 8. Paper presented at the Seminar on the impact of pesticides on the rice agroecosystem in the Muda area, Malaysia. Malaysian Agric. Res. Dev. Inst. (MARDI), Malaysia.

Janiya, J.D., M.A. Dizon, M. Mortimer, C. Piggin, and J.E. Hill. 1999. The impact of cropping practices on rice-weed communities. p. 146–151. Proc. 17th Asian-Pacific Weed Sci. Soc. Conf. Bangkok, Thailand. 22–27 Nov. 1999.

Kahlown, M.A., S.A. Prathapar, and M.A. Gill. 2000. Evaluation of resource-conserving technologies in rice–wheat system of Pakistan. OFWM Mimeograph, Lahore, Pakistan.

Ladha, J.K., K.S. Fischer, M. Hossain, P.R. Hobbs, and B. Hardy. 2000. Improving the productivity and sustainability of rice–wheat systems of the Indo-Gangetic Plains a synthesis of NARS-IRRI partnership research. IRRI Discussion Paper No. 40. IRRI, Los Baños, Philippines.

Luzes, D. 1991. Field leveling and direct seeding methods for rice in Portugal. p. 108–109. *In* Direct seeded flooded rice in the tropics. IRRI, Los Baños, Philippines.

Olk, D.C., K.G. Cassman, G. Simbahan, P.C. Sta. Cruz, S. Abdulrachman, R. Nagarajan, P.S. Tan, S. Satawathananont. 1999. Interpreting fertilizer use efficiency in relation to soil nutrient supplying capacity, factor productivity, and agronomic efficiency. Nutr. Cycl. Agroecosyst. 53:35–41.

Ortiz-Monasterio, J.I., S.S. Dhillon, and R.A. Fischer. 1994. Date of sowing effects on grain yield and yield components of irrigated spring wheat cultivars and relationships with radiation and temperature in Ludhiana, India. Field Crops Res. 37:169–184.

Pandey, S., and L. Velasco. 1999. Economics of direct seeding in Asia: Patterns of adoption and research priorities. Int. Rice Res. Notes 24(2):6–11.

Peng, S., F.V. Garcia, R.C. Laza, A.L. Sanico, R.M. Visperas, and K.G. Cassman. 1996. Increased N-use efficiency using a chlorophyll meter on high-yielding irrigated rice. Field Crops Res. 47:243–252.

Rice–Wheat (Pakistan–Punjab) Interim Report. 1999. Promotion of mechanical rice transplanters for increasing water productivity (rice–wheat cropping system – Punjab). Directorate General Agriculture (Water Management), Punjab, Pakistan, August 1999.

Rickman, J., S. Bunna, and P. Sinath. 1998. Agricultural engineering. p. 142. *In* Program report for 1988. IRRI, Los Baños, Philippines.

Sanchez, P.A. 1973. Puddling tropical soils: 2. Effects on water losses. Soil Sci. 115:303–308.

Saunders, D.A. 1990. Crop management research summary of results. Wheat Research Center (WRC) Monograph 5. WRC, Nashipur, Bangladesh.

Sharma, P.K., and S.K. De Datta. 1985. Physical properties and processes of puddled rice soils. Adv. Soil Sci. 5:139–178.

Turner, F.T., and M.F. Jund. 1994. Assessing the nitrogen requirement of rice crops with chlorophyll meter method. Aust. J. Exp. Agric. 34:1001–1005.

Yamauchi, M., A.M. Aguilar, D.A. Vaughan, and D.V. Seshu. 1993. Rice (*Oryza sativa* L.) germplasm suitable for direct sowing under flooded soil surface. Euphytica 67:177–184.

Yoong, W.C., and L.C. Hui. 1982. Direct seeding of paddy: The Sekingchan experience. BIL/PBLS/K/1 publication. West Shore Selangor Project, Ministry of Agriculture, Malaysia.

7 Resource-Conserving Technologies for Wheat in the Rice–Wheat System

Peter R. Hobbs

Centro Internacional de Mejoramiento de Maiz y Trigo
Cornell University, New York, New York, USA

Raj K. Gupta

Rice–Wheat Consortium for Indo-Gangetic Plains
New Delhi, India

ABSTRACT

Wheat is a major food security crop in South Asia, with 42% of the wheat in this region grown following rice on 13.5 million hectares of land. Resource-conserving technologies, described in this chapter, offer considerable potential for improving the sustainability and productivity of wheat in the rice–wheat system of South Asia. These include raising input use efficiency, cutting costs, providing various environmental benefits, and ultimately improving farmer livelihoods and helping to reduce poverty. The conservation technologies described here are in the form of reduced tillage, zero-tillage, and bed planting for wheat in the rice–wheat system. These technologies effectively attempt to solve the problems of late wheat planting and poor plant stands, which limit yields under rice–wheat rotations. Reduced- and zero-tillage practices are discussed in light of problems that warrant additional research, such as interactions between tillage and fertilizer use (methods and timing of application), residue management, development of improved equipment, pest management strategies, variety, and interactions with crop establishment practices for rice. Farmer response to these technologies has been very favorable, and the area under these practices has risen from a few hectares in 1998 to more than 100 000 ha in 2001. Continued acceleration of adoption of this technology is expected as local manufacturers keep up with equipment demand. We list the large number of benefits explained by farmers for the new technologies. Resource-conserving technologies are key to ensuring sustainable food production in South Asia in the next decade.

Green Revolution technology introduced into South Asia in the mid 1960s has been the basis for sustained food security, rural development, conservation of natural resources, and poverty alleviation for the past 30 yr. The technical change that resulted in the Green Revolution arose from the introduction of improved varieties, fertil-

 Improving the Productivity and Sustainability of Rice–Wheat Systems: Issues and Impacts. ASA Special Publication 65.

izer, irrigation, and other plant protection chemicals. Investments from the public sector were crucial for making these inputs available to farmers. During the past 30 yr, agricultural production has been able to keep pace with population demand for food. This came about through significant area and yield growth. Area growth was a result of new lands being farmed and through increases in cropping intensity, from a single crop to double or even triple crops per calendar year.

Post-Green Revolution or the present-day situation sees area growth slowing, and it is expected that this will not play a major role in future production growth. Yield growth will have to be the mainstay for providing the means for meeting future food demand unless food imports start to play a major role in South Asia. Evidence from some long-term experiments, however, shows that problems of stagnating yields and even yield declines are occurring in the rice–wheat system of South Asia (Regmi et al., 2002; Dawe et al., 2000; Duxbury et al., 2000; Ladha et al., 2002). Total factor productivity is declining, and farmers have to apply more fertilizer to obtain the same yields (Hobbs & Morris, 1996; Murgai et al., 2001). Soil organic matter is also declining; new weeds, pests, and diseases are creating more problems; and irrigation water is less available. Farmers are complaining about high input costs and low prices for their produce. Marketing of excess production is a burden for farmers, and storage is a problem for governments. Therefore, a huge challenge exists in the region to meet future food demand sustainably without damaging the natural resource base on which agriculture depends, producing food at a cost that is affordable by the poor and with incentives to farmers that allow them to improve their livelihoods and ultimately alleviate poverty.

The development and deployment of resource-conserving technologies with farmers in the rice–wheat system of South Asia have been a major success of the Rice–Wheat Consortium (RWC) (Hobbs et al., 2002; Hobbs, 2001). One major hurdle was the changing of the mind-sets of all the partners concerned since the old phrase "the more you till, the more the yield" was adhered to stubbornly. A lot of effort was required to overcome this bias. In this chapter, *resource-conserving technologies* (RCTs) are defined as any practice that will result in improvement of the efficiency of natural resources. The major natural resources discussed in this chapter are water, air, fossil fuels, soils, inputs, and people. Another term, used by FAO for RCTs, is *conservation agriculture*. Globally, these technologies are rapidly gaining popularity among farmers, as they result in higher production at less cost with significant benefits to the environment and more efficient use of natural resources. This ultimately results in higher profits, cheaper food, and improved farmer livelihoods. Crop diversification is also easier because less land is needed to produce staple cereals, freeing up land for other crops.

Another major challenge in the region is how knowledge-intensive technologies are transferred to farmers, especially with weak linkages of existing extension systems. This has been accomplished by the RWC through the promotion of participatory approaches and expanded partnerships and stakeholders. All partners are encouraged to work together to solve the Post-Green Revolution system-level problems at specific geographical sites and targeted to different social groups at the site. In this chapter, we describe the various RCTs being promoted for wheat in the rice–wheat system of South Asia, including the results, responses, and future needs.

A DESCRIPTION OF THE INDO-GANGETIC PLAINS

The Indo-Gangetic Plains (IGP) occupy nearly one-sixth of the total geographical area of the subcontinent and hold nearly 42% of the total population of 1.3 billion of South Asia. The population is increasing at about 2.0% per year, meaning that 24 million more mouths need to be fed each year. The rice–wheat (RW) system is grown on nearly 13.5 million ha (Ladha et al., 2000), with another 10.5 million ha in China. In South Asia, that is almost one-sixth of the cultivated area, and it produces more than 45% of the total food. On the basis of population growth alone, demand for rice and wheat will grow at least 2.0% yr^{-1} in the next 20 yr. The demand will be greater if diets improve and people start eating more animal products (Hobbs & Morris, 1996). At the same time, the per capita rice–wheat growing area has shrunk from 1200 m^2 in 1961 to <700 m^2 in 2001 (based on government statistics for the four South Asian countries). Future food production growth will have to come from yield growth.

The climate of the IGP is subhumid, with a distinct wet monsoon summer season and a dry cool winter season. This allows rice and wheat to be grown in a double cropping pattern in one calendar year, rice in the summer and wheat in the winter. Temperatures can exceed 45°C in the summer and frost occurs in some areas in the winter. Soils are mainly alluvial in nature as a result of the deposition of the Indus and Ganges river systems. Many of the soils are alkaline in pH, although acid soils are also present in the piedmont and some floodplains with acidic geology. Soils range in texture from loamy sands to silty clay loams. The IGP are endowed with extensive canal irrigation systems using water storage reservoirs in the Himalayan mid-hills. Canal irrigation is supplemented with tubewell water, and most rice–wheat areas are irrigated or partially irrigated. The IGP are probably one of the most fertile and productive agricultural areas of the world.

The high population density of the region and the fact that agriculture is the main form of employment and income mean that farm size is relatively small. There are some bigger farms in the area, and some of these farmers use tenants to crop their land. Many farmers crop <1 ha of land, and many of these have to rely on off-farm employment to feed their families. Mechanization has increased gradually from west to east during the last 30 yr (government statistics for the region). In northwestern India and Pakistan, farmers who don't own tractors rent small four-wheel tractors from service providers to do most land preparation. However, in the east, many farmers still rely on animal power for land preparation. This is also changing since it is becoming expensive to keep a pair of bullocks for a year for this purpose. Two-wheel tractors are becoming popular on the small farms and fields in Bangladesh. There is also the question of status, with the use of tractors having a higher status than animals.

Many of the other farm operations, such as rice transplanting, weeding, fertilizer application, harvesting, and threshing, are done manually. However, combine harvesters are becoming popular in northwestern India and Pakistan, and wheat threshers have spread from west to east in the past 20 yr. Labor constraints, especially at key times such as planting and harvesting, are gradually becoming more important for the timeliness of operations as industrialization grows and younger

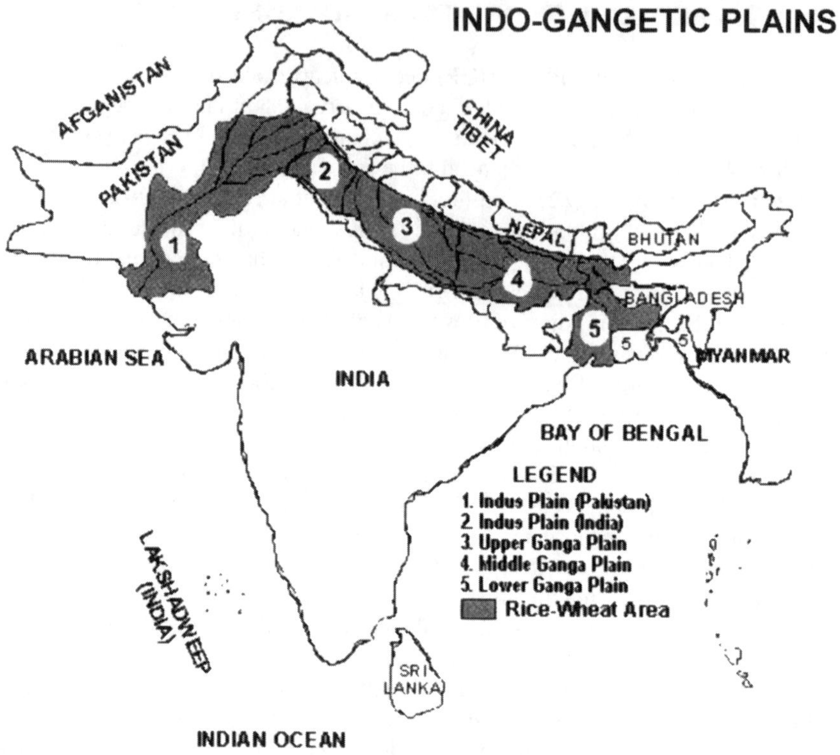

Fig. 7–1. The rice–wheat areas of the Indo-Gangetic Plains by transect.

people shun the drudgery of farming. Therefore, mechanization is bound to increase to fill this power gap.

The IGP are a relatively homogeneous ecological region in terms of vegetation but can be subdivided into five broad transects: the trans (Region 1 in Pakistan and Region 2 in the Indian Punjab and Haryana), upper (Region 3), middle (Region 4), and lower (Region 4 in eastern India and Region 5 in Bangladesh) Indo-Gangetic Plains (Fig. 7–1). These transects have been delineated based on factors such as the increasing demand for food, the development of ground and surface water, infrastructure for inputs and research institutions, variables driving agricultural development, factors that affect the expansion of the rice–wheat area, and productivity constraints. Resource-conserving technologies are being promoted in all five transects, but because of local conditions and farmer circumstances, not all RCTs are possible at all locations. Farmers are given a basket of options that they experiment with and adapt to their own conditions.

THE IMPORTANCE OF TIMELY SOWING

The most common practice for establishing rice in the rice–wheat system of South Asia is puddling—plowing soils when they are saturated—before trans-

Table 7–1. Data on tillage and crop establishment from diagnostic surveys of selected rice–wheat cropping systems, South Asia.

Location†	Area planted late‡	Turnaround time	Number of passes with plow
	%	d	Avg.
Punjab, Pakistan	40	2–10	2–10 (6)
Pantnagar, India	35	15–20	5–12 (8)
Faizabad, India	25	20–45	5–12 (6)
Haryana, India	25	15–35	4–12 (8)
Bhairahawa, Nepal	40	15–35	4–8 (6)

† Source: Data for Punjab, Pakistan, from Byerlee et al. (1984); for Pantangar, India, from Hobbs et al. (1991); for Faizabad, India, from Hobbs et al. (1992); for Haryana, India, from Harrington et al. (1993b); and for Bhairahwa, Nepal, from Harrington et al. (1993a).

‡ Late planting is defined as wheat planted after the first week of December.

planting rice seedlings. Puddling benefits rice by reducing water percolation and controlling weeds. However, puddling also results in degraded soil physical properties, particularly for finer-textured soils, and subsequently creates difficulties when it comes to providing a good soil tilth for wheat (Sharma et al., 2003, Chapter 5 of this publication). This conflicting soil management situation is unique to the rice–wheat and other rice–nonrice cropping systems.

The main implement used for preparing the soil with tractor power is either a nine-tine cultivator or a disk harrow. Deep plowing with a moldboard or disk plow is rare. With animal power, a wooden plow with a metal tip is used. Although tractors allow land to be prepared more rapidly for wheat after rice, farmers still make many passes of plowing and leveling (locally called planking): six to eight passes of the plowing implement are common, usually followed by leveling of the field with a wooden plank. Table 7–1 illustrates the time and number of operations for plowing in several areas of the IGP based on diagnostic surveys conducted at selected locations.

Late planting is a major problem in most rice–wheat areas of South Asia, except for the Indian Punjab (Fujisaka et al., 1994). To improve the productivity of the rice–wheat system, the wheat crop must be planted at the optimal time. A typical response of wheat to date of planting in South Asia shows an optimum yield toward the end of November followed by a linear decline in yield of 1 to 1.5% d^{-1} after that date (Ortiz-Monasterio et al., 1994; Randhawa et al., 1981). Although the slope of the line varies by variety and year, all cultivars show a decline regardless of whether they are short- or medium-duration ones (Fig. 7–2 and 7–3).

Late planting not only reduces yield but also reduces the efficiency of the inputs applied to the wheat crop. Nitrogen response surfaces are much flatter in plots planted late compared with those planted at the optimal time (Saunders, 1990). In other words, applying more N cannot compensate for the decline in yield from late planting. The reasons for late planting of wheat in the rice–wheat system are many (see Fig. 7–4). An obvious cause of late planting is the late harvest of the preceding rice crop or, in some instances, a short-duration third crop planted after rice. Farmers in some parts of the Indo-Gangetic Plains grow long-duration, photosensitive, high-quality basmati rice varieties that mature late. Farmers prefer to grow basmati rice despite its lower yields because of its high market value, good straw

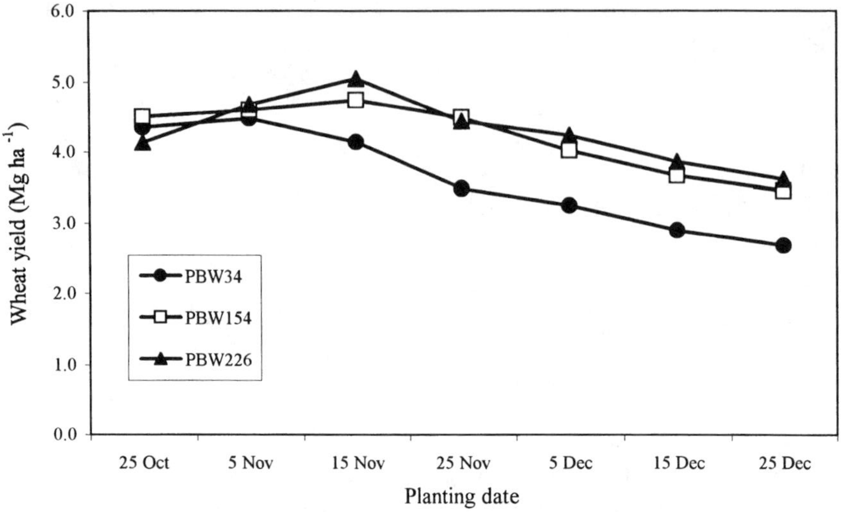

Fig. 7–2. The effect of planting date on wheat yields by variety, Punjab Agricultural University, India.

quality (livestock feed), and lower fertilizer requirements. Basmati varieties cannot be readily replaced by a shorter-duration rice variety, and late rice harvest results. However, in other rice–wheat areas, high-yielding modern rice varieties are grown, and their planting dates can be manipulated so that wheat planting is not delayed. This is the case in the Indian Punjab, where modern rice varieties are planted early (late May to early June) and harvested by early October, and wheat is planted by mid November.

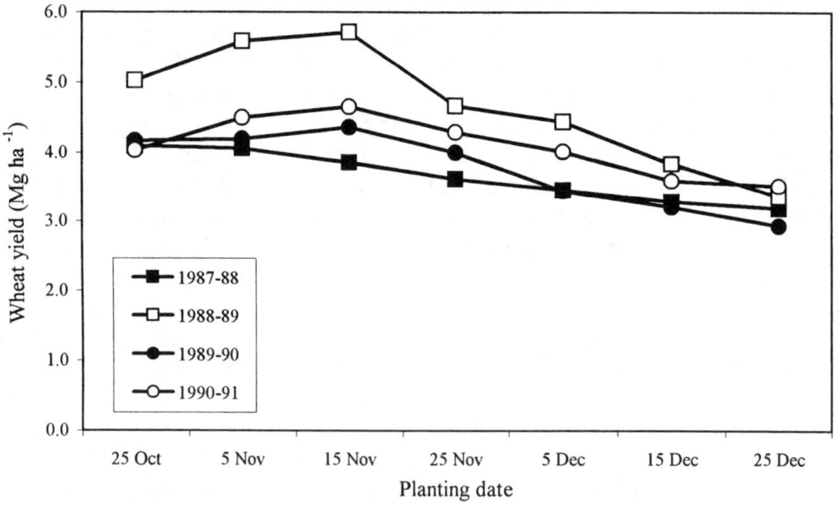

Fig. 7–3. The effect of planting date on wheat yield, Punjab Agricultural University, India from 1987-1988 to 1990-1991.

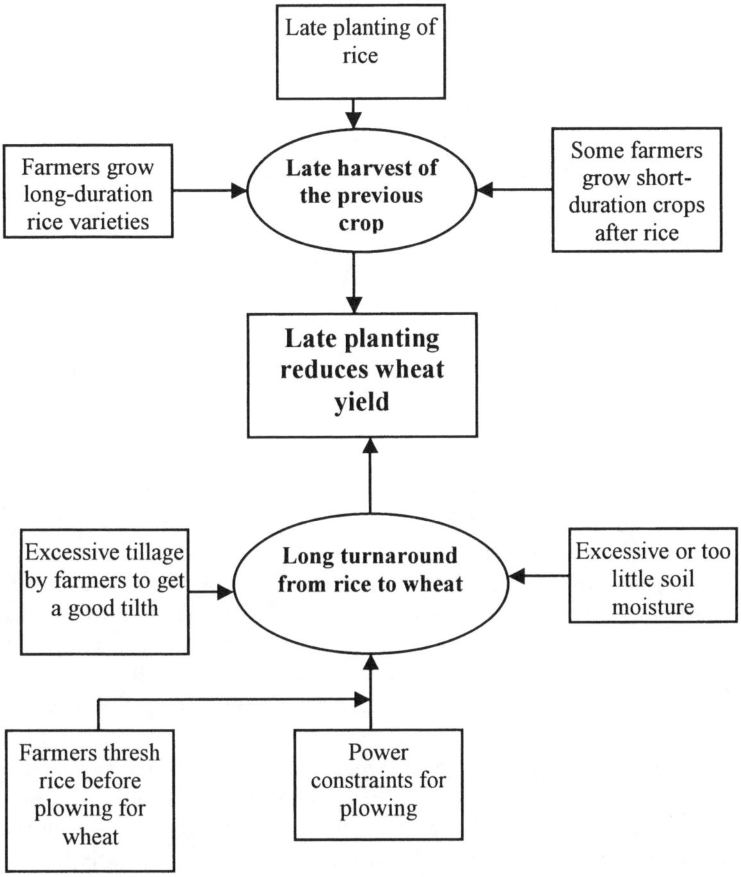

Fig. 7–4. The most common causes of late wheat planting following the rice harvest.

The other major cause of late wheat planting is the long turnaround time after the rice harvest (Fig. 7–4). The long turnaround time can be caused by many factors, such as excessive tillage, soil moisture problems (too wet or too dry), lack of animal or mechanical power for plowing, and the priority farmers place on threshing and handling the rice crop before preparing land for wheat.

Coupled with the problems of late planting of wheat is the problem of poor germination and plant stands. The majority of farmers in the IGP plant wheat by broadcasting the seed into plowed land and incorporating the seed by another plowing and leveling operation. Part of the reason for this is residue management problems in fields following rice. The loose straw and stubbles are raked and clog the seed drills. Broadcast seed results in seed placement at many different depths and into different soil moistures with resulting variable germination. The problems of late planting and poor plant stand have been partially solved in the RWC by promoting various resource-conserving tillage and crop establishment techniques described below.

RESOURCE-CONSERVING TECHNOLOGIES PROMOTED IN THE INDO-GANGETIC PLAINS

Resource-conserving technologies are any practice that results in higher production at less cost. These result in more efficient use of the natural resources used to produce a crop. The following is a brief description of some of the alternative RCT practices being promoted in the region.

Surface Seeding

Surface seeding is the simplest zero-tillage system being promoted. In this tillage option, wheat seed is placed onto a saturated soil surface without any land preparation. This is a traditional farmer practice for wheat, legume, and other crop establishment in parts of eastern India and Bangladesh (Catling et al., 1983). Wheat seed is either broadcast before the rice crop is harvested (relay-planted) or after harvest.

Promotion of surface seeding to plant wheat has been done for several years in areas where the soils are fine-textured and drain poorly and where land preparation is difficult and often results in a cloddy tilth. The key to success with this system is having the correct soil moisture at seeding. Too little moisture results in poor germination and too much moisture can cause seed to rot. A saturated soil surface is best. The roots germinate into the moist soil and follow the saturation fringe as it drains down the soil profile. The high soil moisture reduces soil strength and thus eliminates the need for tillage. Once the root germinates and extends into the soil, the root can follow the saturation fringe and still get sufficient oxygen for growth.

As long as soil moisture can be manipulated, surface seeding is also successful on coarser soils. There must be enough surface moisture to germinate the seed (soaking the seed can help), and soil strength at the root penetration stage must be low (moisture-dependent) to allow root growth. This may require an early light irrigation on coarse-textured soils. Some farmers who relay wheat into the standing rice crop place the cut rice bundles on the ground after harvest. This allows the rice to dry but also acts as mulch, keeping the soil surface moist and ensuring good wheat rooting. In China, where surface seeding is also practiced, farmers apply cut straw to mulch the soil, reduce evaporative losses of moisture, and control weeds (Yonglu et al., 2000; Gupta et al., 2000). The standing stubble also protects the young seedlings from birds. However, relay planting can be done only if the soil moisture is correct for planting at this stage.

One of the major pluses of surface seeding is that no equipment is needed and any farmer can easily adopt this practice. Data presented in Table 7–2 show the benefit of this technology in Nepal, where it is compared with reduced- and conventional-tillage systems. This system is being monitored in farmers' fields to determine whether continuous surface seeding is possible or whether a rotation of tillage systems may be needed to control future weed problems.

Zero-Tillage with Inverted-T Openers

In this RCT, the seed is placed into the soil by a seed drill without prior land preparation. This technology, which has been tested in Pakistan (Sheikh et al., 1993;

Table 7–2. Data from a trial on establishment of wheat following rice, Bhairahwa Agricultural Farm, Nepal, 1993-1994.

Method	Wheat yield	1000-grain	Cost to plow	Net benefit	Extra days needed to plant†
	kg ha^{-1}	g	$ ha^{-1}		
Surface seeding	2775a‡	46.1a	0	287a	0
Chinese seed drill	2831a	45.4b	15	302a	8
Farmers' practice	2314b	40.9c	58	200b	15

† Number of extra days needed for land preparation before seeding compared with the surface seeding.

‡ Numbers followed by the same letter are not significantly different at $P < 0.05$ using Duncan's Multiple Range Test.

Aslam et al., 1993), is presently being tested in other areas of the IGP, including India and Nepal. This technology is more relevant in the higher-yielding, more mechanized areas of northwestern India and Pakistan, where most land preparation is now done with four-wheel tractors.

The basis for this technology is the inverted-T openers (Fig. 7–5) that were developed in and imported from New Zealand. This coulter and seeding system places the seed into a narrow slot made by the inverted-T as it is drawn through the soil by the four-wheel tractor. The coulters can be rigid or spring-loaded depending on the design and cost of the machine. This seed drill works very well in situations where there is little surface residue after the rice harvest. This usually occurs after manual harvesting. The RWC is also promoting the development of animal- and two-wheel tractor-drawn models.

Combine harvesting of wheat is becoming popular among farmers in northwestern India and Pakistan. In this combine system, loose straw and residue are com-

Fig. 7–5. A photograph of the inverted-T coulter used for zero-tillage in the Indo-Gangetic Plains.

Table 7–3. Effects on grain yield of zero-tillage and farmers' practice for establishment of wheat after rice, Punjab, Pakistan, 1985–88.†

Year	Number of locations	Grain yield	
		Zero-tillage	Farmers' practice
		kg ha^{-1}	
1985-1986	15	3600a‡	3516a
1986-1987	13	3791a	3509b
1987-1988	6	4279a	3560b
Pooled data 3 yr	34	3890a	3528b

† Source: Aslam et al. (1993).
‡ Means between zero-till and farmers' practice followed by the same letter do not differ significantly at $P < 0.05$ using Duncan's Multiple Range Test.

monly left after harvest. The inverted-T opener does not work as well where combines are used since the opener acts as a rake for the loose straw. Farmers presently burn residues to overcome this problem of loose stubble whether they use zero-till or the traditional system. Because the RWC wants to discourage this practice, which causes major environmental and air pollution problems, future strategies will look at alternative machinery (disk openers to cut the straw) and techniques to overcome this problem. Leaving the straw as mulch on the soil surface has not been given much thought in Asian agriculture. However, results from rainfed systems and some preliminary results in Asia suggest that this may be very beneficial to the establishment and vigor of crops planted this way (Sayre, 2000). Studies are needed to explore the benefits and longer-term consequences of this idea.

In the late 1980s, 34 zero-tillage trials were conducted in farmers' fields for 3 yr in the rice-growing belt of the Pakistan Punjab (Aslam et al., 1993). The data from these trials are shown in Table 7–3 and show that higher yields are obtained with zero-till because of earlier planting. The implement used in these trials was a tractor-pulled seed and fertilizer drill with inverted T-openers. Any land preparation resulted in loose stubble and a problem of drill clogging. Interestingly, significantly fewer weeds were found under zero-tillage compared with conventional tillage (Verma & Srivastava, 1989, 1994; Singh, 1995), which is the opposite of the experience with zero-tillage systems in developed countries (Kuipers, 1991; Christian & Bacon, 1990). This observation has been confirmed at many other locations. Results from 336 on-farm trials in Haryana (Fig. 7–6) showed that significantly lower weed counts were found in fields with zero-tillage either before or after herbicide applications. This difference can be explained by the nature of the weeds found in the rice–wheat cropping system. Most of the weeds affecting the wheat crop germinate only in the wheat season and, since the soil is disturbed less under zero-tillage, fewer weeds are exposed and fewer germinate. Weed problems typically are more severe under conventional tillage than under zero-tillage. Longer-term research is needed to anticipate future consequences of tillage changes for weed populations (e.g., to quantify likely shifts in weed species and the effects of those shifts on yield stability).

Earlier planting is the main reason for the additional yields obtained under zero-tillage (Table 7–4). Zero-tilled plots were planted as close as possible to 20

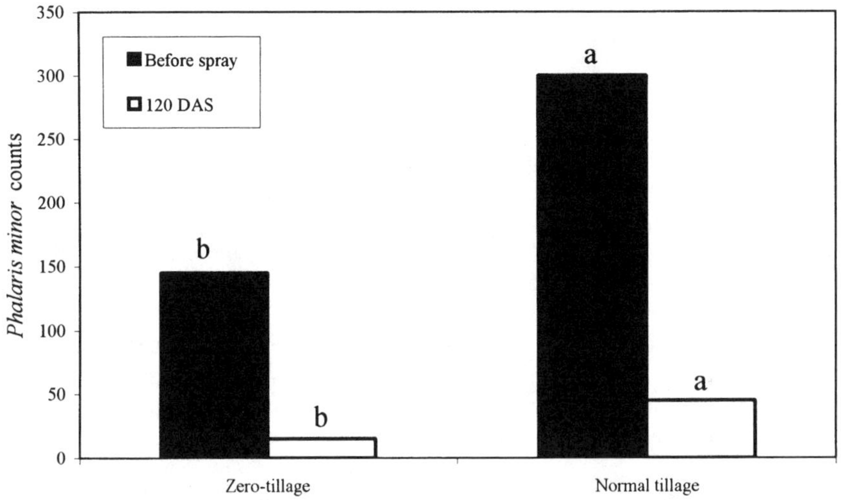

Fig. 7–6. *Phalaris minor* population before herbicide spray and 120 d after seeding (DAS) in zero-till and normal-tilled plots in Haryana, India in 1998-1999. Source: R.K. Malik et al. (2002). Means for zero-tillage and normal tillage followed by the same letter are not significantly different at $P < 0.05$ using Duncan's Multiple Range Test.

November, the optimum date for planting wheat in Pakistan. The results of many trials suggest that the longer the farmer delays planting, the lower the yield. This finding has been confirmed in trials throughout the IGP in the past few years. In Haryana, in surveys and crop cuts in farmers' fields averaged over districts and sowing time, the average yields of wheat were 5.38 and 5.11 Mg ha^{-1} for zero-till and conventional tillage, respectively (Malik et al., 2002; Singh et al., 2002a). This is attributed to earlier timely planting, fewer weeds, better plant stands, and improved fertilizer efficiency because of placement with the seed drill. Some farmers are now

Table 7–4. Comparison of zero-tillage and farmers' practice for establishment of wheat after rice in Punjab, Pakistan, at locations where planting dates for the two methods differ.†

Location	Wheat yield		Days' difference‡
	Zero-tillage	Farmers' practice	
	kg ha^{-1}		
Daska, Site 2	3143	3209	10
Daska, Site 1	3842	2735	13
Ahmed Nagar	4308	3526	20
Maujianwala	2689	2198	22
Mundir Sharif	4245	2660	33
Daska, Site 3	3838	3420	44
Avg.	3677a§	2598b	24

† Source: Aslam et al. (1993).
‡ Difference in days between planting of zero-till and farmers' practice
§ Means for the average yield between zero-tillage and farmers' practice followed by the same letter do not differ significantly at $P < 0.05$ using Duncan's Multiple Range Test.

in their fourth year of adoption of zero-till and find no deleterious effects that would make them revert back to the traditional system (Malik et al., 2002).

Reduced Tillage

The Chinese have developed a seeder for their 12-horsepower, two-wheel diesel tractor that prepares the soil and plants the seed in one operation. This system consists of a shallow rotovator followed by a six-row seeding system and a roller for compaction of the soil. In Bangladesh, farmers are using more than 600 000 hand tractors from China (C. Meisner, personal communication, 2002). Soil moisture was also found to be critical in this reduced-tillage system. The rotovator fluffs up the soil, which then dries out faster than with normal land preparation. The seeding coulter does not place the seed very deep, so soil moisture must be high during seeding to ensure germination and root extension before the soil dries appreciably. Modification of the seed coulter to place the seed a little deeper would help correct this problem.

The main drawback of this technology is that the tractor and the various implements are not easily available, and spare parts and maintenance are a major problem. It would help if the private or public sector in South Asian countries could import this machinery or develop a local manufacturing capability. As it becomes more costly to keep and feed a pair of bullocks for a year, more farmers in the region are turning to significantly cheaper mechanized options of land preparation. One of the benefits of this tractor is that it comes with many options for other farm operations; it includes a reaper, rotary tiller, and a moldboard plow, and it can also drive a mechanical thresher, winnowing fan, or pump. However, most farmers are attracted to the tractor because it can be hitched up to a trailer and used for transportation. For smaller-scale farmers who cannot afford their own tractors, custom hiring is a common alternative.

Table 7–2 shows that this system generates yields equal to those obtained through surface seeding, and the net returns are superior to the farmers' current practice because of the lower land preparation costs. The Chinese tractor can also be used with a rotovator (Bangladesh) to quickly prepare the soil and incorporate the seed after a second pass. This speeds up planting and results in better stands with less cost than traditional methods. However, the Chinese seeder attachment does a better job because the seeds are placed at a uniform depth in the single pass. Engineers are experimenting with removing some of the blades that rototill the soil. In this way, a strip of soil is cultivated rather than the whole area. This reduces the power needs and costs and makes it easier for farmers to manage the tractor. In India, a four-wheel tractor version of this "strip-tillage" machinery is available.

Bed-Planting Systems (Ridge-Till)

In bed-planting systems, wheat or other crops are planted on raised beds. This practice has increased dramatically in the last decade or so in the high-yielding irrigated wheat-growing area of northwestern Mexico (Meisner et al., 1992; Sayre & Ramos, 1997). Bed planting rose from 6% of farmers surveyed in 1981 to 75% in 1994. These data were based on a subsample of 52 growers who were taken from

Table 7–5. Effects of bed size configuration on wheat yield, Punjab Agricultural University, Ludhiana, India, 1994-1995.†

	Sowing method					
	On the flat	75-cm beds		90-cm beds		
Variety	25-cm row	2 rows	2 + 1 rows‡	3 rows	3 + 1 rows‡	Mean
	Wheat yield kg ha^{-1}					
PBW 226	5740	6170	6390	6160	6320	6160a
WH 542	6290	5830	6360	6000	6040	6110a
CPAN 3004	6020	5530	6140	5630	5600	5780b
PBW 154	5460	5110	6000	5930	5880	5680b
HD 2329	5770	4660	6190	5580	5810	5600b
PBW 34	5650	5610	5800	5580	5630	5650b
Mean	5820b§	5490c	6150a	5810b	5880b	

† Source: Dhillon et al. (2000).
‡ 2 + 1 and 3 + 1 mean that an extra row of wheat was planted at the bottom of the furrow.
§ Means of varieties and sowing method followed by the same letter do not differ significantly at $P < 0.05$ using Duncan's Multiple Range Test.

a pool of growers that had been studied intensively by the CIMMYT Economics Section in the past 10 yr. Farmers have given the following reasons for adopting the new system:

- Management of irrigation water is improved.
- Bed planting facilitates irrigation before seeding and thus provides an opportunity for weed control before planting.
- Plant stands are better.
- Weeds can be controlled mechanically, between the beds, early in the crop cycle.
- Wheat seed rates are lower.
- Herbicide dependence is reduced and hand weeding and roguing are easier.
- Less lodging occurs.
- Beds provide better drainage and result in less waterlogging damage to wheat.

This system is now being assessed for suitability in the Asian subcontinent. At Punjab Agricultural University, two bed widths (75 and 90 cm, Table 7–5) and two or three rows of wheat planted per bed were compared with conventional flat-bed planting. Two of the major constraints to higher yields in northwestern India and Pakistan are weeds and lodging. Both can be reduced in bed-planting systems. The major weed species affecting wheat, *Phalaris minor*, is normally controlled using the herbicide isoproturon, which is not always effective. Farmers do not always apply isoproturon well or on time. In addition, recent reports have confirmed that *P. minor* has developed isoproturon resistance (Malik & Singh, 1995; Malik 1996; Malik et al., 1998; Yadav et al., 2002). Alternative integrated weed strategies must be developed to overcome this problem. Preliminary observations made by the author indicate that *P. minor* is less prolific on dry tops of raised beds than on the wetter soil found in conventional flat-bed planting. Cultivating between the beds can also reduce weeds. Thus, bed planting provides farmers with additional options for controlling weeds.

Lodging is also less of a problem on raised beds (Meisner et al., 1992). Additional light enters the canopy and strengthens the straw and the soil around the base of the plant is drier. Reduced lodging can have a significant effect on yield, since many farmers in the Punjab do not irrigate after heading precisely because they want to avoid lodging. As a result, water can become limiting during grain filling, resulting in lower yields. On raised beds, this irrigation need not be avoided for the reasons stated.

Results show that there is no significant difference between flat-bed and 90-cm-bed planted systems (Table 7–5) (Dhillon et al., 2000). The 75-cm beds gave higher yields when an additional row was planted in the furrow compared with flat-bed planting, but less yield when this row was left out. This reduction in yield for two rows on 75-cm beds was a result of low yields with two varieties, PBW154 and HD2329, with this planting system. These two varieties have an upright plant structure and cannot compensate for the gap between the beds. Other varieties such as PBW 226 have a more spreading architecture and can compensate for the gap between rows. This suggests that varieties need to be selected for this planting system. Figure 7–7 shows results from Pantnagar and Haryana for variety PBW 343, a well-adapted variety for bed planting that does better on beds than flat-bed planting. Farmers have expressed that water savings are a major advantage they obtain from bed systems (Singh et al., 2002b).

An additional advantage of bed planting becomes apparent when beds are "permanent," that is, when they are maintained over the medium term and not broken down and reformed for every crop. In this system, wheat is harvested and straw is left on the surface or burned. Passing a shovel down the furrows reshapes the beds.

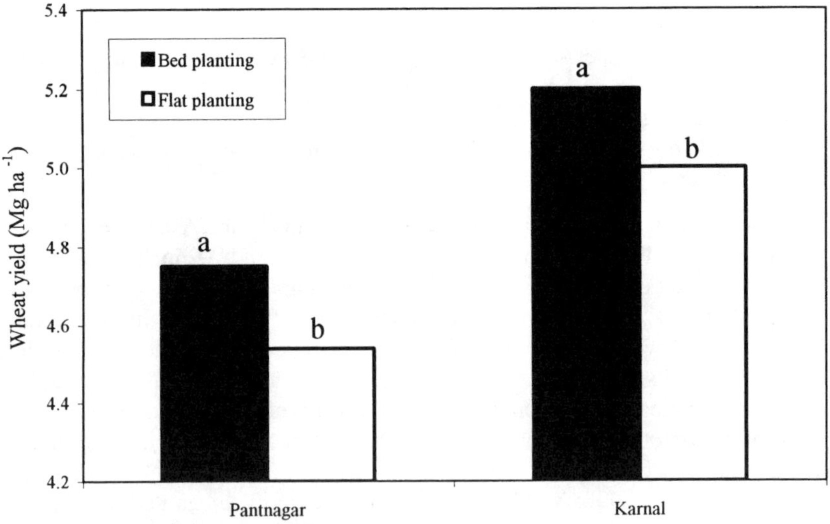

Fig. 7–7. Comparison of bed vs. flat planting in farmers' fields in Pantnagar ($n = 3$) and Karnal ($n = 8$), India, in 1998-1999. Means within sites followed by the same letter are not significantly different at $P < 0.05$ using Duncan's Multiple Range Test.

The next crop (e.g., soybean, maize, sunflower, cotton, or rice) can then be planted into the stubble in the same bed. Research in farmers' fields has also shown that rice can be grown on beds, either transplanted or direct-seeded, making this system feasible in the rice–wheat pattern (Balasubramanian et al., 2003, Chapter 6 of this publication; Gupta et al., 2003, Chapter 1 of this publication). Rice can be grown on beds by either transplanting seedlings or direct seeding. At the moment, transplanting on beds is best since the normal herbicides used for transplanted rice can be used to control weeds. As dry-seeded herbicides become available and weeds can be managed, dry-seeded rice on beds will become more attractive. One farmer in the Punjab reported to the author that he obtained 9 Mg ha^{-1} of rice transplanted on beds and saved more than 50% of the water he normally applies on flat-surface transplanted rice. His flat-surface planted rice yields from 7 to 8 Mg ha^{-1}, slightly less than beds because it lodges. This is being confirmed through monitoring water use in bed-planted plots this year at several RWC sites. Data are not available at this time from these sites.

The use of beds also provides a way for improving fertilizer use efficiency. This is achieved by placing a band of fertilizer in the bed at planting or topdress. Using slow-release formulations and experimenting with urea supergranules can make further improvements. Both can be applied in the bed at the time of planting with the seed-cum-fertilizer drill.

Benefits Obtained in a Systems Perspective

One subject that will not be covered in detail here concerns the benefits that can accrue when zero-till or bed planting is combined with rice establishment without puddling the soil. Data presented in Fig. 7–8 show that wheat yields are significantly better (300 kg ha^{-1}) when wheat is planted with zero-till after non-

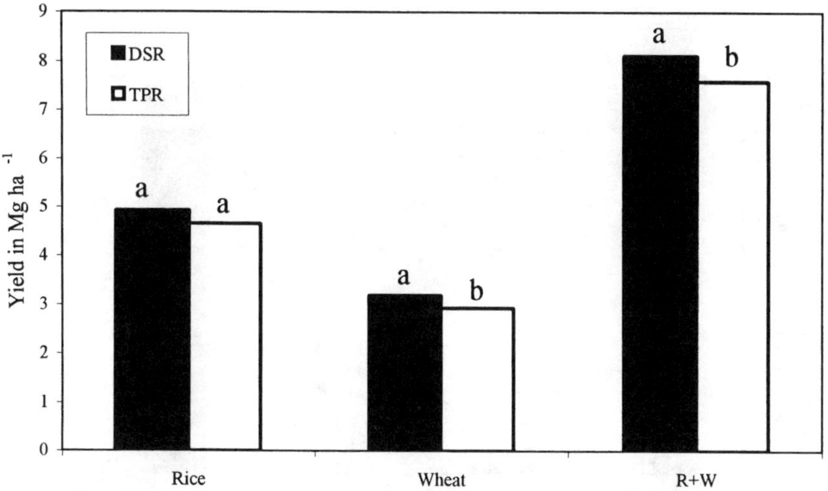

Fig. 7–8. Rice and wheat yields with (TPR) and without (DSR) puddling in Nepal in 1998–1999. Means followed by the same letter are not significantly different at $P < 0.05$ using Duncan's Multiple Range Test.

puddled rice than after puddled rice. The data also show that rice yields remain similar under puddled and nonpuddled situations. Weeds were controlled properly in these experiments and so did not hamper yield. This shows that RCTs need to be assessed on a systems basis and not on a single crop. This is being promoted strongly in the RWC.

ADOPTION OF RESOURCE-CONSERVING TECHNOLOGIES IN SOUTH ASIA

The adoption of RCTs in South Asia has been rapid in the past few years, especially for zero-till with the inverted-T planter. Figure 7–9 shows data from India and Pakistan (data from Malik et al., 2002 and Gill et al., 2000). This success was possible because of the application of participatory approaches to accelerate adoption. The traditional extension system that was so effective in the early years of the Green Revolution was based on the development of recommendations and packages and then having the extension service demonstrate the technology to farmers. Seed and fertilizer were easily packaged, and it was possible to lay out many trials at low cost. When this traditional extension system was used for extending RCTs, problems arose. The first problem was the availability of the machinery to conduct the demonstrations, although the main constraint was convincing farmers, extension workers, and at times scientists that this technology had any benefit. Success came once partners were allowed to work together and experiment with the technology. Local manufacturers had to be involved in the development and manufacture of the equipment. Machinery had to be of high quality, yet at a cost within the budgets of farmers. Farmers had to be shown how the drill worked and then allowed to experiment with the equipment before they could be convinced to accept this radical technology. Stories abound of how the farmers who first tried the technology

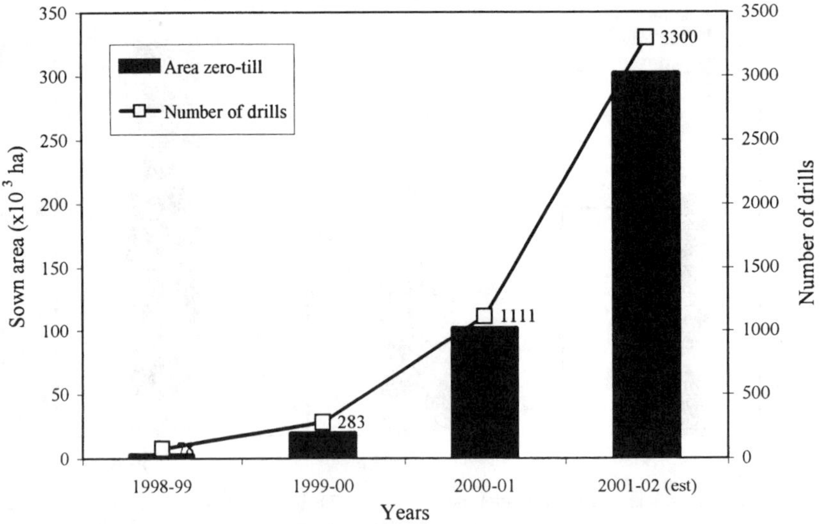

Fig. 7–9. Trends in zero-tillage area for wheat in rice–wheat systems in India and Pakistan.

were ridiculed by their neighbors for trying something so alien, but once the seed germinated, farmers begged the innovator to help them sow their fields. It is now thought that zero-till wheat is an acceptable technology and will be part of the recommendations for planting wheat. Similarly, other RCTs, such as bed planting, will become accepted practices as machinery is made available and more farmers experiment with the system.

One question that is often asked is, Who can benefit from this technology? Is it just for the large commercial farmer? The answer from a survey done by Haryana Agricultural University appears to indicate that this technology is scale-neutral and that farmers from all social classes can benefit from the many advantages that this system brings to wheat cultivation (Malik et al., 2002; Punia et al., 2002). The survey conducted in Haryana in 2001 of 20 villages and 91 farmers showed that 24% of the zero-till adopters owned a tractor, whereas the rest used service providers. The average farm size of adopters ranged from 0.8 to 20.2 ha. Twenty-two percent had <2 ha and 37% from 2- to 4-ha farms. All but one farmer agreed that zero-till was highly profitable, and data show that zero-till resulted in \$75 ha^{-1} more cost reduction than with conventional practices. Yields from zero-till were 5.4 Mg ha^{-1} vs. 5.1 Mg ha^{-1} for conventional practices. Resource-poor farmers and farmers without tractors are using contract services to plow their fields now. It is becoming too expensive to keep a pair of bullocks just for land preparation and so using a service provider is more economical. When this technology is applied to zero-till or bed planting, the benefits are even more pronounced. In this case, farmers have only to rent the service once and their fields are planted. This saves them money and time to do other activities. Data from socioeconomic and impact assessment surveys in India (Malik et al., 2002; Punia et al., 2002) and Pakistan (Anwar et al., 2001) show this to be true. The first innovators are larger, better endowed tractor owners. Later, less-endowed farmers adopt the technology as they see the benefits and obtain the services for this technology.

Farmer responses and feedback on the RCTs and especially zero-tillage provide valuable feedback to scientists in the RWC for improving the technology. At the same time, scientists have been monitoring the fields where these technologies are being adopted and collecting data on soil, biotic, and resource use. Table 7–6 lists some of this feedback for different variables. These data show that zero-tillage provides many benefits to farmers and explains why farmers have adopted this technology so quickly. This adoption rate will increase further as more machinery is manufactured and made available to farmers and more farmers become aware of the benefits to their farming.

ENVIRONMENTAL ISSUES

Apart from the benefits of greater yields at less cost, there are several environmental benefits from adopting RCTs (Grace et al., 2003, Chapter 2 of this publication). Foremost among these are the savings in fossil fuel and the reduction in greenhouse gas emissions. Data in Table 7–6 show that 60 L of diesel can be saved per hectare in zero-till. If eventually used on even 1 million ha of rice–wheat area, zero-tillage would reduce diesel use by 60 million liters. Using a conversion fac-

Table 7–6. Benefits of zero-tillage over conventional tillage for sowing wheat after rice in Haryana, India.

Item	Farmers' perceptions	Researchers' findings
Sowing	Wheat sowing earlier by 5–8 d (smaller-to-medium farms) to 2 wk (large farms)	On average, wheat sowing can be advanced by 5–15 d
Fuel savings	Not available	On average 60 L diesel per ha
Cost of cultivation	\$42–92 ha^{-1}	\$37–62 ha^{-1}
Plant population	20–30% more plants in zero-till fields	13.5% more plants in zero-tillage fields
Weed infestation	20% less and weaker weeds in zero-till fields	43% less weeds in zero-till fields
Irrigation	Saves 30–50% water in the first irrigation and 15–20% in subsequent irrigations	36% less water used, on average
Rice stem borer	Less, because of less stubble sprouting	Winter coolness impairs sprouting and thus borer development. Beneficial insects in stubble help control borers.
Rice stubbles	Decayed faster	Decayed faster
Fertilizer use efficiency	High	Higher because of placement
Wheat yields	Higher than under conventional system depending on days planted earlier	420–530 kg more ha^{-1}

tor of 2.6 kg CO_2 produced per liter of diesel burned, this would represent a reduction of more than 156 000 Mg yr^{-1} in CO_2 emissions. Other greenhouse gas emissions, including methane and nitrous oxides, have an even greater effect on global warming. Changing how rice is grown and promoting more nonpuddled rice cultivation on beds or on flat surfaces can reduce methane emissions (Grace et al., 2003, Chapter 2 of this publication). Improving fertilizer efficiency can also reduce nitrous oxide emissions. This has very important implications for global warming.

No-tillage practices also result in less oxidation of soil organic matter and so it is hypothesized that this would help with sequestering of C. Data are being collected on this important problem by monitoring farmers who have adopted RCTs. The management of residues left over after harvesting fields, especially combine-harvested fields, is also important in terms of C management, air pollution, and greenhouse gas emissions. Many farmers now burn these residues. This creates severe air pollution and a health hazard to people in the area. It also provides short-term carbon dioxide emissions. If burning residues can be eliminated, this problem can be solved. Research is under way to find a solution where the residue can be left as mulch on the surface and crops sown into the residue without burning or incorporation.

Water will become a major constraint to agriculture in the future as competition for water for domestic and industrial uses increases. At the moment, water use efficiency in the IGP is poor, and ways to increase this efficiency are needed. Resource-conserving technologies are one way to achieve this. Practices being promoted to achieve this are leveling of fields, lining of distribution channels, and adoption of zero-tillage and bed planting. Because zero-tillage takes immediate advantage of residual moisture from the previous rice crop, as well as cutting down on subsequent irrigation requirements, water use is reduced by about 10 cm or approximately 1 million L ha^{-1}. Farmer feedback indicates that application of the first irrigation is much faster in zero-till since the water flows faster across the field and farmers take one-third to one-half less time to irrigate their fields (Malik et al., 2002). In bed planting, farmers have also said that it takes half the time to irrigate their fields than on the flat surface. Sayre and Moreno Ramos (1997) and Sayre (2000) put a figure of 35% as savings in water with bed-planted wheat compared with flat-planted wheat in Mexico. Similar data are reported by Malik et al. (2002). Monitoring is presently being done in farmers' fields to better quantify this savings scientifically. Data are also being collected at the command and basin level to ascertain whether savings at the field level translate into savings at the basin level.

Resource-conserving technologies also have environmental implications in terms of reducing the use of organic chemicals for weed and insect control. Zero-tillage results in less weed germination because less weed seed is exposed than with normal tillage (Malik et al., 2002; Fig. 7–6). In bed planting, weeds can be controlled mechanically, although this increases fuel costs, instead of using expensive herbicides that have detrimental effects on soil biology. This is particularly important in areas where the main grassy weed in wheat cultivation, *Phalaris minor*, has developed resistance to the herbicide isoproturon (Malik et al., 1998; Yadav et al., 2002). Data collected from monitoring of farmers' fields also suggest that leaving the stubbles in the field and not burning them in zero-tillage results in a favorable habitat for beneficial insects and spiders and thus less of a problem with stem borers in the next rice crop (Jaipal et al., 2002). The data also suggest that the diversity of beneficial insects was higher in zero-till fields than in conventionally tilled ones.

ISSUES FOR FUTURE RESEARCH

More and Better Machinery. Zero-till drills are currently manufactured in the Indian and Pakistan Punjabs and must be shipped to other places that want to adopt zero-till and bed planting. Also, current drills become plugged or do not always sow seed at uniform depths. Fertilizer distribution is not uniform and machinery does not handle well the larger urea supergranules. If machinery is to be used for intercultivation and control of weeds, appropriate attachments are needed to do this efficiently. The RWC will be working with local manufacturers to produce prototypes that will be tested with farmer innovators in their fields.

Residue Management. Some way to manage the large amounts of residues produced in rice–wheat rotations, besides burning them, is needed. Options include developing systems to plant into residue, baling and removal for use as animal feed or for industry, and possibly the use of microbial sprays that speed decomposition. Planting into the residue will require possible chopping and spreading of residue from the combine and the use of specialized coulters that can plant into the residue without clogging. Star-wheel planters and double-disk openers with various cutting disks are being evaluated (Gupta & Rickman, 2002).

New Varieties and Crops. Researchers need to identify new varieties and crops, including wheat, rice, legumes, oilseeds, fodders, and vegetables, that will grow well under zero-till and on beds. Data do suggest that a variety × tillage interaction occurs and that wheat varieties with a more spreading canopy do better than upright varieties. In rice, "aerobic" rice varieties that do well under nonpuddled direct-seeded conditions and are more competitive with weeds are urgently needed. Crops that fit the new tillage systems are also needed.

Land Leveling. As machinery becomes more available, land leveling of fields to improve the efficiency of water use is being promoted. Use of these systems at the field level also needs to be translated into benefits at the basin level.

Water Distribution Improvements. Water distribution losses from canals can be reduced by introducing water-course improvements, including lining of canals.

Long-Term Consequences. Monitoring of soils and biotic factors and socioeconomic surveys are needed to assess the longer-term benefits of these new systems. Quantitative data are needed on the benefits of RCTs for C sequestration, water savings, and improved input efficiency. Are there longer-term problems of weeds, diseases, and insects that need to be anticipated? Who benefits from the RCTs and do the benefits lead to improved livelihoods and poverty alleviation?

Policy. Policy issues are also very important for developing efficient agricultural systems anywhere in the world. Issues of marketing of extra produce are already a major concern of farmers (Pingali & Shah, 1999; Kumar et al., 1999). Problems of input costs, water pricing, and the competitiveness of South Asian farmers in relation to global trade will become more important with the introduction of the General Agreement on Tariffs and Trade. Should there be subsidies to promote the adoption of RCTs? Research on policy issues has been weak from RWC partners and will need to be given importance in future planning.

CONCLUSIONS

In this chapter we have tried to demonstrate the benefits to the region when farmers adopt various resource-conserving technologies in wheat in terms of improved production at a lower cost, improving the efficiency of natural resources, benefits to the environment, improved livelihoods of farmers, and ultimately help in alleviating poverty. The success of this technology depends on rapid adoption by farmers. Accelerated adoption was mainly the result of the change in paradigm for extending the technology. Instead of the linear approach to extension commonly found in the Green Revolution era, the importance of partnerships, expanded stakeholders, and participatory approaches in which farmers could experiment and feed back information soon became apparent. Resource-conserving technologies are a key to ensuring sustainable food production in South Asia in the next decade. Overcoming mind-sets that hold traditional beliefs about excessive tillage and providing the enabling factors that allow exposure of the technology to all those involved in agriculture will be key factors for future success. This technology revolution is seen as one way to sustainably increase food production to meet future demand while conserving natural resources, improving farmer livelihoods, and reducing negative effects on the environment. Of course, all of these benefits will be of little use unless the nations in South Asia control their population growth.

REFERENCES

Anwar, M.Z., M.A. Khan, and M. Aslam. 2001. Impact of zero tillage wheat sowing technology in the rice–wheat farming system of Pakistan's Punjab. PARC & Rice–Wheat Consortium Paper No. 2001-1. Agricultural Economic Research Unit, SSI, NARC, Islamabad, Pakistan.

Aslam, M., A. Majid, N.I. Hashmi, and P.R. Hobbs. 1993. Improving wheat yield in the rice–wheat cropping system of the Punjab through zero tillage. Pakistan J. Agric. Res. 14:8–11.

Balasubramanian, V., J.K. Ladha, R.K. Gupta, R.K. Naresh, R.S. Mehla, Bijay-Singh, and Yadvinder-Singh. 2003. Technology options for rice in the rice–wheat systems in Asia. p. 115–148. *In* J.K. Ladha et al. (ed.) Improving the productivity and sustainability of rice–wheat systems: Issues and impacts. ASA Spec. Publ. 65. ASA, CSSA, SSSA, Madison, WI.

Byerlee, D., A. Sheikh, M. Aslam, and P.R. Hobbs. 1984. Wheat in the rice-based farming system of the Punjab: Implications for research and extension. PARC/CIMMYT Wheat Paper No. 84. Pakistan Agricultural Research Council (PARC) and CIMMYT, Islamabad, Pakistan.

Catling, H.D., P.R. Hobbs, Z. Islam, and B. Alam. 1983. Agronomic practices and yield assessments of deepwater rice in Bangladesh. Field Crops Res. 6:109–132.

Christian, D.G., and E.T.G. Bacon. 1990. A long-term comparison of ploughing, tyne cultivation and direct drilling on the growth and yield of winter cereals and oilseed rape on clayey and silty soils. Soil Tillage Res. 18:311–331.

Dawe, D., A. Dobermann, P. Moya, S. Abdulrachman, Bijay-Singh, P. Lal, S.Y. Li, B. Lin, G. Panaullah, O. Sariam, Y. Singh, A. Swarup, P.S. Tan, and Q.X. Zhen. 2000. How widespread are yield declines in long-term rice experiments in Asia? Field Crops Res. 66:175–193.

Dhillon, S.S., P.R. Hobbs, and J.S. Samra. 2000. Investigations on bed planting system as an alternative tillage and crop establishment practice for improving wheat yields sustainably. *In* Proc. 15th Conf. Int. Soil Tillage Res. Org. [CD-ROM] Fort Worth, TX. 2–7 July 2000.

Duxbury, J.M., I.P. Abrol, R.K. Gupta, and K.F. Bronson. 2000. Analysis of long-term soil fertility experiments with rice–wheat rotations in South Asia. p. 7–22. *In* I.P. Abrol et al. (ed.) Long-term soil fertility experiments with rice–wheat rotations in South Asia. Rice–Wheat Consortium Paper Ser. 6. Rice–Wheat Consortium for the Indo-Gangetic Plains, New Delhi, India.

Fujisaka, S., L.W. Harrington, and P.R. Hobbs. 1994. Rice–wheat in South Asia: System and long-term priorities established through diagnostic research. Agric. Syst. 46:169–187.

Gill, M.A., M.A. Kahlown, M.A. Choudhary, and P.R.Hobbs. 2000. Evaluation of resource conservation technologies in rice–wheat system of Pakistan. Report of the OFWM Division, Punjab Government, Lahore, Pakistan.

Grace, P.R., M.C. Jain, L. Harrington, and G.P. Robertson. 2003. Long-term sustainability of the tropical and subtropical rice–wheat system: An environmental perspective. p. 27–44. *In* J.K. Ladha et al (ed.) Improving the productivity and sustainability of rice–wheat systems: Issues and impact. ASA Spec. Publ. 65. ASA, CSSA, and SSSA, Madison, WI.

Gupta, R.K., P.R. Hobbs, M. Salim, N.H. Chowdhary, and S.I. Bhuiyan (ed.) 2000. Study of research and extension issues in the Sichuan Province of China for farm-level impact on productivity of the rice–wheat system. Rice–Wheat Consortium Traveling Seminar Report Ser. 2. Rice–Wheat Consortium, New Delhi, India.

Gupta, R.K., R.K. Naresh, P.R. Hobbs, Zheng Jiaguo, and J.K. Ladha. 2003. Sustainability of post-Green Revolution agriculture: The rice–wheat cropping system of the Indo-Gangetic Plains and China. p. 1–26. *In* J.K. Ladha et al. (ed.) Improving the productivity and sustainability of rice–wheat systems: Issues and impacts. ASA Spec. Publ. 65. ASA, CSSA, and SSSA, Madison, WI.

Gupta, R.K., and J. Rickman. 2002. Design improvements in existing zero-till machines for residue conditions. Rice–Wheat Consortium Traveling Seminar Report Ser. 3. Rice–Wheat Consortium, New Delhi, India.

Harrington, L.W., S. Fujisaka, P.R. Hobbs, C. Adhikary, G.S. Giri, and K. Cassaday. 1993a. Rice–wheat cropping systems in Rupandehi District of the Nepal Terai: Diagnostic surveys of farmers' practices and problems, and needs for further research. CIMMYT, Mexico.

Harrington, L.W., S. Fujisaka, M.L. Morris, P.R. Hobbs, H.C. Sharma, R.P. Singh, M.K. Chaudhary, and S.D. Dhiman. 1993b. Wheat and rice in Karnal and Kurukshetra Districts, Haryana, India: Farmers' practices, problems and an agenda for action. Indian Council for Agricultural Research (ICAR), Haryana Agricultural University (HAU), CIMMYT, Mexico, and IRRI, Los Baños, Philippines.

Hobbs, P.R. 2001. Tillage and crop establishment in South Asian rice–wheat systems: Present and future options. J. Crop Prod. 4:1–23.

Hobbs, P.R., E.P. Hettel, R.P. Singh, Y. Singh, L.W. Harrington, and S. Fujisaka (ed.) 1991. Rice–wheat cropping systems in the Terai areas of Nainital, Rampur, and Pilibhit Districts in Uttar Pradesh, India: Sustainability of the rice–wheat system in South Asia. Diagnostic surveys of farmers' practices and problems, and needs for further research. CIMMYT, Mexico, ICAR, and IRRI, Los Baños, Philippines.

Hobbs, P.R., E.P. Hettel, R.K. Singh, R.P. Singh, L.W. Harrington, V.P. Singh, and K.G. Pillai. 1992. Rice–wheat cropping systems in Faizabad District of Uttar Pradesh, India: Exploratory surveys of farmers' practices and problems, and needs for further research. ICAR, NDUAT, CIMMYT, Mexico, and IRRI, Los Baños, Philippines.

Hobbs, P.R., and M.L. Morris. 1996. Meeting South Asia's future food requirements from rice–wheat cropping systems: Priority issues facing researchers in the post Green Revolution era. NRG Paper 96-01. CIMMYT, Mexico.

Hobbs, P.R., Y. Singh, G.S. Giri, J.G. Lauren, and J.M. Duxbury. 2002. Direct seeding and reduced tillage options in the rice–wheat systems of the Indo-Gangetic Plains of South Asia. p. 201–205. *In* S. Pandey et al. (ed.) Direct seeding in Asian rice systems: Strategic research issues and opportunities. IRRI, Manila, Philippines.

Jaipal, Saroj, Samar Singh, A. Yadav, R.K. Malik, and P.R. Hobbs. 2002. Species diversity and population density of macro-fauna of rice–wheat cropping habitat in semi-arid subtropical north-west India in relation to modified tillage practices of wheat sowing. p. 166–171. *In* R.K. Malik et al. (ed.) Herbicide resistance management and zero-tillage in the rice–wheat cropping system. Proc. Int. Workshop. Hissar, India. 4–6 Mar. 2002. CCSHA Univ., Hissar, India.

Kuipers, H. 1991. Agronomic aspects of ploughing and non-ploughing. Soil Tillage Res. 21:167–176.

Kumar, P., P.K. Joshi, C. Johansen, and M. Ashokan. 1999. Sustainability of rice–wheat based cropping systems in India: Socio-economic and policy issues. p. 61–77. *In* P.L. Pingali (ed.) Sustaining rice–wheat production systems: Socio-economic and policy issues. Rice–Wheat Consortium Paper Ser. 5. Rice–Wheat Consortium for the Indo-Gangetic Plains, New Delhi, India.

Ladha, J.K., D. Dawe, H. Pathak, A.T. Padre, R.L. Yadav, Bijay-Singh, Yadvinder-Singh, Y. Singh, P. Singh, A.L. Kundu, R. Sakal, N. Ram, A.P. Regmi, S.K. Gami, A.L. Bhandari, R. Amin, C.R. Yadav, E.M. Bhattarai, R.K. Gupta, and P.R. Hobbs. 2002. How extensive are yield declines in long-term rice–wheat experiments in Asia? Field Crops Res. 4149:1–22.

Ladha, J.K., K.S. Fischer, M. Hossain, P.R. Hobbs, B. Hardy. 2000. Improving the productivity of rice–wheat systems of the Indo-Gangetic Plains: A synthesis of NARS–IRRI partnership research. IRRI Discussion Paper 40. IRRI, Los Baños, Philippines.

Malik, R.K. 1996. Herbicide resistant weed problems in developing world and methods to overcome them. p. 665–673. *In* Proc. 2nd Int. Weed Science Congress. Copenhagen, Denmark. FAO, Rome.

Malik, R.K., G. Gill, and P.R. Hobbs. 1998. Herbicide resistance: A major issue for sustaining wheat productivity in the rice–wheat cropping system in the Indo-Gangetic plains. Rice–Wheat Consortium Research Paper Ser. 3. Rice Wheat Consortium for the Indo-Gangetic Plains, New Delhi, India.

Malik, R.K., and S. Singh. 1995. Littleseed canarygrass resistance to isoproturon in India. Weed Technol. 9:419–425.

Malik, R.K., A. Yadav, S. Singh, R.S. Malik, R.S. Balyan, R.S. Banga, P.K. Sardana, S. Jaipal, P.R. Hobbs, G. Gill, S. Singh, R.K. Gupta, and R. Bellinder. 2002. Herbicide resistance management and evolution of zero-tillage: A success story. Research Bulletin. Haryana Agricultural University, Hisar, India.

Meisner, C., E. Acevedo, D. Flores, K. Sayre, I. Ortiz-Monasterio, and D. Byerlee. 1992. Wheat production and grower practices in the Yaqui valley, Sonora, Mexico. Wheat Special Rep. 6. CIMMYT, Mexico.

Murgai, R., Mubarik Ali, and D. Byerlee. 2001. Productivity growth and sustainability in post-Green Revolution agriculture: The case of the Indian and Pakistan Punjabs. World Bank Observer 16(2):199–218.

Ortiz-Monasterio, J.I., S.S. Dhillon, and R.A. Fischer. 1994. Date of sowing effects on grain yield and yield components of irrigated spring wheat cultivars and relationships with radiation and temperature in Ludhiana, India. Field Crops Res. 37:169–184.

Pingali, P.L., and M. Shah. 1999. Rice–wheat cropping systems in the Indo-Gangetic plains: Policy redirections for sustainable resource use. p. 1–12. *In* P. Pingali (ed.) Sustaining rice–wheat production systems: Socio-economic and policy issues. Rice–Wheat Consortium Paper Ser. 5. Rice–Wheat Consortium, New Delhi, India.

Punia, S.S., R.K. Malik, A. Yadav, S. Singh, and A.S. Nehra. 2002. Acceleration of zero-tillage technology in Haryana. p. 146–148. *In* R.K. Malik et al. (ed.) Herbicide resistance management and zero-tillage in rice–wheat cropping system. Proc. Int. Workshop. Hissar, India. 4–6 Mar. 2002. CCSHA Univ., Hissar, India.

Randhawa, A.S., S.S. Dhillon, and D. Singh. 1981. Productivity of wheat varieties, as influenced by the time of sowing. J. Res. Punjab Agric. Univ. 18:227–233.

Regmi, A.P., J.K. Ladha, E.M. Pasuquin, H. Pathak, P.R. Hobbs, M.L. Shrestha, D.B. Ghauti, and E. Duveiller. 2002. Potassium in sustaining yields in a long-term rice–wheat experiment in the Indo-Gangetic Plains of Nepal. Biol. Fert. Soils. 36:240–247.

Saunders, D.A. 1990. Crop management research summary of results. Monogr. 5. Wheat Research Centre, Nashipur, Bangladesh.

Sayre, K.D. 2000. Effects of tillage, crop residue retention and nitrogen management on the performance of bed-planted, furrow-irrigated spring wheat in northwest Mexico. *In* Proc. 15th Conf. Int. Soil Tillage Res. Org. [CD-ROM] Fort Worth, TX. 2–7 July 2000.

Sayre, K.D., and O.H. Moreno Ramos. 1997. Applications of raised-bed planting systems to wheat. Wheat Program Special Rep. 31. CIMMYT, Mexico.

Sharma, P.K., J.K. Ladha, and L. Bushan. 2003. Soil physical effects of puddling in rice–wheat cropping systems. p. 97–114. *In* J.K. Ladha et al. (ed.) Improving the productivity and sustainability of rice–wheat systems: Issues and impacts. ASA Spec. Publ. 65. ASA, CSSA, and SSSA, Madison, WI.

Sheikh, A.D., M.A. Khan, M. Aslam, Z. Ahmad, and M. Ahmad. 1993. Zero tillage wheat cultivation technology in the rice–wheat areas of the Punjab. Mimeograph. National Agricultural Research Council (NARC), Islamabad, Pakistan.

Singh, Samar, A. Yadav, R.K. Malik, and Harpal Singh. 2002a. Long term effect of zero-tillage sowing technique on weed flora and productivity of wheat in rice–wheat cropping zones of Indo-Gangetic Plains. p. 155–157. *In* R.K. Malik et al. (ed.) Herbicide resistance management and zero-tillage in rice–wheat cropping system. Proc. Int. Workshop. Hissar, India. 4–6 Mar. 2002. CCSHA Univ., Hissar, India.

Singh, Samar, A. Yadav, R.K. Malik, and Harpal Singh. 2002b. Furrow irrigated raised bed planting system: A resource conservation technology for increasing wheat productivity in rice–wheat sequence. p. 198–200. *In* R.K. Malik et al. (ed.) Herbicide resistance management and zero-tillage in rice–wheat cropping system. Proc. Int. Workshop. Hissar, India. 4–6 Mar. 2002. CCSHA Univ., Hissar, India.

Singh, Y. 1995. Rice–wheat cropping system NARP basic research sub-project. Progress Report 1994-95. G.B. Pant University of Agriculture and Technology, Pantnagar, U.P., India.

Verma, U.N., and V.C. Srivastava. 1989. Weed management in wheat under zero and optimum tillage conditions. Indian J. Agron. 34:176–179.

Verma, U.N., and V.C. Srivastava. 1994. Production technology of non-tilled wheat after puddled transplanted rice. Indian J. Agric. Sci. 64:277–284.

Yadav, A., R.K. Malik, B.S. Chauhan, and G. Gill. 2002. Present status of herbicide resistance in Haryana. p. 15–23. *In* R.K. Malik et al. (ed.) Herbicide resistance management and zero-tillage in rice–wheat cropping system. Proc. Int. Workshop. Hissar, India. 4–6 Mar. 2002. CCSHA Univ., Hissar, India.

Yonglu, T., Huang Gang, Yu Yao, and Yuan Lixun. 2000. High yielding cultivation techniques for wheat under the rice–wheat cropping system in Sichuan Province of China. *In* P.R. Hobbs and R.K. Gupta (ed.) Soil and crop management practices for enhanced productivity of the rice–wheat cropping system in Sichuan Province of China. Rice–Wheat Consortium Paper Ser. 9. Rice–Wheat Consortium, New Delhi, India.

8 Managing Crop Residues in the Rice–Wheat System of the Indo-Gangetic Plain

Jagir S. Samra

Indian Council of Agricultural Research
New Delhi, India

Bijay-Singh

Punjab Agricultural University
Ludhiana, India

Kuldip Kumar

University of Minnesota
St. Paul, Minnesota, USA

ABSTRACT

Large quantities of crop residues produced in the rice–wheat system spanning 13.5 million hectares in South Asia are a potential source for improving soil organic matter dynamics, nutrient cycling, and the soil physical environment. Burning of plant residues is causing sustainability concerns. The incorporation of residues into the soil improved soil properties in the rice–wheat system. Crop residue management produced greater effects on yields in fine-textured soils than in coarse-textured soils. The incorporation of wheat residues, in general, increased the yield of rice and also had a positive residual effect on the yield of subsequent wheat. However, when 50% of the inorganic N applied to wheat was replaced by wheat residue N, rice yield declined significantly. Recycling of rice residues constitutes more potential problems than recycling of wheat straw. The incorporation of rice straw in wheat and wheat straw in rice also caused yield reductions. However, yield reductions can be overcome by either allowing some time between incorporating residues and sowing the next crop or by composting rice straw or adding farmyard manure (FYM). High crop yields were obtained from the combined long-term use of fertilizers and crop residues. Control of weeds, diseases, and pests is important for the success of residue management practices. Residues can be incorporated into the soil, left on the soil surface, or partially burned. A moldboard plow or disk can incorporate residue into the soil, but appropriate machinery for the remaining two options is still in the development stage.

Crop residues—the parts of plants left in the field after crops have been harvested and thrashed—have at times been regarded as waste materials that require disposal,

 Improving the Productivity and Sustainability of Rice–Wheat Systems: Issues and Impacts. ASA Special Publication 65.

Table 8–1. Rice and wheat crop residues available in different states of India in the Indo-Gangetic Plain and usable nutrients contained in residues (adapted from Sarkar et al., 1999).

State	Residues available†			Residues available for recycling‡	Nutrient (N + P + K) potential§		
	Rice	Wheat	Total		Total¶	Available for direct recycling	Fertilizer replacement value#
	Tg						
Punjab	10.0	18.2	28.2	9.40	0.462	0.154	0.077
Haryana	2.5	9.7	12.2	4.07	0.194	0.065	0.032
Uttar Pradesh	14.0	27.5	41.5	13.83	0.677	0.226	0.113
Bihar	9.6	5.3	14.9	4.97	0.257	0.086	0.043
West Bengal	16.7	0.1	16.8	5.60	0.308	0.103	0.051
Total	52.8	60.8	113.6	37.87	1.898	0.634	0.316

† Calculated by dividing the economic yield by grain straw ratio of 1:1.5 as used by Bhardwaj (1995).
‡ Assuming that two-thirds of the crop residues are used as feed for animals and other purposes and one-third is available for direct recycling.
§ Based on amount of residues available for recycling.
¶ Using average oven-dry nutrient contents as given by Tandon (1995).
# Assuming that one-half of the total N + P + K is mineralized in a season.

but it is increasingly realized that these are a tremendous natural resource, not a waste. In the Indo-Gangetic Plain of South Asia spread over 13.5 million hectares in India, Pakistan, Nepal, and Bangladesh, rice followed in sequence by wheat is a cropping system dating back thousands of years (Ladha et al., 2000; Woodhead et al., 1994). Traditionally, wheat and rice straws have been removed from the fields for use as cattle feed and for several other purposes such as livestock bedding, thatching material for houses, and fuel. Recently, because of the advent of mechanized harvesting, farmers prefer to burn large quantities of crop residues left in the field in situ as these interfere with tillage and seeding operations for the next crop. This practice has important implications for the recycling of nutrients contained in crop residues and for organic matter content of the soil. According to Narang and Virmani (2001), who analyzed the factors responsible for signs of a yield plateau in the rice–wheat system of the Indo-Gangetic Plain of India, one important priority task is to develop agrotechnology to recycle the vast biomass of rice and wheat residues. These residues are now burned annually, causing a serious waste of precious nutrient resource and contributing to intense air pollution.

Nutrient cycling in the soil–plant ecosystem is an essential component of sustainable agriculture. Although during the last three decades fertilization practices have started playing a dominant role in the rice–wheat cropping system, crop residues still play an essential role in the cycling of nutrients. Estimates of directly recyclable residues of rice and wheat in different states of the Indo-Gangetic Plain region of India and the nutrient potential associated with these are given in Table 8–1. With availability of 37.87 Tg of rice and wheat residues for recycling, associated nutrient (N + P + K) potential was 0.634 Tg. According to Sarkar et al. (1999), the rice–wheat system accounts for nearly one-fourth of the total crop residue production in India. One tonne of rice residues contains approximately 6.1 kg N, 0.8

kg P, and 11.4 kg K, while one tonne of wheat residues contains 4.8 kg N, 0.7 kg P, and 9.8 kg K.

The incorporation of crop residues alters the soil environment, which in turn influences the microbial population and activity in the soil and subsequent nutrient transformations (Kumar & Goh, 2000). Through this chain of events, management of crop residues regulates the efficiency with which fertilizer, water, and other reserves are used in a cropping system. A major problem encountered in the profitable use of cereal crop residues is the occurrence of microbial immobilization of soil and fertilizer N in the short term (Mary et al., 1996). Crops sown immediately after the incorporation of residues of cereal crops suffered because of a deficiency of plant available N. The addition of fertilizer N to the decomposing residues could only partially offset the immobilization process. Suitable manipulations of processes such as nutrient immobilization are an important component of an efficient crop residue management program. For example, allowing adequate time for the decomposition of crop residues before planting of the next crop can be beneficial in alleviating adverse effects caused by N immobilization. Efficient crop residue management strives for improved crop yields by increasing soil nutrient availability, decreasing erosion, improving soil structure, and increasing soil water-holding capacity (Kumar & Goh, 2000). In the long term, crop residues can contribute to the formation of soil organic matter depending on climatic conditions. It seems that future increases in the productivity of the rice–wheat system will greatly depend upon improvements in soil productivity through proper management and use of crop residues and other agricultural wastes.

A distinct feature of the rice–wheat cropping system in the Indo-Gangetic Plain is the inherent conflict between maximizing short-term production at minimum cost vs. providing for the sustainable health and long-term productivity of the soil. Many soils under the rice–wheat system in the Indo-Gangetic Plain are low in organic matter and nutrient supplies. The exploitative agriculture and decreasing inputs of organic materials have led to severe degradation of the soil, resulting in a reduction in soil organic matter, soil fertility, and productivity. Crop residues are the primary source of C inputs, and the ways in which these are managed have a significant effect on soil physical, chemical, and biological properties (Kumar & Goh, 2000).

Prasad and Power (1991) and more recently Kumar and Goh (2000) emphasized that no single residue management practice is superior under all conditions. Therefore, it is important to determine the benefits and adverse effects of residue management options before these are recommended to farmers for adoption. Considerable research has been conducted in the last few decades relating residue management to soil chemical, physical, and biological properties and consequent fertilizer management practices in the rice–wheat cropping system. In this chapter, we have tried to use this knowledge to make recommendations and conclusions for crop residue management in this cropping system. We have not attempted to review all available literature; only pertinent data sets have been used to substantiate different conclusions. We have focused our attention on the in situ incorporation of crop residues left naturally in the field under the rice–wheat cropping system in the Indo-Gangetic Plain of South Asia.

Table 8–2. Greenhouse gas emissions (CO_2 equivalent) from burning of rice straw in northwestern India (adapted from Kalra, unpublished data, 2001).

State	CO_2	CO	CH_4	N_2O
			Tg	
Punjab	13.60	0.869	0.047	0.175
Haryana	1.56	0.099	0.005	0.013
Uttar Pradesh	11.10	0.706	0.038	0.090
Total	26.26	1.674	0.090	0.278

IN SITU BURNING OF CROP RESIDUES

A 1996 survey of 237 farmers selected by stratified random sampling from rice–wheat cropping area in the 11 districts of the Indian Punjab revealed that 87% of rice and 52% of wheat were combine-harvested and burned. After 4 yr, the combine-harvested area under rice increased to 91%, but in the case of wheat, it increased dramatically to 82% from 52% 4 yr earlier (Indian Council of Agricultural Research, 1999).

The gaseous emissions from burning of rice straw were 70% CO_2, 7% CO, 0.66% CH_4, and 2.09% N_2O. Estimated emissions of greenhouse gases (Table 8–2) caused by the burning of rice straw in three states in the Indo-Gangetic Plain of India are noteworthy. However, little is known about the effects of burning on nutrient losses and dynamics in the rice–wheat cropping system. It has been reported from the United Kingdom that 40 to 80% of wheat crop residue N is lost as ammonia when it is burned in the field (Lee & Atkins, 1994). The same study reported that emissions of ammonia declined from 20 Gg N yr^{-1} in 1981 to 3.3 Gg N yr^{-1} in 1991 as a result of changes in agricultural practices because of an imposed ban on the burning of crop residues in the UK. These data suggest that emissions of ammonia resulting from residue burning cause a significant N loss that has not been quantified in the rice–wheat cropping system. Kumar et al. (2001) recently reported from New Zealand that, for every tonne of wheat residues burned, 2.4 kg of N was lost. Likewise, S losses from the burning of high-S and low-S rice crop residues in Australia were 60 and 40% of S content, respectively (Lefroy et al., 1994). Alkaline ash left on the soil surface after the burning of crop residues causes an increase in urease activity and may cause N losses from soil and applied fertilizer because of ammonia volatilization (Bacon & Freney, 1989). There is a need to quantify the nutrient losses from the burning of crop residues in the rice–wheat cropping system.

SOIL PROPERTIES AS INFLUENCED BY IN SITU INCORPORATION OF CROP RESIDUES

Crop residues, including roots, are the primary source of organic material added to soil in many cropping systems in the tropics. For a given climatic region and soil type, the rate of addition of C inputs is an important factor determining the amount of organic matter that can be maintained in the soil (Kumar & Goh, 2000). Soils tend to reach equilibrium if farming techniques and crop residue management

Table 8–3. Effect of crop residue management on organic C and total N content of soil under the rice–wheat cropping system at different locations in the Indo-Gangetic Plain in India and Pakistan.

Reference	Type of crop residue	Duration of study	Residue management	Organic C	Total N
		yr		%	
Beri et al. (1995)	Rice straw in wheat and wheat straw in rice	10	Removed	0.38	0.051
			Burned	0.43	0.055
			Incorporated	0.47	0.056
Sharma et al. (1987)	Rice straw in wheat and wheat straw in rice	6	Removed	1.15	0.144
			Incorporated	1.31	0.159
Zia et al. (1992)	Rice straw in rice	3	Removed	0.53	-
			Incorporated	0.63	-
Yadvinder-Singh et al. (2000)	Wheat straw, green manure (GM), and wheat straw + GM in rice	6	Removed	0.38	-
			Incorporated	0.49	-
			GM	0.41	-
			Straw + GM	0.47	-

practices stay the same for a long enough period. Only a few studies exist on the long-term effect of crop residue management on organic matter and N contents of soils under the rice–wheat cropping system in the Indo-Gangetic Plain (Table 8–3). Field experiments conducted in India on the rice–wheat cropping system showed that the incorporation of crop residues compared with burning or removal increased soil organic C and total N contents (Table 8–3). The incorporation of crop residues along with green manure in rice increased soil organic C compared with straw removal, but the increase was similar to that when crop residues were applied alone (Table 8–3). In these studies, increases in organic matter content because of crop residue recycling were relatively small compared with those reported from temperate regions (Prasad & Power, 1991). Even then, the incorporation of crop residues increased organic C by 14 to 29% over residue removal treatments in 3 to 10 yr of experiments. Tillage, soil texture, C inputs, and climatic conditions are the primary factors that control the stabilization of soil organic matter (Anderson, 1995; Kumar & Goh, 2000).

It is important to study the factors that are responsible for the accumulation of organic C or losses of organic C from the system under different management. Particulate organic C (POC) has been found to be the fraction preferentially lost when soils are cultivated (Cambardella & Elliott, 1992), and it is a more sensitive indicator of changes in organic C in soils (Chan et al., 2002). Chan et al. (2002) showed that tillage and residue burning preferentially removed different organic C fractions. Tillage preferentially removed POC, and residue burning significantly reduced incorporated organic C (<53 μm). Consequences of this preferential removal were that tillage and residue burning lowered the water stability of aggregates >2 mm, whereas residue burning caused a reduction in the water stability of aggregates <50 μm. The preferential effect of tillage, residue burning, or incorporation on different fractions of organic C in the soil may have more effect on the productivity and sustainability of the rice–wheat cropping system where soils are extensively tilled and puddled. Also, there is a strong correlation between POC and N availability of soils (Wilson et al., 2001; Chan et al., 2002). Therefore, it is important to maintain or improve the level of POC in the soils under rice–wheat crop-

Table 8–4. Effect of crop residue management on soil fertility of a loamy sand soil during 11 yr of the rice–wheat cropping system at Ludhiana, India (adapted from Beri et al., 1995).

	Crop residue management		
Soil property	Burned	Removed	Incorporated
Total P, mg kg^{-1}	390	420	612
Total K, g kg^{-1}	17.1	15.4	18.1
Olsen P, mg kg^{-1}	14.4	17.2	20.5
Available K, mg kg^{-1}	58	45	52
Available S, mg kg^{-1}	34	55	61

ping, and this can be achieved by proper tillage and residue management. There is a need to ascertain the minimum amount of organic C (either through aboveground crop residues or roots) that should be recycled annually in this system to either maintain or improve the soil organic C levels so that the system remains sustainable. This can be achieved by using some process-based simulation models.

In an 11-yr field experiment on a loamy sand soil in the Indian Punjab, the incorporation of residues of both crops in the rice–wheat cropping system increased the total P, available P, and K contents in the soil over the removal of residues (Table 8–4). Total P, available P, and available S were in the order: residue incorporation > residue removal > residue burning. In another, 5-yr study on a silt loam soil at Palampur in Himachal Pradesh having a relatively cooler climate than the Punjab, the incorporation of rice straw in wheat caused a slight increase in the availability of P, Mn, and Zn and a marked increase in the availability of K (Verma & Bhagat, 1992). The incorporation of crop residues on a long-term basis increased the DTPA-extractable Zn, Cu, Fe, and Mn content in the soil (Yadvinder-Singh et al., 2000). In a long-term study carried out in the eastern part of the Indo-Gangetic Plain (Bihar), Misra et al. (1996) observed increases in available N, P, and K in the soil with the incorporation of crop residues in the rice–wheat rotation. In general, both nutrient contents and their availability increased with the incorporation of crop residues compared with their removal or burning.

Crop residues also play an important role in maintaining soil physical conditions. In most climates, removal or burning of crop residues deteriorated soil physical properties (Prasad & Power, 1991). Puddling of soil by cultivation in standing water before transplanting of rice adversely affected soil structure by destroying aggregates and peds (Sharma & De Datta, 1985) and formed a pan of low permeability immediately below the cultivated layer. This hardpan can be detrimental to the productivity of wheat that followed rice (Sur et al., 1981). Recycling of crop residues helped reduce the adverse effects of hardpan formation in rice-based cropping systems.

Several researchers have reported an improvement in soil aggregation after the incorporation of crop residues into the soil under rice-based cropping systems (Oh, 1984; Liu et al., 1990; Liu & Shen, 1992; Meelu et al., 1994). In a loamy sand soil under the rice–wheat cropping system, the incorporation of wheat straw into the soil before transplanting of rice during a 5-yr period promoted the formation of soil aggregates, particularly 1 to 2 mm in size, and increased the mean weight diameter of the aggregates (Table 8–5). The mixed application of green manure and crop residues was more effective than their separate applications. The incorpora-

Table 8–5. Effect of incorporating green manure and crop residues for 5 yr on soil aggregation and bulk density in the rice–wheat cropping system on a loamy sand soil at Ludhiana, India (adapted from Meelu et al., 1994).

Treatment	Water-stable aggregates >2 mm	1–2 mm	0.5–1 mm	0.1–0.5 mm	Mean weight diameter	Bulk density 0–10 cm	10–20 cm
	%				mm	mg m^{-3}	
Residue removed	9.8	10.0	5.6	11.3	1.42	1.59	1.72
Residue incorporated	11.7	15.0	5.5	11.3	1.56	1.49	1.72
Green manure (GM)	11.1	15.5	6.1	12.0	1.58	1.51	1.71
Crop residue + GM	17.1	11.1	6.9	9.1	1.68	1.48	1.68

tion of residues alone or in conjunction with green manure also lowered the bulk density of surface soil (Table 8–5). Verma and Singh (1974) observed that wheat straw incorporation markedly improved soil aggregation in four different soils varying in texture. Maximum aggregation occurred in the sandy loam soil; alkali soil showed the poorest aggregation. The application of rice straw to alkali clayey soil significantly increased water stable aggregates >0.25 mm.

In long-term experiments in the rice–wheat cropping system in sandy loam soil, the incorporation of both rice and wheat straw compared with their burning or removal increased both the infiltration rate and cumulative infiltration (Walia et al., 1995; Singh et al., 1996). Similarly, in another 5-yr study on the rice–wheat cropping system on a loamy sand soil, Meelu et al. (1994) observed increased rates of infiltration on soil amended with green manure and crop residues. The combined application of green manure and crop residues was more effective not only in increasing infiltration but also in reducing bulk density, penetration resistance, and crusting of surface soil layers compared with their separate application (Verma & Singh, 1974; Meelu et al., 1994). In general, the incorporation of crop residues into paddy soils reduced bulk density, penetration resistance, and compaction in soils under the rice–wheat cropping system (Walia et al., 1995; Singh et al., 1996).

Soil microorganisms constitute a driving force for soil organic matter buildup and maintenance and several nutrient transformations in the soil. The availability of nutrients such as N, P, and S is particularly dependent upon microbial biomass and microbial activity, which in turn depend on the supply of organic substrates in the soil (Kumar & Goh, 2000). The regular addition of sufficient amounts of organic materials such as crop residues to the soil thus leads to the maintenance of microbial biomass and improvement of soil fertility. Malik et al. (1998) found that the application of wheat straw and green manure in a rice–wheat cropping system caused a large increase in microbial biomass throughout the growing season of rice and synchronized between N release and N uptake. Patra et al. (1992) found more biomass C in wheat straw than in cowpea [*Vigna unguiculata* (L.) Walp.] residue–amended soil, but the amount of biomass N was significantly higher in the cowpea residue–amended soil. In a rice–wheat rotation, Beri et al. (1992) and Sidhu et al. (1995) observed that soil treated with crop residues held 5 to 10 times more aerobic bacteria and 1.5 to 11 times more fungi than soil for which residues were either burned or removed (Table 8–6).

Table 8–6. Microbial population in the soil as influenced by residue management in the rice–wheat system at Ludhiana, India (adapted from Sidhu et al., 1995).

Residue management	After rice		After wheat	
	Bacteria	Fungi	Bacteria	Fungi
	× 10^5	× 10^3	× 10^5	× 10^3
Burning	59.8†	69.0†	16.2	34.5
Removal	15.6	13.5	13.9	33.2
Incorporation	115.4	143.0	126.6	133.6

† Rice residue never burned completely because of its high moisture content and therefore relatively better microbial activity.

In general, residue incorporation produced more microbial activity than residue removal or burning. It is also well known that the chemical composition of crop residues is an important determinant of both the size and composition of microbial (Broder & Wagner, 1988) and faunal (Parmele et al., 1989) communities. It is generally believed that fungi dominated the colonization of nonleguminous residues and bacteria dominated the colonization of leguminous residues (Campbell et al., 1991; Kumar, 1998) and that the C/N ratio of fungal communities is wider than that of bacterial communities (Beare et al., 1992).

Most studies on crop residue management have ignored the contribution of roots in terms of C and N additions and their effect on soil properties. Some earlier field estimates based on root excavations suggested that root C and N represented only 10 to 15% of total plant C and N (Bergersen et al., 1989; Unkovich et al., 1994) and were considered to play only a minor role. However, recent ^{15}N feeding studies now suggest that only a proportion of roots may be physically recovered as intact roots (Russell & Fillery, 1996). Lynch and Whipps (1990) reported that 30 to 60% of the net fixed C is transferred to roots in annual plants; thus, the amount of N and C in roots can be considerable. The data from Kumar et al. (2001) show that a similar amount of N was added through wheat roots and aboveground crop residues. The contributions of C and N additions through roots in the rice–wheat cropping system have so far been ignored.

AGRONOMIC RESPONSES TO CROP RESIDUE MANAGEMENT

Recycling of crop residues is an important component in achieving sustainability in crop production systems. The evaluation of crop residues in terms of fertilizer effect is complicated by the variable nutrient contents of the materials and a host of other factors. In some cases, straw incorporation may lead to reduced crop yields because of the release of phytotoxic compounds during decomposition or the immobilization of soil and fertilizer N, particularly when the crop is planted immediately after straw incorporation (Stirzaker & Bunn, 1996; Nelson, 1996; Bhogal et al., 1997). It is impossible to monitor all the effects of organic matter on nutrient availability. The evaluation of the fertilizer effects of an organic resource (e.g., for N) requires that the material be assessed both at an equal-N application and on an equal-mass (or C) application, preferably in each case for a range of application

Table 8–7. Effect of wheat straw management in rice and its residual effect in the following wheat crop in the rice–wheat cropping system at different locations in the Indo-Gangetic Plain in India.

Experimental details	Crop	Grain yield of rice or wheat with management of wheat straw in rice as: Straw removed	Straw burned	Straw incorporated	Straw + GM†	Reference
		Mg ha^{-1}				
Haryana, 3-yr study, clay loam soil	Rice	6.97	7.23	7.01 (+0.6)‡	7.17§	Agrawal et al. (1995)
	Wheat	4.65	4.84	4.43 (–4.7)	4.52	
Punjab, 6-yr study, loamy sand soil	Rice	6.15	--	5.76 (–6.3)	6.35	Yadvinder-Singh et al. (1996)
	Wheat	4.30	--	4.30 (0.0)	4.30	
Madhya Pradesh, 2-yr study	Rice	2.21	--	2.82 (+27.6)	--	Pandey et al. (1985)
	Wheat	4.48	--	5.59 (+24.8)	--	
West Bengal, acidic silty clay loam soil, 2-yr study, wheat straw incorporated 10 d before rice planting	Rice	3.74	--	4.17 (+11.5)	--	Sharma & Mittra (1992)
	Wheat	1.80	--	2.00 (+11.1)	--	
Uttar Pradesh, clay loam soil (pH 8.6), wheat straw (10 Mg ha^{-1}) incorporated 30 d before rice transplanting	Rice (100% NPK)	4.10	--	4.45 (+2.5)	--	Rajput (1995)
	Rice (50% NPK)	--	--	4.08	--	
	Wheat	2.29	--	2.72 (+18.8)	--	

† Green manure.
‡ Numbers in parentheses represent percentage increase or decrease in grain yield over residue removal.
§ 30 kg ha^{-1} extra fertilizer N.

rates. The effect of residue incorporation on the succeeding crop depends on the amount of residues and time and method of incorporation. Though the long-term effects of crop residue incorporation are generally expected to be beneficial, the effect on crop yields in the short term is often unpredictable. Thus, despite a large body of literature on the recycling of crop residues, little information enables the proper evaluation of organic residues for their fertilizer value.

Effect on Crop Yields

Results from field experiments that examined the effects of crop residue management practices are conflicting. Several factors such as residue quality, edaphic factors, and the health of the previous crop and their complex interactions with various management practices are involved in determining the ultimate yield of crops (Kumar & Goh, 2000). Crop residue management as practiced in the rice–wheat cropping system is of three types:

Wheat Straw Management in the Rice Crop and Its Residual Effect on the Following Wheat Crop

The incorporation of wheat straw into the soil has pronounced but variable effects on the growth and yield of the subsequent crop of rice at different locations in the Indo-Gangetic Plain (Table 8–7). In a field experiment on a clay loam soil in Haryana, rice yields with removal or incorporation of wheat straw were similar

(Agrawal et al., 1995). Interestingly, green manuring in conjunction with wheat straw helped to mitigate the adverse effect of wheat straw on rice. Incorporation of wheat straw in an acidic clay loam soil in West Bengal significantly increased the grain yield of rice, and a residual effect was also observed on the succeeding wheat crop (Sharma & Mittra, 1992). Similar observations were made by Pandey et al. (1985) in Madhya Pradesh. A beneficial effect of wheat straw on rice grain yield in the first year of study was reported by many workers (Sarkar, 1997; Alam & Azmi, 1989; Zia et al., 1992). However, on a loamy sand soil in Punjab, the incorporation of wheat straw adversely affected rice yield, compared with plots receiving no residues (Yadvinder-Singh et al., 1996). In general, the incorporation of wheat straw in rice in heavier soils increased rice grain yields by 0.5 to 28% and also increased the yields of subsequent wheat by 10 to 25%. However, on lighter soils, rice yields declined because of the incorporation of wheat residues.

In a long-term experiment carried out from 1984 to 1994, Rattan et al. (1996) observed that both rice and wheat yielded the highest with inorganic fertilizers alone rather than with any crop residue management treatment in the first year. After 2 to 3 yr, the combination of wheat straw and inorganic fertilizers produced yields similar to those observed with inorganic fertilizers. After 3 to 4 yr, the combined use of inorganic fertilizers and wheat straw started giving higher yields than the inorganic fertilizer treatment. In another study conducted by Yadav et al. (2000) at seven locations, rice grain yield in the treatment receiving 50% of the recommended N dose through wheat residues (Table 8–8) was, in general, significantly lower by 4 to 18% than the yield obtained with the recommended fertilizer NPK dose at most of the locations studied. Also, the residual effect of wheat residues applied to rice was visible through the grain yield of subsequent wheat. Six of eight locations showed a marginal decline in wheat yield (Table 8–8). Possibly, the wide C/N ratio and high lignin content of wheat residues impeded decomposition and

Table 8–8. Grain yield of rice and wheat (6-yr average) as influenced by the application of wheat residues in the rice–wheat rotation at seven locations in India (adapted from Yadav et al., 2000).

		Rice grain yield		Wheat grain yield	
Location	Duration of experiment	100% NPK†	50% NPK + wheat residues‡	100% NPK†	50% NPK + wheat residues‡
		Mg ha^{-1}			
Ludhiana	1983-84 to 1997-98	6.08	5.06 (−16.8)§	4.46	4.40 (−1.3)
Pantnagar	1983-84 to 1997-98	4.56	3.84 (−15.8)	3.75	3.45 (−8.0)
Kanpur	1985-86 to 1997-98	4.47	3.66 (−18.1)	4.53	4.38 (−3.3)
Faizabad	1984-85 to 1997-98	4.21	3.78 (−10.1)	3.45	3.31 (−4.1)
Sabour	1984-85 to 1997-98	4.11	3.87 (−5.8)	3.13	3.15 (+ 0.6)
Kalyani	1986-87 to 1998-99	3.33	3.47 (+ 4.2)	2.67	3.09 (+15.7)
Jabalpur	1985-86 to 1997-98	4.78	4.37 (−8.6)	2.55	2.48 (−2.7)
Mean		4.24	4.01	3.51	3.47

† 100% NPK to rice or wheat consisted of 120 kg N, 26 kg P, and 33 kg K ha^{-1} through urea, single superphosphate, and muriate of potash.

‡ 50% NPK through fertilizer plus 50% replacement of N by wheat crop residues to rice, 100% NPK through fertilizer to wheat.

§ Numbers in parentheses represent percentage increase or decrease in grain yields over 100% NPK treatment.

immobilized native soil N and also only 50% of the N was applied and the rest was replaced by wheat residue-N, which may not be available. As a consequence, the productivity of rice that followed residue incorporation was adversely affected. In general, ^{15}N studies have shown that the recovery of nonleguminous residue N in the subsequent crops is 10 to 15% lower than that for fertilizer N (35–55%) (Kumar & Goh, 2000). Recently, Kumar et al. (2001) showed that the wheat crop recovered 29 to 37% of N applied through leguminous residues [white clover (*Trifolium repens* L.) and pea] and only 11 to 13% of N applied through nonleguminous residues (wheat and ryegrass *Lolium* spp.).

Rice Straw Management in the Wheat Crop and Its Residual Effect on the Following Rice Crop

In a field experiment conducted using ^{15}N-labeled urea, grain yields of wheat and the following rice were not adversely affected by incorporation of rice straw at least 20 d before sowing (Table 8–9). In another 6-yr study, in situ incorporation of rice straw 10, 20, or 40 d before sowing wheat did not affect the wheat yield significantly (Table 8–9). Also, rice straw incorporated in wheat did not show a residual effect on the succeeding rice crop. Several reports show similar rice and wheat yields under different residual management practices such as burning, removal, or incorporation (Walia et al., 1995; Singh et al., 1996; Bijay-Singh et al., 2001).

At Faislabad in Pakistan, the incorporation of rice straw into the soil produced significantly higher yields of wheat (3.51 Mg ha^{-1}), compared with when rice

Table 8–9. Effect of time of rice straw incorporation and fertilizer N (120 kg N ha^{-1}) management in wheat on losses of applied N, wheat yield, and residual effect in the following rice crop in the rice–wheat cropping system at Ludhiana, India.

	Exp. 1: Bijay-Singh et al. (2001)				Exp. 2: Yadvinder-Singh et al. (2000)		
	Grain yield of				Grain yield of		
Treatment	Wheat	Rice	^{15}N recovery in crop	^{15}N losses	Wheat (1993–1996)	Wheat (1996–1999)	Rice (1994–1998)
	Mg ha^{-1}		%		Mg ha^{-1}		
Straw removed	5.06a†	4.90a	40.9a	31.7b	5.04	4.58	5.93
Straw burned	5.11a	5.13a	40.8a	31.3b	--	4.65	5.65
Straw incorporated (40 DBS‡)	4.89a	4.87a	36.1ab	36.6ab	5.05	4.82	5.81
Straw incorporated (20 DBS)	5.00a	4.97a	34.4bc	34.0ab	--	4.86	5.82
Straw incorporated (20 DBS) and 25% N applied at incorporation	4.79a	5.02a	30.4c	45.2a	--	4.62	5.66
Straw incorporated (10 DBS)	--	--	--	--	--	4.70	5.68

† Numbers followed by the same letter in a column are not significantly different at $P < 0.05$ according to Duncan's Multiple Range Test.

‡ Days before sowing of wheat.

Table 8–10. Effect of rice straw management on wheat and its residual effect in the following rice crop in the rice–wheat cropping system at different locations in northwestern India.

Experimental details	Crop	Grain yield of rice and wheat with rice straw management as: Straw removed	Straw burned	Straw incorporated	Reference
		Mg ha^{-1}			
Himachal Pradesh, data averaged for 4 yr, acidic clay loam soil, rice straw chopped and incorporated 4 wk before wheat sowing	Wheat	2.8	--	2.8 (0.0)†	Sharma et al. (1985, 1987)
	Rice	2.4	--	2.5 (+4.2)	
Himachal Pradesh, 5-yr study, acidic clay loam soil, rice straw chopped and incorporated 4 wk before wheat sowing	Wheat (1984–1987)	2.6	2.6	2.2 (−15.4)	Verma & Bhagat (1992)
	Rice	3.7	3.6	3.7 (0.0)	
	Wheat (1987–1989)	2.4	2.4	2.4 (0.0)	
	Rice	3.8	3.7	4.0 (+5.3)	
Punjab, sandy loam soil. Data are reported for the 4th cropping cycle	Wheat	--	4.88	5.18	Beri & Sidhu (unpublished data, 2000)
	Rice	--	6.05	6.00	

† Numbers in parentheses represent percentage increase or decrease in grain yields over straw-removed treatment.

straw was removed (2.91 Mg ha^{-1}) (Salim, 1995). Singh et al. (1996) reported that the incorporation of rice straw 3 wk before wheat sowing significantly increased wheat yield on clay loam soil but not on sandy loam soil. Studies conducted by Sharma et al. (1985, 1987) showed no adverse effect of straw incorporation on the grain yield of wheat and the following rice (Table 8–10). The application of FYM (5 Mg ha^{-1} dry weight) along with rice straw increased average wheat yields by 40% over the rice straw treatment and exhibited a significant residual effect on the following rice crop. Most of the studies on rice straw management were conducted at recommended N rates, and thus it is difficult to quantify the contribution of rice straw in supplying N to plants in the cropping system. On the other hand, in Himachal Pradesh (India), the incorporation of rice straw (5 Mg ha^{-1} dry weight) 30 d before the sowing of wheat produced significantly lower wheat yields than the removal or burning of straw in the first two years. The incorporation of straw over longer periods produced wheat yields similar to those with straw removal or burning (Table 8–10) (Verma & Bhagat, 1992).

Thus, it is clear that, in general, rice straw can be managed in situ successfully either by allowing sufficient time between the incorporation and sowing of the wheat crop or by adding FYM with straw.

Wheat Straw in the Rice Crop and Rice Straw Management in the Following Wheat Crop

In a long-term field experiment carried out at Ludhiana in northwestern India, Beri et al. (1995) found that the incorporation of crop residues in rice and wheat crops significantly lowered yields compared with the removal or burning of

Table 8–11. Effect of rice straw management in wheat and wheat straw management in rice on crop yields in the rice–wheat cropping system at Ludhiana, India (adapted from Beri et al., 1995).

Experimental details	Crop	N rate	Grain yield of rice and wheat with rice straw management as: Straw removed	Straw burned	Straw incorporated	Mean
			Mg ha^{-1}			
Sandy loam soil, wheat straw incorporated in rice and rice straw in wheat. Data pooled over 10 yr for rice and 11 yr for wheat	Rice	60	4.46	4.65	3.62	4.24
		120	5.50	5.65	4.63	5.26
		180	6.04	6.42	5.26	5.91
		Mean	5.33	5.57	4.51	
	LSD ($P = 0.05$)		Residue = 0.55, N = 0.16, Residue × N = NS†			
	Wheat	60	3.48	3.46	2.94	3.29
		120	4.14	4.26	3.87	4.10
		180	4.42	4.64	4.34	4.47
		Mean	4.02	4.12	3.72	
	LSD ($P = 0.05$)		Residue = 0.25, N = 0.22, Residue × N = NS			

† NS, Not significant.

residues (Table 8–11). The decline in wheat yield because of straw incorporation was 0.54 Mg ha^{-1} at 60 kg N ha^{-1} and 0.08 Mg ha^{-1} at 180 kg N ha^{-1}. This suggested that the depression in rice yield was caused by N immobilization. In a calcareous sandy loam soil in Bihar, India, Prasad and Sinha (1995) studied the effect of incorporation of rice and wheat residues soaked in 2% urea solution after chopping (2-cm size) and inoculating with cellulytic culture (*Aspergillus* spp.) to hasten decomposition. At recommended fertilizer levels, the incorporation of treated crop residues increased yields of wheat and rice by 7.2 and 8.5% (3-yr average), respectively. Kavinandan et al. (1987) observed that the incorporation of wheat straw 10 d before transplanting rice and rice straw 3 wk before sowing wheat increased grain yield by 0.25 and 0.42 Mg ha^{-1} of rice and wheat, respectively, over the incorporation of wheat straw 3 d before transplanting rice and rice straw 2 wk before sowing wheat; differences, however, were not significant.

The application of rice straw in wheat and wheat straw in rice may pose some problems in the short term; however, allowing the proper time between straw incorporation and the sowing of crops or enhancing the decomposition of residues by additional inorganic N, GM, and FYM or by using cellulytic bacteria may alleviate these problems.

Fertilizer Nitrogen Management in Straw-Amended Soils

A few studies have been carried out on improved fertilizer management practices for the rice–wheat cropping system on soils amended with crop residues. In a 3-yr study carried out in Uttar Pradesh by Misra et al. (1996), total grain yield of rice and wheat increased because of the incorporation of both straws along with a starter dose of 20 kg N ha^{-1} over burning and straw incorporation without a starter dose. Similarly, Bali et al. (1986) and Sharma and Mittra (1992) found that incorporating rice straw 15 to 20 d before sowing wheat decreased grain yield, while incorporating wheat straw 15 to 20 d before transplanting increased rice yields. The application of 15 kg N ha^{-1} as a starter dose with straw application increased the yields of both rice and wheat.

Using ^{15}N-labeled urea in field experiments and applying a part of recommended N (25%) at the time of straw incorporation (to hasten straw decomposition) showed large N losses and low wheat yields (Bijay-Singh et al., 2001). The recovery of ^{15}N by wheat was maximum (40.9%) when rice straw was removed or burned and minimum (30.4%) when 30 kg out of a total of 120 kg N ha^{-1} fertilizer was applied along with straw incorporation at 20 d before sowing wheat (Table 8–9). In a 6-yr, two-site study conducted by Yadav (1997), when crops of a rice–wheat system were grown on recycled residues of the preceding crop, the application of 20 kg N ha^{-1} at the time of residue incorporation was found to be essential at the site with subhumid, moderate climate conditions. On the other hand, in the warmer climate at Kanpur in the heart of the Indo-Gangetic Plain, the application of N at the time of residue incorporation was not beneficial. However, at this site, raising the recommended level of N by 20 kg ha^{-1} for both rice and wheat produced more than 16% higher yield from the total system compared with the yield from the prevalent recommended N dose. Thakur and Pandya (1997) reported that preconditioning of urea with rice straw and soil at the ratio of 1:3:1 was significantly superior to urea alone in increasing grain yield and N uptake of wheat.

It is clear that incorporated straw causes rapid immobilization of soil and fertilizer N during the early stages of decomposition, thus causing N deficiency in the succeeding crop. Proper management of fertilizer N should reduce immobilization of N by crop residues and produce higher N use efficiency. Improved fertilizer management practices may include optimum method, time, and rate of fertilizer N application. One obvious solution to the N immobilization problem would be to place the fertilizer below the C-enriched surface soil that results from surface placement of crop residues (Doran & Smith, 1987). For example, in soils amended with crop residues, band placement of urea prills and deep placement of large urea granules or nest placement of urea prills immobilized significantly less fertilizer N than the mixed application of commercial urea granules (Yadvinder-Singh et al., 1994). Malik and Jaiswal (1993) observed that the application of 58 kg N ha^{-1} as urea super granules plus 28 kg N ha^{-1} through wheat straw produced significantly higher rice yields than the recommended practice of applying 87 kg N ha^{-1} as commercial urea granules. The limited contact between fertilizer N and the decomposing microorganisms was the main reason for the low rates of N immobilization with large urea granules. The adverse effect of N immobilization on crop growth can also be avoided by applying additional fertilizer N (15–20 kg) as a starter dose at the time of residue incorporation to minimize the adverse effects of short-term N immobilization or delayed planting.

Integrated Management of Fertilizers and Crop Residues

Integrated nutrient management aims to maintain adequate nutrient cycling on a long-term basis while supplying sufficient quantities of nutrients to crops in the short term. Organic matter management is an essential component of this strategy as organic inputs can help in ameliorating a large number of other constraints besides contributing to the nutrient supply. Pathak and Sarkar (1997) found that a very high dose of urea-N along with rice straw incorporation was necessary to obtain respectable rice yields. The beneficial effect of straw could not be obtained in the succeeding wheat crop. At the recommended fertilizer N level, rice straw incorporation produced lower rice yields than urea alone. In another study, the application of 30 kg extra N ha^{-1} above the recommended fertilizer increased rice yields only slightly (Agrawal et al., 1995) (Table 8–7). Thakur and Singh (1987) estimated optimum N rates of 140 and 115 kg ha^{-1} for rice on fields with and without wheat straw (5 Mg ha^{-1}) incorporation. Thus, a higher fertilizer N dose may be required for crops grown on soils amended with crop residues to get maximum benefits. Jha et al. (1992) obtained higher grain yield of rice when rice straw and green manure (mungbean) along with 60 kg N ha^{-1} were applied in three equal split doses than rice straw or green manure applied alone. Grain yield of the following wheat was also higher when rice straw and green manure were incorporated. In a 2-yr field experiment at Modipuram (northwestern India) the incorporation of one-half of the crop residues along with recommended fertilizers consistently produced high yields of both crops of rice and wheat (Sarkar, 1997). In a rice–wheat rotation on a calcareous soil, the application of crop residues along with FYM produced the highest yields, followed by FYM, crop residues, and no amendment (Prasad & Sinha, 1995). The grain yield recorded with 50% NPK+ FYM + crop residues was higher than that of the 100% recommended dose of NPK alone, indicating that FYM + crop residues replaced 50% of NPK for both wheat and rice production.

Singh and Singh (1995) found that incorporating rice straw (10 Mg ha^{-1}; C/N = 75.5) 3 wk before planting rice integrated with 50% of recommended NPK fertilizers produced the highest rice yield and improved the profitability of the system. Rajput (1995) found that the incorporation of wheat straw (10 Mg ha^{-1}) resulted in up to 50% savings in the recommended NPK fertilizers (60 kg N + 13.1 kg P + 25 kg K ha^{-1}). A higher yield potential of rice was achieved when wheat straw was applied along with recommended NPK fertilizers (Table 8–7). The residual effects of wheat straw incorporation on the following wheat were also substantial. However, several other long-term studies showed that it is not possible to replace a part of the fertilizer N requirement of rice with N added through wheat straw. Katyal et al. (1998) reported results from long-term field experiments conducted at five different sites in India during 1983 to 1991. At Kanpur in the heart of the Indo-Gangetic Plain, 50% recommended NPK fertilizers plus 50% N through wheat straw in rice followed by 100% recommended NPK fertilizers in wheat stabilized the yields of rice and wheat. However, at the other three locations, it was not possible to replace a part of fertilizer N (25% N) with crop residues. In a field experiment carried out in West Bengal, the application of semidecomposed wheat straw (0.78% N, dry weight basis) at 3 Mg ha^{-1} along with 25 or 50% of the recommended NPK fertilizers (80 kg N + 26 kg P + 33 kg K ha^{-1}) 30 d before sowing rainfed rice gave the highest yields. Wheat straw applied to rice had little effect on the grain yield of the succeeding wheat crop (Kundu et al., 1994). Verma and Mathur (1990) found that the incorporation of rice straw (11.2 Mg ha^{-1}) along with cellulytic microorganisms and rock phosphate 15 d before sowing wheat produced a significant increase in the yield of wheat over recommended fertilizer management practices.

The equivalent replacement of inorganic fertilizer nutrients with nutrients supplied from crop residues is not possible, as discussed above and also shown by ^{15}N studies. This is because of large differences in the recovery of N from different sources (nonleguminous residues, leguminous residues, and inorganic fertilizers) by different crops (Kumar & Goh, 2000; Kumar et al., 2001). In some studies reported here, 25 to 50% of fertilizer N was replaced by wheat or rice crop residue N and produced similar crop yields. The effects may not be due to direct nutrient supply or replacement of these residues, but due to indirect effects as discussed by Tian et al. (1993).

It is clear from these studies that the integrated management of nutrients from various sources, including inorganic fertilizers, organic manures, and crop residues, can improve the productivity of the rice–wheat cropping system.

TILLAGE, MECHANIZATION, AND CROP RESIDUE INCORPORATION

Three options exist for managing crop residues in the rice–wheat system in South Asia:

1. Crop residues incorporated into the soils by using conventional tillage systems.
2. Crop residues retained at or near the soil surface by using zero or conservation tillage systems.
3. A proportion of crop residues removed from the soil surface for fodder, fuel, or other purposes or partially burned.

Tillage operations vary with each option, and these exert a profound influence on the soil environment. The incorporation of residues into the soil helps crop residues decompose faster. This is generally achieved by disking or using a moldboard plow. Obviously, incorporated residues come in contact with soil having hydrothermal regimes favorable for the microbial decomposition of organic material. On the other hand, residues left on the soil surface are subject to large fluctuations in moisture and temperature. In many cases of surface-applied residues, the moisture becomes a limiting factor, and many times the temperature becomes either too low or too high for the decomposers to act efficiently (Kumar & Goh, 2000). In the rice–wheat system, wheat can be sown using zero-tillage after rice straw is left on the soil surface. However, appropriate machinery is still in the development stage. Some of the factors directly affecting this residue management option are the uneven distribution of residues in the field, residue load, and rolling of rice residues while using drills. Beneficial effects of conservation tillage over conventional tillage have been observed in a Typic Ustochrept (Jat et al., unpublished data, 2001). Growth of wheat was better when sown with a zero-and strip-tillage drill in standing residues of rice than when rice residues were incorporated using conventional tillage. A third management option is the partial burning of residues, leaving behind about 60% of total residues, or the partial removal of residues from the field. The advantage of partial burning or partial removal is that the zero-till drill can be efficiently used in standing stubble for sowing wheat. Not enough information is available yet on the comparative advantages of different residue management options in the rice–wheat system.

In relatively cold hilly regions under the rice–wheat cropping system in the Himalayas, rice straw is used for mulching wheat but subsequently mixed with the soil during puddling before transplanting the following rice crop. In a study carried out by Verma and Sharma (2000), while productivity of wheat because of mulching with rice straw was always higher than with residue incorporation, productivity of rice was not influenced by mulching and the incorporation of rice straw.

WEED CONTROL AND HERBICIDE EFFICIENCY

Weeds pose a major problem in the productivity of the rice–wheat cropping system. In a recent experiment on a sandy loam soil in Punjab, Gajri et al. (1999) showed that rice yield decreased exponentially with an increase in weed biomass. They suggested that, for effective weed control and high rice yields, the field must receive prepuddling tillage. Most past studies in weed control in rice cropping systems have been confined to evaluating the effects of herbicide, tillage, water, and their interactions (Baltazar & De Datta, 1992; Bhagat et al., 1999). Few studies have been conducted on the dynamics of weed populations and the productivity of the rice–wheat cropping system because of crop residue management practices. Such information is needed in weed control strategies for the rice–wheat cropping system to improve its productivity.

There are many ways by which crop residues and their appropriate management can suppress weeds (Kumar & Goh, 2000), for example, (i) through their physical presence on the soil surface as mulch and by restricting solar radiation reaching below the mulch layer, (ii) by direct suppression caused by allelopathy, (iii) by controlling N availability, and (iv) burning residues for effective removal of weeds and weed seeds.

Narwal (2000) recently showed that weeds could be managed in the rice–wheat rotation by using allelopathy. Narwal's studies showed that growing the early summer (April–June) fodder crops of sorghum and pearl millet [*Pennisetum glaucum* (L.) R. Br.] smothered the weed population and biomass and provided satisfactory weed control in the subsequent rice crop. This may help in reducing the use of herbicides in the rice crop. Likewise, the replacement of wheat by winter fodder crops of oat (*Avena sativa* L.) or berseem once in 2 to 3 yr of the rice–wheat rotation may also help control winter weeds that affect the wheat crop.

In addition to influencing the weed growth and population, crop residue management and tillage practices also influence the efficiency of soil-applied preemergence herbicides (Kumar & Goh, 2000). Because preemergence herbicides are applied to the soil, the amount and quality of residues and also ash left behind after residue burning might affect their activity. Consequences of this may be that higher rates of herbicide application may be required to achieve weed control (Kumar & Goh, 2000).

Further studies on these aspects of weed control and herbicide efficiency under different residue management practices are needed to improve the productivity of the rice–wheat cropping system.

PESTS AND DISEASES

The effects of crop residue management on the pests and diseases in the rice–wheat cropping system have not received much attention. Since the introduction of stubble mulch farming, it has been reported from most parts of the world that crop yields are generally depressed when the crop is planted in a field with residues of the previous crop lying on the surface of the soil (Kumar & Goh, 2000). The burning of residues is thought to be a good option for controlling soilborne fungal diseases, and this can destroy pests such as aphids present in residues. However, the effects of burning depend on the burning intensity and temperatures reached during burning. Even a small amount of inoculum left in the soil may be sufficient to carry the disease over to the next crop (de Boer et al., 1993).

The effects of the incorporation of infected residues on pest and disease incidence are controversial. The fungi that affect cereals may have a limited or wide range of hosts. Barley (*Hordeum vulgare* L.), for example, will not contract mildew or rust from volunteer plants of wheat or from a wheat crop growing in the adjacent fields and vice versa. However, the pathogen causing take-all (whiteheads) (*Gaeumannomyces graminis* var. *tritici*; syn. *Ophiobolus graminis* var. *tritici*) can affect both crops and carry disease over from one to the other (Gair et al., 1987). Colbach and Meynard (1995) tried to explain disease incidence on the basis of crop succession and soil tillage management. For example, when the previous crop was a host crop preceded by a nonhost crop, soil inversion buried host residues, thus decreasing the risk of primary infection. In another scenario, where the previous crop was a nonhost crop preceded by a host crop, soil inversion carried the infected host residues back to the soil surface, thus increasing the primary infection risk. This phenomenon may play an important role in disease incidence in the rice–wheat cropping system in the long term. For example, the buried disease-infected wheat residues may not cause any problem to the subsequent rice crop, but tillage of soil before sowing the wheat crop following rice may bring the infection to the soil surface and cause disease or pest problems in the wheat crop.

It has been shown that retention of crop residues in the field increases the population of soil micro- and meso-fauna (Kumar & Goh, 2000). These can play an important role in suppressing various plant pathogens and they offer a potential for biological disease control as numerous species of nematodes and collembola feed on pathogenic fungi such as *Rhizoctonia*, *Botrytis*, and *Fusarium*, (Curl, 1988). This aspect of crop residue management needs more field studies and could be exploited for biological control of diseases.

The system becomes very complex with few definitive principles for management. Thus, decisions on crop residue management should be made keeping in view the health of previous crops and the potential susceptibility of the subsequent crops, cultivar selection, and planting date (Kumar & Goh, 2000).

FUTURE RESEARCH NEEDS

To interpret soil organic matter dynamics and nutrient cycling in the rice–wheat system in the Indo-Gangetic Plain, more reliable data on yields with both above- and below-ground crop residues are needed. More information needs to be collected on the effect of management (including residue quality, application rate, application method, timing, interactions with inorganic fertilizers, temperature, soil moisture, soil type, and tillage) on decomposition and nutrient release rates and the production of phytotoxic compounds from rice and wheat residues. Limited information is available on the interactions between crop residue management and tillage practices under the rice–wheat cropping system. More studies are needed to provide a better understanding of how the different crop residues and their management can be exploited to control weeds, diseases, and pests so that the productivity of this system can be increased. At present, proper technology and machinery for the in situ incorporation of crop residues are not available. Development of better technology for the in situ incorporation and rapid decomposition of crop residues by appropriate tillage, irrigation, fertilizer management, and inoculation of suitable microorganisms, particularly for the intensive rice–wheat cropping system commonly followed in South Asia, is needed.

Many of the changes in soil properties and nutrient cycling may become apparent only after several years (10 yr or more) of residue management, and long-term results may differ from those obtained in the short term. Thus, there is a strong need to establish long-term experiments at sites carefully selected for variations in temperature, moisture regimes, soil mineralogy, and agricultural management covering the rice–wheat cropping system in the Indo-Gangetic Plain. Because of several variables and the complexities of interactions among soils, crops, and climate, an accurate prediction of the effect of crop residue management practices on nutrient cycling, crop growth and yield, and soil properties will be possible only by developing realistic process-based computer simulation models. Several models are available to simulate the effects of management practices on soil organic matter and crop yield, but most of them have not been evaluated and tested for the rice–wheat system in the Indo-Gangetic Plain.

ACKNOWLEDGMENTS

The authors are grateful to Dr. S.S. Pal, Project Directorate, Cropping Systems Research, Modipuram, India, for providing useful information on the subject.

REFERENCES

Agrawal, R.P., V.K. Phogat, T. Chand, and M.S. Grewal. 1995. Improvements of soil physical conditions in Haryana. p. 69–75. *In* Research Highlights. Dep. of Soil Sci., C.C.S. Haryana Agric. Univ., Hisar, India.

Alam, S.M., and A.R. Azmi. 1989. Influence of wild plant and crop residues on growth and nutrient content of wheat. Pakistan J. Scientific Indust. Res. 32:749–751.

Anderson, D.W. 1995. Decomposition of organic matter and carbon emissions from soils. p. 165–175. *In* R. Lal et al. (ed.) Advances in soil science: Soils and global change. CRC Press, Boca Raton, FL.

Bacon, P.E., and J.R. Freney. 1989. Nitrogen loss from different tillage systems and the effect on cereal grain yield. Fert. Res. 20:59–66.

Bali, S.V., S.C. Modgal, and R.D. Gupta. 1986. Effect of recycling of organic wastes on rice–wheat rotation under Alfisol soil conditions of North-western Himalayas. Himachal J. Agric. Res. 12:98–107.

Baltazar, A.M., and S.K. De Datta. 1992. Weed management in rice. Weed Abstr. 41:495–508.

Beare, M.H., R.W. Parmelee, P.F. Hendrix, W. Cheng, D.C. Coleman, and D.A. Crossley. 1992. Microbial and faunal interactions and effects on litter nitrogen and decomposition in agroecosystems. Ecol. Monogr. 62:569–591.

Bergersen, F.J., J. Brockwell, R.R. Gault, L.J. Morthrope, M.B. Peoples, and G.L. Turner. 1989. Effects of available soil nitrogen and rates of inoculation on nitrogen fixation by irrigated soybeans and evaluation of $\delta^{15}N$ methods for measurement. Aust. J. Agric. Res. 40:763–780.

Beri, V., B.S. Sidhu, G.S. Bahl, and A.K. Bhat. 1995. Nitrogen and phosphorus transformations as affected by crop residue management practices and their influence on crop yield. Soil Use Manage. 11:51–54.

Beri, V., B.S. Sidhu, A.K. Bhat, and Bhupinder Pal-Singh. 1992. Nutrient balance and soil properties as affected by management of crop residues. p. 133–135. *In* M.S. Bajwa et al. (ed.) Nutrient management for sustained productivity. Proc. Int. Symp. Vol. II. Dep. Soils, Punjab Agric. Univ., Ludhiana, India.

Bhagat, R.M., S.I. Bhuiyan, and K. Moody. 1999. Water, tillage and weed management options for wet seeded rice in the Philippines. Soil Tillage Res. 52:51–58.

Bhardwaj, K.K.R. 1995. Recycling of crop residues, oil cakes and other plant production agriculture. p. 9–30. *In* H.L.S. Tandon (ed.) Recycling of crop, animal, human and industrial wastes in agriculture. Fertiliser Development and Consultation Organisation, New Delhi, India.

Bhogal, A., S.D. Young, and R. Sylvester-Bradeley. 1997. Straw incorporation and immobilization of spring-applied nitrogen. Soil Use Manage. 13:111–116.

Bijay-Singh, K.F. Bronson, Yadvinder-Singh, T.S. Khera, and E. Pasuquin. 2001. Nitrogen-15 balance and use efficiency as affected by rice residue management in a rice wheat system in northwest India. Nutr. Cycl. Agroecosyst. 59:227–237.

Broder, M.W., and G.H. Wagner. 1988. Microbial colonization and decomposition of corn, wheat, and soybean residue. Soil Sci. Soc. Am. J. 52:112–117.

Cambardella, C.A., and E.T. Elliott. 1992. Particulate soil organic matter changes across a grassland cultivation sequence. Soil Sci. Soc. Am. J. 56:777–783.

Campbell, C.A., V.O. Beiderbeck, R.P. Zentner, and G.P. Lafond. 1991. Effect of crop rotations and cultural practices on soil organic matter, microbial biomass and respiration on a thin Back Chernozem. Can. J. Soil. Sci. 71:363–376.

Chan, K.Y., D.P. Heenan, and A. Oates. 2002. Soil carbon fractions and relationship to soil quality under different tillage and stubble management. Soil Tillage Res. 63:133–139.

Colbach, N., and J.M. Meynard. 1995. Soil tillage and eyespot: Influence of crop residue distribution on disease development and infection cycles. Eur. J. Plant Pathol. 101:601–611.

Curl, E.A. 1988. The role of soil micro-fauna in plant-disease suppression. Crit. Rev. Plant Sci. 7:175–196.

de Boer, R.F., G.R. Steed, J.F. Kollmorgen, and B.J. Macauley. 1993. Effects of rotation, stubble retention and cultivation on take-all eyespot of wheat in northeastern Victoria, Australia. Soil Tillage Res. 25:263–280.

Doran, J.W., and M.S. Smith. 1987. Organic matter management and utilization of soil and fertility nutrients. p. 51–70. *In* R.F. Follet et al. (ed.) Soil fertility and organic matter as critical components of production systems. SSSA Spec. Publ. 19. ASA and SSSA, Madison, WI.

Gair, R., J.E. Jenkins, E. Lester, and P. Bassett. 1987. Cereal pests and diseases. Farming Press, Ipswich, England.

Gajri, P.R., K.S. Gill, R. Singh, and B.S. Gill. 1999. Effect of pre-planting tillage on crop yields and weed biomass in a rice–wheat system on a sandy loam soil in Punjab. Soil Tillage Res. 52:83–89.

Indian Council of Agricultural Research. 1999. Enhancing the biodegradation of rice residue in rice based cropping systems to sustain soil health and productivity. Terminal Proj. Rep. Indian Council of Agric. Res., New Delhi, India.

Jha, J.N., B. Roy, and S.P. Singh. 1992. Effect of combined-application of organic manures and inorganic fertilizer on rice and their residue on wheat. J. Appl. Biol. 2:55–59.

Katyal, V., S.K. Sharma, and K.S. Gangwar. 1998. Stability analysis of rice (*Oryza sativa*)–wheat (*Triticum aestivum*) cropping system in integrated nutrient management. Indian J. Agric. Sci. 68:51–53.

Kavinandan, S.K., J.P. Gupta, and I.C. Mahapatra. 1987. Studies on residue management and biofertilizers in rice–wheat sequence. Indian J. Agron. 32:278–279.

Kumar, K. 1998. Effects of different crop residues and management practices on soil nitrogen dynamics, yield and recovery of nitrogen by wheat. Ph.D. diss. Lincoln University, Canterbury, New Zealand.

Kumar, K., and K.M. Goh. 2000. Crop residue management: Effects on soil quality, soil nitrogen dynamics, crop yield, and nitrogen recovery. Adv. Agron. 68:197–319.

Kumar, K., K.M. Goh, W.R. Scott, and C.M. Frampton. 2001. Effects of ^{15}N-labelled crop residues and management practices on subsequent winter wheat yields, nitrogen benefits and recovery under field conditions. J. Agric Sci. (Cambridge) 136:35–53.

Kundu, A.L., R.K. Ghosh, and T.K. Mondal. 1994. Effect of crop waste on production in a rice–wheat crop sequence under rainfed condition in the Gangetic plains of West Bengal. Adv. Plant Sci. 7:228–234.

Ladha, J.K., K.S. Fischer, M. Hossain, P.R. Hobbs, and B. Hardy. 2000. Improving the productivity and sustainability of rice–wheat systems of the Indo-Gangetic Plains: A synthesis of NARS-IRRI partnership research. p. 1–31. Discussion Paper No. 40. IRRI, Los Baños, Philippines.

Lee, D.S., and D.H.F. Atkins. 1994. Atmospheric ammonia emissions from agricultural waste combustion. Geophys. Res. Lett. 21:281–284.

Lefroy, R.D.B., W. Chaitep, and G.J. Blair. 1994. Release of sulfur from rice residues under flooded and non-flooded soil conditions. Aust. J. Agric. Res. 45:657–667.

Liu, J.R., and R.P. Shen. 1992. The effects of fertilization on the properties and role of soil organic matter. p. 246–251. *In* Proc. Int. Symp. on Paddy Soils. Nanjing, China.

Liu, J.R., D.Y. Zhang, and W. Zhou. 1990. The effects of mixed application of organic and inorganic fertilizers to paddy soil (third report). Acta Agric. Universitatis Jiangxiensis 12:37–42.

Lynch, J.M., and J.M. Whipps. 1990. Substrate flow in the rhizosphere. Plant Soil 129:1–10.

Malik, N., and L.M. Jaiswal. 1993. Integrated use of organic and inorganic nitrogen sources and levels of N in wetland rice (*Oryza sativa*) in eastern Uttar Pradesh. Indian J. Agron. 38:641–643.

Malik, V., B. Kaur, and S.R. Gupta. 1998. Soil microbial biomass and nitrogen mineralization in straw incorporated soils. p. 557–565. *In* Ecological agriculture and sustainable development. Proc. Int. Conf. on Ecological Agriculture: Towards Sustainable Development. Vol. 1. 15–17 Nov. 1997. Chandigarh, India.

Mary, B., S. Recous, D. Darwis, and D. Robin. 1996. Interactions between decomposition of plant residues and nitrogen cycling in soil. Plant Soil 181:71–82.

Meelu, O.P., Yadvinder-Singh, Bijay-Singh, T.S. Khera, and K. Kumar. 1994. Recycling of crop residues and green manuring for soil and crop productivity improvement in rice–wheat cropping system. p. 605–613. *In* E. Humphreys et al. (ed.) Temperate rice: Achievements and potential. Vol. 2. NSW Agriculture, Griffith, NSW, Australia.

Misra, R.D., D.S. Pandey, and V.K. Gupta. 1996. Crop residue management for increasing the productivity and sustainability in rice–wheat system. p. 42. *In* Abstract of poster sessions. 2nd Int. Crop Sci. Congr. National Academy of Agricultural Sciences and ICAR, New Delhi, India.

Narang, R.S., and S.M. Virmani. 2001. Rice–wheat cropping systems of the Indo-Gangetic plain of India. Rice–Wheat Consortium Pap. Ser. 11. Rice–Wheat Consortium for the Indo-Gangetic Plains, New Delhi, India.

Narwal, S.S. 2000. Weed management in rice:wheat rotation by allelopathy. Crit. Rev. Plant Sci. 19:249–266.

Nelson, C.J. 1996. Allelopathy in cropping systems. Agron. J. 88:991–996.

Oh, W.K. 1984. Effects of organic matter on rice production. p. 477–488. *In* Organic matter and rice. IRRI, Los Baños, Philippines.

Pandey, S.P., H. Shankar, and V.K. Sharma. 1985. Efficacy of some organic and inorganic residues in relation to crop yield and soil characteristics. J. Indian Soc. Soil Sci. 33:178–181.

Parmele, R.W., M.H. Beare, and J.M. Blair. 1989. Decomposition and nitrogen dynamics of surface weed residues in no-tillage agroecosystems under drought conditions: Influence of resource quality on the decomposer community. Soil Biol. Biochem. 21:97–103.

Pathak, H., and M.C. Sarkar. 1997. Nitrogen supplementation with rice straw in an Ustochrept. J. Indian Soc. Soil Sci. 45:103–107.

Patra, D.D., J.C. Bhandari, and A. Misra. 1992. Effect of plant residues on the size of microbial biomass and nitrogen mineralization in soil: Incorporation of cowpea and wheat straw. Soil Sci. Plant Nutr. 38:1–6.

Prasad, B., and S.K. Sinha. 1995. Nutrient recycling through crop residues management for sustainable rice and wheat production in calcareous soil. Fert. News. 40(11):15–23.

Prasad, R., and J.F. Power. 1991. Crop residue management. Adv. Soil Sci. 15:205–239.

Rajput, A.L. 1995. Effect of fertilizer and organic manure on rice (*Oryza sativa*) and their residual effect on wheat (*Triticum aestivum*). Indian J. Agron. 40:292–294.

Rattan, K.R., M.P. Singh, R.O. Singh, and U.S.P. Singh. 1996. Long term effect of inorganic and organic–inorganic nutrient supply system on yield trends in rice–wheat cropping system. J. Appl. Biol. 6:56–58.

Russell, C.A., and I.R.P. Fillery. 1996. Estimates of lupin below-ground biomass nitrogen, dry matter, and nitrogen turnover to wheat. Aust. J. Agric. Res. 47:1047–1059.

Salim, M. 1995. Rice crop residue use for crop production. p. 99. *In* Organic recycling in Asia and the Pacific. RAPA Bulletin Vol. 11. Regional Office for Asia and the Pacific, FAO, Bangkok, Thailand.

Sarkar, A. 1997. Energy-use patterns in sub-tropical rice–wheat cropping under short-term application of crop residue and fertilizer. Agric. Ecosyst. Environ. 61:59–67.

Sarkar, A., R.L. Yadav, B. Gangwar, and P.C. Bhatia. 1999. Crop residues in India. Tech. Bull. Project Directorate for Cropping System Research, Modipuram, India.

Sharma, A.R., and B.N. Mittra. 1992. Integrated nitrogen management in rice (*Oryza sativa*)–wheat (*Triticum aestivum*) cropping system. Indian J. Agric. Sci. 62:70–72.

Sharma, H.L., S.C. Modgal, and M.P. Singh. 1985. Effect of applied organic manure, crop residues and nitrogen in rice–wheat cropping system in north-western Himalayas. Himachal J. Agric. Res. 11:63–68.

Sharma, H.L., C.M. Singh, and S.C. Modgal. 1987. Use of organics in rice–wheat crop sequence. Indian J. Agric. Sci. 57:163–168.

Sharma, P.K., and S.K. De Datta. 1985. Effects of puddling on soil physical properties and processes. p. 217–234. *In* Soil physics and rice. IRRI, Los Baños, Philippines.

Sidhu, B.S., V. Beri, and S.R. Gosal. 1995. Soil microbial health as affected by crop residue management. p. 45–46. *In* Proc. Natl. Symp. on Developments in Soil Sci. Ludhiana, India. 2–5 Nov. 1995. Indian Soc. Soil Sci., New Delhi, India.

Singh, H., and K.P. Singh. 1995. Effect of plant residue and fertilizer on grain yield of dryland rice under reduced tillage cultivation. Soil Tillage Res. 34:115–125.

Singh, Y., D. Singh, and R.P. Tripathi. 1996. Crop residue management in rice–wheat cropping system. p. 43. *In* Abstracts of poster sessions. 2nd International Crop Science Congress, National Academy of Agricultural Sciences, New Delhi, India.

Stirzaker, R.J., and D.G. Bunn. 1996. Phytotoxicity of ryegrass and clover cover crops, and a lucerne alley crop for no-till vegetable production. Biol. Agric. Hortic. 13:83–101.

Sur, H.S., S.S. Prihar, and S.K. Jalota. 1981. Effect of rice–wheat and maize–wheat rotations on water transmission and wheat root development in a sandy loam of the Punjab, India. Soil Tillage Res. 1:361–371.

Tandon, H.L.S. 1995. Fertilisers, organic manures, recyclable wastes and biofertilisers. p. 36–51. *In* H.L.S. Tandon (ed.) Recycling of crop, animal, human and industrial wastes in agriculture. Fertiliser Development and Consultation Organisation, New Delhi, India.

Thakur, D.S., and K.S. Pandya. 1997. Response of rice (*Oryza sativa* L.) to nitrogen levels and pretreated urea in an Inceptisol. Fert. News 42(8):37–40.

Thakur, K.S., and M.N. Singh. 1987. Effect of organic wastes and N levels on transplanted rice. Indian J. Agron. 32:161–164.

Tian, G., B.T. Kang, and L. Brussaard. 1993. Mulching effect of plant residues with chemically contrasting compositions on maize growth and nutrient accumulation. Plant Soil 153:179–187.

Unkovich, M.J., J.S. Pate, and J. Hamblin. 1994. The nitrogen economy of broadcare lupin in southwest Australia. Aust. J. Agric. Res. 45:149–164.

Verma, S., and R.S. Mathur. 1990. The effects of microbial inoculation on the yield of wheat when grown in straw-amended soil. Biol. Wastes 33:9–16.

Verma, S.M., and N.T. Singh. 1974. Effect of some indigenous inorganic materials on soil structure. J. Indian Soc. Soil Sci. 22:220–225.

Verma, T.S., and R.M. Bhagat. 1992. Impact of rice straw management practices on yield, nitrogen uptake and soil properties in a wheat–rice rotation in northern India. Fert. Res. 33:97–106.

Verma, T.S., and P.K. Sharma. 2000. Effect of organic residue management on the productivity of the rice–wheat cropping system. p. 163–171. *In* I.P. Abrol et al. (ed.) Long-term soil fertility experiments in rice–wheat cropping systems. Rice–Wheat Consortium for the Indo-Gangetic Plains, New Delhi, India.

Walia, S.S., S.S. Brar, and D.S. Kler. 1995. Effect of management of crop residues on soil properties in rice–wheat cropping system. Environ. Ecol. 13:503–507.

Wilson, T.C., E.A. Paul, and R.R. Hardwood. 2001. Biological active soil organic matter fractions in sustainable cropping systems. Appl. Soil Ecol. 16:63–76.

Woodhead, T., R. Huke, and E. Huke. 1994. Areas, locations, and ongoing collaborative research for the rice–wheat system of Asia. p. 68–96. *In* R.S. Paroda et al. (ed.) Sustainability of rice–wheat production systems in Asia. FAO, Bangkok, Thailand.

Yadav, R.L. 1997. Urea-N management in relation to crop residue recycling in rice–wheat cropping system in North-Western India. Bioresour. Technol. 61:105–109.

Yadav, R.L., B.S. Dwivedi, Kamta Prasad, O.K. Tomar, N.J. Shurpali, and P.S. Pandey. 2000. Yield trends and changes in soil organic-C and available NPK in a long-term rice–wheat system under integrated use of manures and fertilizers. Field Crops Res. 68:219–246.

Yadvinder-Singh, Bijay-Singh, and O.P. Meelu. 1996. Organic manuring and crop residue recycling for sustained soil and crop productivity in rice–wheat cropping system. p. 73–83. *In* P. S. Pathak and B. Gopal (ed.) Studies in Indian agroecosystems. National Institute of Ecology, New Delhi, India.

Yadvinder-Singh, Bijay-Singh, O.P. Meelu, and C.S. Khind. 2000. Long-term effects of organic manuring and crop residues on the productivity and sustainability of rice–wheat cropping system in Northwest India. p. 149–162. *In* I.P. Abrol et al. (ed.) Long-term soil fertility experiments in rice–wheat cropping systems. Rice–Wheat Consortium for the Indo-Gangetic Plains, New Delhi, India.

Yadvinder-Singh, S.S. Malhi, M. Nyborg, and E.G. Beauchamp. 1994. Large granules, nests or bands: Method of increasing efficiency of fall applied urea for small grains in North America. Fert. Res. 38:61–87.

Zia, M.S., M. Munsif, M. Aslam, and M.A. Gill. 1992. Integrated use of organic manures and inorganic fertilizers for the cultivation of lowland rice in Pakistan. Soil Sci. Plant Nutr. 38:331–338.

9 Prospects for Permanent Beds for the Rice–Wheat System

David J. Connor

The University of Melbourne
Victoria, Australia

Jagadish Timsina and Elizabeth Humphreys

Commonwealth Scientific and Industrial Research Organization
Griffith, Australia

ABSTRACT

The ability of the rice–wheat (RW) system of the Indo-Gangetic Plains (IGP) to continue to meet the food needs of an increasing population is threatened by yield stagnation, resource sustainability, and environmental pollution. An important question is whether prolonged puddling and ponding for rice and the flat, poorly drained layouts required for ponding are inimical to high aggregate productivity and efficient resource use of the RW system. The evidence for the adverse effects of the puddled, flat layout on wheat productivity in the RW system, however, is scant, inconsistent, and inconclusive. It may depend on soil type and site-specific management, and needs more comprehensive investigation. Even so, issues of resource use warrant research on a radical redesign of the system, such as the use of permanent beds. Permanent raised beds may offer farmers significant advantages, such as reduced tillage, increased opportunity for crop diversification, mechanical weeding and placement of fertilizers, opportunities for relay cropping and intercropping, and water savings. Such a radical change in the system, however, will also have implications for a host of other production and environmental factors, including cultivar characteristics; irrigation and nutrient management; pest, disease, weed, and residue management; soil structure; deep drainage; groundwater depletion and pollution; and atmospheric pollution. At present, there seem to be good prospects for permanent beds to help address the sustainability constraints of the RW system, although overcoming them will require considerable investment in detailed research on all aspects of the system.

The rice–wheat system in South and East Asia has developed quickly in the past three decades with the introduction of rice into traditional wheat areas (e.g., northwest India) and wheat into traditional rice areas (e.g., Bangladesh, eastern India, and Nepal). Consequently, wide variation exists in the growing conditions, principally in climate and soil properties, across the RW region that encompasses more than 13 Mha in the IGP of South Asia and a further 13 Mha in China (Timsina &

 Improving the Productivity and Sustainability of Rice–Wheat Systems: Issues and Impacts. ASA Special Publication 65.

Connor, 2001). The initial expansion of the RW area was driven by the need to intensify cropping to meet the increasing demand for food, and was made possible by the development of short- and medium-duration cultivars of both rice and wheat, and by the expansion of irrigation. For example, the irrigated area in the Indian Punjab doubled from 1960 to 1997, mostly because of the increased use of groundwater (Hira & Khera, 2000).

The IGP is a great crescent of alluvial soils that stretches from the delta of the Indus River in the west to the Ganga-Brahmaputra River delta in the east. The RW system in the IGP is practiced on a range of soil types. Without appropriate data, it is difficult to estimate the area under various soils where RW is practiced. However, some approximations can be made, extrapolating from Huke and Huke (1992) and Narang and Virmani (2001). Thus, in the Indus or Trans-Gangetic Plains (areas in Pakistan and parts of Punjab and Haryana in India), soils are generally sandy, loamy sands, and loams, representing approximately 25% of the RW area in the IGP. In the Upper Gangetic Plains (most of Uttar Pradesh and parts of Bihar and Nepal), soils are alluvial and deep, well-drained, with coarse to medium texture (about 30% of the RW area). In the Middle Gangetic Plains (most of Bihar and parts of Nepal), they are medium- to fine-textured (about 25% of the RW area), while in the Lower Gangetic Plains (parts of Bihar and West Bengal and all of Bangladesh), they are mostly fine-textured (about 20% of the RW area). Thus, the RW system is practiced on significant areas of each soil type, coarse-, medium-, and fine-textured, with perhaps a slightly larger area on coarse-textured soils than on the other two types.

Two aspects of traditional rice culture may disadvantage the productivity of wheat and other nonrice crops grown after rice in many parts of the IGP. The first is puddling that impedes seedling establishment and restricts root penetration in those soils in which hard pans are formed (Meelu et al., 1979; Sur et al., 1981; International Board for Soil Research Management [IBSRAM], 1991; Kirchhof & So, 1996; Aggarwal et al., 1995; Kirchhof et al., 2000). The second is that the bunded paddies, required to hold standing water for rice, accumulate water and impede surface drainage during the production periods of other crops (IBSRAM, 1991; Kirchhof & So, 1996; Kirchhof et al., 2000). Although the nonrice season across the RW region is one of low rainfall, heavy premonsoonal rain can have disastrous effects on the third crop, such as maize or mungbean grown after wheat or before rice, both during establishment because of waterlogging and during grain filling because of waterlogging and mechanical damage (Timsina & Connor, 2001; Quayyum et al., 2002).

Pressure now exists to continue expanding the area under the RW system and also to increase its productivity in the IGP. Expansion and intensification already impose major threats to groundwater depletion, groundwater and atmospheric pollution, waterlogging, salinization (Pingali & Shaha, 1999; Singh, 2000), and declining soil organic matter (SOM), the latter in soils with high initial SOM (Yadav et al., 2000). Significant atmospheric pollution results from stubble burning and from fuel used for performing tillage operations and pumping water. For example, 10.7 and 2.7 million megagrams, respectively, of rice residues are burned annually in Punjab and Haryana Districts of India (P.R. Gajri, personal communication, 2002). Modified production systems are needed to rectify these problems rather than have them exacerbated by further expansion and intensification of existing systems. An important question is whether prolonged inundation and puddling for rice are inimical to high aggregate productivity of the RW system. If so, what are the prospects for suitably modified or alternative and feasible intensive food production systems for the region?

PUDDLING AND THE WET–DRY TRANSITION

Land preparation for rice by puddling and then continuing inundation until maturity have significant effects on the physical, chemical, and biological status of soils that influence growing conditions for all crops in the system (Sharma & de Datta, 1986; Kleinhenz & Schnitzler, 1996). Puddling is the repeated tillage of saturated (inundated) soil in preparation for a rice crop, and it is undertaken because it offers significant advantages to rice production (Sánchez, 1976; Ghildyal, 1978). It softens soil, facilitates transplanting of rice seedlings, promotes root growth, aids weed control, and reduces water and nutrient losses by leaching. Further, it mobilizes P and holds N in the ammonium form, which is readily available for rice. The procedure, however, disperses soil aggregates, destroys soil structure, and, particularly in fine-textured soils, forms a massive topsoil that sets hard and often cracks widely when dry (Ghildyal, 1978; Sharma & de Datta, 1986; Kirchhof et al., 2000). When repeated from year to year, puddling may form a compacted layer up to 5 to 10 cm thick to a depth of 10 to 40 cm (Sur et al., 1981; Sharma & de Datta, 1986; Aggarwal et al., 1995). In continuous rice culture, the formation and maintenance of the compacted layer can be interpreted as an investment in infrastructure. In the rice–nonrice system, however, the destruction of the structure of the surface soil and the hardpan that forms as the compacted layers dry might be expected to be serious liabilities to the establishment and performance of wheat and other crops grown after rice.

There is surprisingly little published data on the direct effect of puddling for rice on the performance of succeeding crops for both irrigated and rainfed situations. The data in Table 9–1 reveal variable effects on wheat, maize, and legumes grown after rice for a range of soil types for both irrigated and rainfed situations. The majority of the data for the RW system come from the irrigated coarse-textured soils of the Indian Punjab, and even then the effects are inconsistent. Furthermore, many of those reports compare yields of wheat grown after a nonpuddled crop (e.g., maize) with yields after puddled rice, and are therefore not a true comparison of the effects of puddling for rice on wheat yield in a RW rotation. For example, better soil physical conditions and greater productivity of wheat were achieved from maize–wheat and cotton–wheat systems than from a (puddled) rice–wheat system. Data from long-term field experiments across Asia (Dawe et al., 2000) and across India (Duxbury et al., 2000) also suggest little or no decline in wheat yield following puddled rice, but these studies do not make comparisons with puddled rice. The degree to which current soil management for rice constrains the productivity of other crops in the rotation, and aggregate system productivity, requires serious investigation.

Table 9–1. Effect of puddling for rice on soil physical properties and growth and yield of succeeding crops.

Location/country	Crop	Soil	Effects of puddling on.... Yield	Effects of puddling on.... Other	Reference
			Irrigated		
Punjab, India	Wheat	Sandy loam	Increased, Year 1 Decreased, Year 2		Singh, 1998
Punjab, India	Wheat	Silty clay loam	Increased, Year 1 No effect, Year 2		
Punjab, India	Wheat	Sandy loam	Decreased growth and yield	Increased bulk density, decreased shoot and root growth, and decreased leaf water potential	Aggarwal et al., 1995
Punjab, India	Wheat	Sandy loam	Decreased after rice compared with after maize		Boparai et al., 1992
Punjab, India	Wheat	Sandy loam	Decreased after rice compared with after maize	Decreased rooting density after rice compared with after maize	Sur et al., 1981
Punjab, India	Wheat	Sandy loam	Decreased after rice compared with after maize	Increased bulk density, modulus of elasticity, dispersion ratio, restricted root proliferation	Meelu et al., 1979
Bhairahwa, Nepal	Wheat	Silty clay	Decreased after TPR† (by 7%) compared with after DSR‡	Hobbs et al., 2002	
Pantnagar, India		Sandy loam and silty loam	Decreased after TPR (by 13%) compared with after DSR		
Southern NSW, Australia	Wheat and canola Wheat	Clay Clay loam	No effect on growth and yield No effect on growth and yield		Humphreys et al., 1996

East Java, Indonesia	Maize	Clay	Increased by 1%	Increased length and width of cracks, decreased penetration resistance and tensile strength	Willatt & Tranggono, 1987; Tranggono & Willatt, 1988
East Java, Indonesia	Maize	Loam	Decreased by 5%	Increased width but decreased the length of cracks, increased penetration resistance and tensile strength	
Los Baños, Philippines	Mungbean	Clay loam	Increased		IRRI, 1988 (p. 405–406.)
		Sandy loam	Decreased		
Punjab, Pakistan	Wheat	Loam and sandy loam	Decreased after rice compared with after maize or cotton		Altaf, 1994
			Rainfed		
East Java, Indonesia, and Luzon, Philippines	Mungbean and peanut		Clay	No effect	Kirchhoff et al., 2000
	Soybean		Decreased	Decreased root growth and water uptake	
Bangladesh, China, Pakistan	Wheat	Clay	Decreased	Impaired soil structure, hardpans	IBSRAM, 1991
Philippines, Thailand Indonesia	Maize and mungbean		Clay	Decreased	
	Maize	Clay	Increased		
	Loam	Decreased			

† TPR, transplanted rice.
‡ DSR, direct-sown rice.

PRODUCTIVITY AND NUTRIENT BALANCE

Rice–wheat is a highly extractive system at the levels of yield that are being sought. The design and management of the system, therefore, present a challenge increasingly facing modern agronomy, the effort to establish and maintain productive systems that are also sustainable. In the current RW system, high nutrient extraction is exacerbated by the frequent harvest or burning of straw, and by the limited supply of organic matter that can be returned as farmyard manure because of competing use as fuel for cooking.

The annual productivity of the RW system in the IGP is small (5–7 Mg ha^{-1}) compared with currently attainable (8–10 Mg ha^{-1}) and site-potential (12–19 Mg ha^{-1}) yields (Aggarwal et al., 2000). Consideration of yield responses to nutrients, nutrient extraction in harvest, and nutrient input by fertilizer reveals that greater yields will require different, and more carefully targeted, fertilizer practices. There are two parts to this.

The first is attention to the macronutrients needed to sustain the system without depleting soil fertility. Nutrient extraction for target yields gives a clue to the magnitude and composition required, even though it does not take into account other inputs, such as irrigation water or biological N_2 fixation, or losses by volatilization or leaching. Thus, based on average nutrient concentrations in grain and straw (Timsina & Connor, 2001), a yield of 10 Mg ha^{-1}, equally divided between wheat and rice, removes 163, 34, and 47 kg N, P, and K ha^{-1} in the grain alone. The quantity would rise to 222, 43, and 182 kg ha^{-1} in those many cases in which stubble is also removed for fuel or animal fodder. In many areas, such as in Bangladesh, fertilizer application is much lower than the requirement (Morris et al., 1997). In other areas, it may be excessive and/or unbalanced. For example, in the Indian Punjab, the recommended doses of nutrients are 120, 0, and 25 and 120, 26, and 25 kg ha^{-1} N, P, and K for rice and wheat, respectively. Farmers, however, tend to overapply N but ignore K. The average application rates are 150, 10, and 0 and 140, 26, and 0 kg ha^{-1} N, P, and K for rice and wheat, respectively (Yadvinder-Singh & Bijay-Singh, 2001). This reflects the fact that the Indian government has withdrawn fertilizer subsidies for P and K, but not for N.

The second is the need to pay attention to the possibility in some regions, depending predominantly on soil type, of micronutrient deficiencies induced by long-term extraction from these highly productive systems or, on coarse-textured soils, by leaching. Further, evidence of deficiencies of Zn, Fe, Mn, and Cu in some long-term experiments in India, especially where recommended amounts of N, P, and K have been applied for a long time (Yadvinder-Singh et al., 2000), suggests micronutrient deficiency as a result of nutrient interactions. In contrast, although Fe and Zn deficiency in rice is sometimes observed on the coarse-textured soils in the Indian Punjab, the levels of those nutrients are now generally increasing in soils because of the greater use of those fertilizers (Nayyar et al., 2001). There, most farmers apply about 11 kg Zn ha^{-1} to wheat and rice and some spray rice crops with 1 to 2% $FeSO_4$. There is, however, increasing evidence of Mn deficiency in wheat in the RW system on those soils because of the leaching of Mn during the rice season (Nayyar et al., 2001). Likewise, Cu levels are also declining in the Indian Punjab and B deficiency in wheat is widespread in Nepal and eastern India (Kataki et al.,

2001). The declines in Mn and Cu in the Indian Punjab are due to antagonistic effects of Zn and Fe fertilizers (Arora et al., 1993) and are particularly associated with the expansion of rice to coarse-textured soils. Unless the micronutrient deficiencies are remedied, responses to macronutrients will be, at best, muted. It is also important to realize that the export of micronutrients increases with crop yields such that, without remediation, currently chronic deficiencies will become acute.

The inclusion of legumes is seen by many as a means of supplying N to the system. Nutrient balances have shown, however, that legumes can make only small contributions to N relative to the high yields required of the system (Timsina & Connor, 2001). Further, legume N is not free, but is accumulated at a cost of land, time, labor, water, fertilizer, and the opportunity costs of more profitable crops. It is not surprising that, on N balance alone, few farmers find legumes to be economical at the current price of N fertilizer. There are, however, reasons to include legumes and other crops in the system to assist soil, disease, and weed management, improve human nutrition, and increase or stabilize economic returns. In the subtropical environment of South and East Asia where the RW system is practiced, the range of choices is unfortunately limited (Johansen et al., 2000), and there is little information on productivity and appropriate management practices within such a system. Further, because of the low economic return from legumes and the need to produce cereals as major staple foods and associated price support, the RW system has largely replaced both summer and winter legumes that otherwise were grown instead of rice in the summer and wheat in the winter (Morris et al., 1997; Johansen et al., 2000).

A RESEARCH PARADIGM

The need for greater and more carefully balanced nutrient input to sustain high yields of the RW system is evident, but crops will only respond to improved nutrition as other constraints are removed—at each step concentrating on the (most) limiting factor. A detailed review (Timsina & Connor, 2001) has shown many constraints, such as poorly adapted cultivars; suboptimal irrigation and drainage practices; pressure from weeds, pests, and diseases; and impaired root growth.

One direction for research and investment in technology development is to seek gradual improvement across the many constraints to production in current systems. This means a continuation of work on cultivar improvement and crop management so that phenologically adapted cultivars with improved resistance to biotic and abiotic stresses become available to farmers. It is critical that this research be accomplished region by region and site by site to identify and remove the constraints to fertilizer response. The essential outcome for farmers would then be test-based approaches to fertilizer management on individual farms so that as yield increases, the responses remain economic. In that way, farmers will be able to continue to increase yield. With this scenario, the system continues in its present form and research seeks, in a coordinated fashion, to remove the most serious limitations to greater productivity and resource use efficiency.

There is, however, another option that by redesign would seek to remove the major conflicts (or constraints) that currently exist within many parts of the system. These issues derive from the deleterious effects of traditional rice culture involving puddling, prolonged inundation and flat layouts, and the major limitations they present to the productivity of wheat and other crops in the system (Meelu et al., 1979; Sur et al., 1981; IBSRAM, 1991; Aggarwal et al., 1995; Kirchhof & So, 1996; Kirchhof et al., 2000). If these could be resolved, larger yield responses to fertilizer applications would be achieved and, most likely, achieved more quickly.

As discussed above, puddling offers significant advantages to rice by reducing percolation of water and nutrients, especially on coarse-textured soils, but it may not be necessary or could be reduced in fine-textured soils (Utomo et al., 1996). Alternatives to puddling for rice, especially for vast areas of the IGP with fine-textured soils, include dry seeding on flat surfaces (De Datta, 1986) or on beds (Borrell et al., 1997; Cooten & Borrell, 1999). Dry seeding is less labor-intensive than puddled transplanted rice and suffers from serious weed problems, but can improve soil structure to the benefit of the ensuing wheat crop. Recent experiments on fine-textured soils at RW sites in Bhairahawa (Nepal) and Pantnagar (India) showed that rice yields were not reduced by direct seeding in nonpuddled fields, and yields of the following direct-seeded wheat increased (Hobbs et al., 2002). Dry seeding, either on beds or on flat surfaces, facilitates mechanization but would require different cultivars and nutrient-, water-, and weed-management strategies than for transplanted rice.

The radical change in management by the cultivation of crops on beds could not only reduce the current deleterious effects of waterlogging and restrictions to the exploration of the soil profile by root systems of nonrice crops, but could provide other benefits also.

THE POTENTIAL OF RAISED BEDS

Changing from flat to bed layouts alters the geometry and hydrology of the system and offers greater control of irrigation, drainage, and their effects on the transportation and transformation of nutrients, and possibly better capture and use of rainfall.

During irrigation of bed and furrow systems, water moves horizontally from furrows into beds (*subbing*), pulled upward toward the soil surface by evaporation, transpiration, and capillarity, and downward largely by gravity. Irrigation and fertilizer management must be considered together and in relation to rainfall and groundwater conditions to avoid creating problems. These include salinization in the seed zone or reduced nutrient availability because of the transportation of dissolved solutes to the surface of the beds (Wright et al., 1985) with evaporative flow, or to below the root zone by leaching. Management must also seek to minimize the loss of N to the atmosphere. During irrigation, air-filled porosity decreases in the furrows and in the ridges below the level of ponded water, and, when water is ponded for long periods in the furrows, denitrification can cause a significant loss of N (Muirhead et al., 1985).

Preliminary studies on both experiment stations and in farmers' fields have shown that both wheat and rice crops can be grown successfully on beds in areas of the IGP. For wheat, it has been established that

1. A bed width of 67 cm (37-cm bed top and 30-cm furrow) is optimum on coarse- to medium-textured soils.
2. Various cultivars perform differently on beds.
3. The seed rate can be reduced by up to 30% compared with flat surfaces.
4. Yields with two and three rows per bed are comparable, but lodging is greater with three rows per bed.
5. Irrigation water requirement is reduced by up to 50% (based on pumping time) as reported by farmers.
6. There is reduced incidence of the grass weed *Phalaris minor*.
7. Mechanical cultivation is possible as an alternative to chemical weed control.
8. Fertilizer can be banded below the soil surface before and during the crop season (Hobbs et al., 1997; Malik et al., 1998; S.S. Dhillon, personal communication, 2002).

Furthermore, there are indications that yield of wheat on beds can be further increased through higher N application rates and later irrigation because of the reduced risk of lodging on beds (S.S. Dhillon, personal communication, 2002). Studies with other nonrice crops, such as cotton and rapeseed, in Punjab, India, have also shown that bed planting can save 20 to 40% of irrigation water without reducing yield (Aujla et al., 1991, 1992).

The performance of direct-seeded and transplanted rice has also been evaluated during the last two years on new or reshaped beds in the IGP and elsewhere. Experience of farmers and researchers in western Uttar Pradesh, Punjab, and Haryana, India, suggests that rice on beds uses less water (by 25–50% based on pumping time) and that growth and yield of transplanted rice on beds can be comparable with those of traditional culture (R. Gupta, personal communication, 2001). However, dry-seeded rice on beds has generally been inferior to date. Results from Bangladesh suggest that the performance of rice on beds depends upon soil type. For example, rice used less water and had better growth on beds compared with the traditional puddled system on a fine-textured clay soil at Joydebpur, while its performance was poor on a coarse-textured soil at Nashipur (R. Karim, BRRI, and Sufian, BARI, personal communication, 2001). Other evidence from the United States and Australia also indicates water savings for rice grown on beds compared with flooded rice. For example, on fine-textured soils in Missouri in the USA, furrow-irrigated rice used approximately one-half the water of flooded rice (Tracy et al., 1993). Also on fine-textured soils in subtropical Queensland, Australia, rice grown on beds with saturated soil culture used 32% less water than that on flat surfaces with permanent flood established from the three-leaf stage (Borrell et al., 1997; Ockerby & Fukai, 2001).

Evidently, significant steps have been made in demonstrating the potential benefits of raised beds in the IGP and in other places. With few exceptions (e.g., Maynard, 1991; Kleinhenz et al., 1996; Limon et al., 2000), the beds are heavily cultivated and reshaped or completely reformed each year, at considerable expense. Few detailed measurements have been made, however, to understand the soil, water, and crop processes, and the potential application of permanent beds in the

RW system has not yet been investigated in detail, including opportunities for residue retention or incorporation.

It can also be hypothesized that bed systems may not be suited to coarse-textured soils because of excessive loss by percolation, or to sodic soils that form crusts on ridge tops and disperse quickly, resulting in poor emergence and infiltration. Similarly, they may be unsuited to saline soils, shallow saline water table situations, and irrigation with saline water. Some soils would require amelioration and careful management for beds to function; for example, sodic soils could be improved by the addition of gypsum and/or organic matter. Research is required to determine the suitability of soils for the RW system on permanent beds.

FORCES FOR CHANGE TO BED SYSTEMS

The change from RW cultivation on flat surfaces to beds offers diverse advantages to production, resource conservation, resource use efficiency, and profitability, especially because of the intensive nature of the system and the high aggregate yields required of it.

Farmers now have much interest in bed systems, driven by many factors, including those mentioned above. Some of the main factors appear to be opportunities for relay cropping (e.g., wheat–sugarcane [*Saccharum bicolor* L.]) mechanical cultivation of weeds, reduced irrigation time, and improved yields on fine-textured soils prone to waterlogging. However, we hypothesize that many other potential benefits will lead to increased productivity and profitability through permanent beds, including

- Reduced tillage and direct seeding with permanent beds that reduce costs of labor, diesel, and machinery when offset against the costs of initial bed formation and maintenance.
- Increased opportunities for crop diversification in both the wet and dry seasons that allow response to market opportunities as well as to the reduced supply of water—for example, research at AVRDC in tropical Taiwan has concluded that vegetables could be grown more profitably than rice (Kleinhenz & Schnitzler, 1996; Kleinhenz et al., 1996) and arrest soil structural degradation and hardpan formation from intensive rice.
- Mechanical weeding and interculture operations in wheat and other non-rice crops that reduce herbicide or labor costs—these advantages may also extend to rice on beds.
- Mechanical placement of fertilizers below the soil surface, leading to improved fertilizer use efficiency.
- Reduced seed requirement of a range of crops compared with flat surfaces.
- Reduced cost of irrigation—this includes the digging, maintenance, and deepening of wells as well as power for pumping and labor time for irrigation. (These savings also accrue to governments in those areas, such as Punjab, where farmers are not charged for electricity for pumping.)
- Greater yields of all crops that can be obtained in response to better soil structure and surface drainage, more timely sowing because of direct seeding and good drainage, and other management operations.

There are additional public benefits, including solutions to

- Current and impending water scarcity—bed culture promotes the possibility (i) to grow rice with intermittent irrigation to save water and (ii) to diversify crops to other less water-demanding crops.
- Environmental pollution—it can be hypothesized that both groundwater pollution and methane emissions from nonpuddled intermittently irrigated rice on beds would decrease compared with those from puddled rice on flat surfaces. (Conversely, leaching of pollutants in the absence of puddling and emissions of nitrous oxides may increase.)

DISCUSSION AND CONCLUSIONS

There is an urgent need to investigate new systems of RW culture to overcome the major productivity constraints, resource depletion, and pollution problems of existing systems. Particularly urgent problems are yield stagnation (or possibly decline in some situations), groundwater depletion and pollution, and atmospheric pollution from stubble burning. Much new knowledge and technology can be considered for that quest, such as direct and surface seeding, zero or reduced tillage, stubble management, bed planting with furrow irrigation, herbicides, and fertilizers. Mechanization can be of great assistance not just in reducing the drudgery of many traditional practices and in retention of stubble to the benefit of soil organic matter and air quality, but also in contributing to timely sowing and harvesting, which are important requirements for success under tight cropping schedules.

Management of soil for high and sustainable productivity requires attention to maintaining soil structure, rooting depth, and aeration, as appropriate to the individual crops concerned. Improved conditions for all crops in the RW system could be provided by a change in production strategy to raised-bed culture. Rice can be grown satisfactorily by furrow irrigation, as can the other crops in the system. Raised beds not only facilitate irrigation but also drainage, and therein lies their potential to increase the yield of crops other than rice in the system. Further, beds are best left in place to provide controlled traffic to maintain soil structure and provide a healthy and deep-rooting environment. Under this system, there would be new problems relating to weed control and new issues concerning nutrient and residue management, especially when rice and wheat are grown on zero-tillage conditions on beds. There is good reason to believe, however, that the greater efficiency of water and nutrient use found elsewhere for other cropping systems on raised-bed systems (e.g., Tracy et al., 1993; Borrell et al., 1997; Limon et al., 2000) could also apply to RW production systems of South Asia. These problems and the identification of areas where raised-bed systems would be appropriate are arguably a valuable focus for major investment in future research in the RW system.

Crop diversification may also be an important contributor to environmental sustainability and economic viability of RW areas. Permanent beds provide the opportunity for increased crop diversification, especially during the wet season, and higher productivity for crops traditionally grown on flat surfaces because of less waterlogging. With permanent beds, farmers will be able to rapidly change crop choice, for example, from rice to vegetables, in response to market opportunities (Kleinhenz & Schnitzler, 1996; Kleinhenz et al., 1996). The ability to diversify away from rice will also reduce demand for water or leave more water available for other uses.

Last but not least, research on the permanent-bed RW technique will need to tackle many interacting problems across a range of disciplines. These include production problems such as plant type, planting geometry, zero- or reduced-tillage options, integrated weed management (especially for dry-seeded rice), residue management (especially for rice stubble), irrigation and nutrient management (especially N, P, Zn, and Fe for rice), soil management, water use efficiency and deep drainage, and environmental problems of groundwater and air pollution.

ACKNOWLEDGMENTS

Our research on the rice–wheat system is supported by the Australian Centre for International Agricultural Research. We thank S.S. Dhillon, Yadvinder-Singh, P.R. Gajri, and Bijay-Singh, Punjab Agricultural University, for their valuable comments and suggestions that have improved the manuscript.

REFERENCES

Aggarwal, G.C., A.S. Sidhu, N.K. Sekhon, K.S. Sandhu, and H.S. Sur. 1995. Puddling and N management effects on crop response in a rice–wheat cropping system. Soil Tillage Res. 36:129–139.

Aggarwal, P.K., K.K. Talukdar, and R.K. Mall. 2000. Potential yields of rice–wheat system in the Indo-Gangetic Plains of India. Rice–wheat Consortium Paper Ser. 10. Rice–Wheat Consortium for the Indo-Gangetic Plains, New Delhi, India.

Altaf, Z. 1994. Rice–wheat systems in Pakistan: The sustainability questions. p. 197–203. *In* Paroda et al. (ed.) Sustainability of rice–wheat production systems in Asia. Regional Office for Asia and the Pacific. FAO, Bangkok, Thailand.

Arora, C.L., M.S. Brar, H.S. Hundal, and P.N. Takkar. 1993. Changing nutrient status of crop plants in intensive agriculture. Fert. News 38:47–51.

Aujla, M.S., C.J. Singh, K. Kumar, and B.S. Sandhu. 1992. Efficiency of different methods of irrigation in Indian rape. J. Arid Environ. 33:374.

Aujla, M.S., C.J. Singh, K.K. Vashist, and B.S. Sandhu. 1991. Evaluation of methods for irrigation of cotton in canal-irrigated areas of south-west Punjab. Arid Soil Res. Rehab. 5:225–234.

Boparai, B.S., Yadvinder-Singh, and B.D. Sharma. 1992. Effect of green manuring with *Sesbania aculeata* on physical properties of soil and on growth of wheat in rice–wheat and maize–wheat cropping systems in a semi-arid region of India. Arid Soil Res. Rehab. 6:135–143.

Borrell, A., A. Garside, and S. Fukai. 1997. Improving efficiency of water use for irrigated rice in a semi-arid tropical environment. Field Crops Res. 52:231–248.

Cooten, D.E. van, and A.K. Borrell. 1999. Enhancing food security in semi-arid eastern Indonesia through permanent raised-bed cropping: A review. Aust. J. Exp. Agric. 39:1035–1046.

Dawe, D., A. Dobermann, P. Moya, S. Abdulrachman, Bijay-Singh, L.P. Singh, S.Y. Li, B. Lin, G. Panaullah, O. Sarim, Y. Singh, A. Swarup, P.S. Tan, and Q. Zhen. 2000. How widespread are yield declines in long-term rice experiments in Asia? Field Crops Res. 66:175–193.

De Datta, S.K. 1986. Technology development and the spread of direct-seeded rice in Southeast Asia. Exp. Agric. 22:417–426.

Duxbury, J.M., I.P. Abrol, R.K. Gupta, and K.F. Bronson. 2000. Summary: Analyses of long-term soil fertility experiments with rice–wheat rotations in South Asia. p. vii–xxii. *In* I.P. Abrol et al. (ed.) Proc. Long-term soil fertility experiments in rice–wheat cropping systems. RWC Paper Series 6. Rice–Wheat Consortium for the Indo-Gangetic Plains, New Delhi, India.

Ghildyal, B.P. 1978. Effect of compaction and puddling on soil physical properties and rice growth. p. 317–336. *In* Soils and rice. IRRI, Los Baños, Philippines.

Hira, G.S., and K.L. Khera. 2000. Water resource management in Punjab under rice–wheat production system. Res. Bull. 2/2000. Punjab Agricultural University, Ludhiana, Punjab.

Hobbs, P., G.S. Giri, and P. Grace. 1997. Reduced and zero tillage options for the establishment of wheat after rice in South Asia. Rice–Wheat Consortium Paper Ser. 2. Rice–Wheat Consortium for the Indo-Gangetic Plains, New Delhi, India.

Hobbs, P., Y. Singh, G.S. Giri, J.G. Lauren, and J.M. Duxbury. 2002. Direct-seeding and reduced-tillage options in the rice–wheat systems of the Indo-Gangetic Plains of South Asia. p. 201–215. *In* S. Pandey et al. (ed.) Direct seeding: Research issues and opportunities. Proc. of the International Workshop on Direct Seeding in Asian Rice Systems: Strategic Research Issues and Opportunities. Bangkok, Thailand. 25–28 Jan. 2000. IRRI, Los Baños Philippines.

Huke, R., and E. Huke. 1992. Rice–wheat atlas of South Asia. IRRI, Los Baños, Philippines.

Humphreys, E., W.A. Muirhead, B.J. Fawcett, J.T. Townsend, and E.A. Murray. 1996. Puddling in mechanized rice culture: Impacts on water use and the productivity of rice and post-rice crops. p. 213–218. *In* G.S. Kirchhof and H.B. So (ed.) Management of clay soils for rainfed lowland rice-based cropping systems. ACIAR Proc. No. 70. Aust. Cent. for Int. Agric. Res., Canberra, Australia.

International Board for Soil Research Management. 1991. The management of lowland claey soils after rice in Asia. IBSRAM Proc. No. 11. IBSRAM, Bangkok, Thailand.

IRRI. 1988. IRRI annual report for 1987. IRRI, Los Baños, Philippines.

Johansen, C., J.M. Duxbury, S.M. Virmani, C.L.L. Gowda, S. Pande, and P.K. Joshi (ed.) 2000. Legumes in rice and wheat cropping systems of the Indo-Gangetic Plains: Constraints and opportunities. ICRISAT, Hyderabad, India, and Cornell University, Ithaca, NY.

Kataki, P.K., S.P. Srisvastava, M. Saifuzzaman, and H.K. Upreti. 2001. Sterility in wheat and response of field crops to applied boron in the Indo-Gangetic Plains. J. Crop Prod. 4:133–165.

Kirchhof, G., S. Priyono, W.H. Utomo, T. Adisarwanto, E.V. Dacanay, and H.B. So. 2000. The effect of soil puddling on the soil physical properties and the growth of rice and post-rice crops. Soil Tillage Res. 56:37–50.

Kirchhof, G.S., and H.B. So. 1996 (ed.) Management of clay soils for rainfed lowland rice-based cropping systems. ACIAR Int. Workshop, Bureau of Soil and Water Management. Quezon City, Manila. 20–24 Nov. 1995. ACIAR Proc. No. 70. Aust. Cent. for Int. Agric. Res., Canberra, Australia.

Kleinhenz, V., D.J. Midmore, and W.H. Schnitzler. 1996. High bed systems for off-season vegetable production in the tropics and subtropics. Agric. Res. 1/96:60–62.

Kleinhenz, V., and W.H. Schnitzler. 1996. Diversification and transformation of Asian paddy fields to upland vegetable production. Plant Res. Dev. 43:81–94.

Limon, O.A., K.D. Sayre, and C.A. Francis. 2000. Wheat and maize yields in response to straw management under a bed planting system. Agron. J. 92:295–302.

Malik, R.K., G. Gill, and P.R. Hobbs. 1998. Herbicide resistance: A major issue for sustaining wheat productivity in rice–wheat cropping systems in the Indo-Gangetic Plains. Rice–Wheat Consortium Paper Ser. 3. Rice–Wheat Consortium for the Indo-Gangetic Plains, New Delhi, India.

Maynard, M. 1991. Permanent beds: Their potential role in soil management for the future. Farmers' Newsl. 137:14–18.

Meelu, O.P., V. Beri, K.N. Sharma, S.K. Jalota, and B.S. Sandhu. 1979. Influence of paddy and corn in different rotations on wheat yield, nutrient removal, and soil properties. Plant Soil 51:51–57.

Morris, M.L., N. Chowdhury, and C. Meisner. 1997. Wheat production in Bangladesh: Technological, economic, and policy issues. International Food Policy Research Institute, Washington, DC.

Muirhead, W.A., F.M. Melhuish, R.J.G. White, and M.L. Higgins. 1985. Comparison of several N fertilizers applied in surface irrigation systems. II. Nitrogen transformations. Fert. Res. 8:49–65.

Narang, R.S., and S.M. Virmani. 2001. Rice–wheat cropping systems of the Indo-Gangetic Plain of India. RWC Paper Series 11. Rice–Wheat Consortium for the Indo-Gangetic Plains, New Delhi, India.

Nayyar, V.K., C.L. Aroara, and P.K. Kataki. 2001. Management of soil micronutrient deficiencies in the rice–wheat cropping systems. J. Crop Prod. 4:87–132.

Ockerby, S., and S. Fukai. 2001. Rice genotypes grow differently on raised beds with continuous furrow irrigation compared with paddy. Field Crops Res. 69:215–226.

Pingali, P.L., and M. Shaha. 1999. Rice–wheat cropping systems in the Indo-Gangetic plains: Policy re-directions for sustainable resource use. p. 1–13. *In* Sustaining rice–wheat production systems: Socio-economic and policy issues. Rice–Wheat Consortium Paper Ser. 5. Rice–Wheat Consortium for the Indo-Gangetic Plains, New Delhi, India.

Quayyum, M.A., J. Timsina, M.A.H.S. Jahan, R.A. Begum, and D.J. Connor. 2002. Growth, yield and system productivity for rice–wheat–mungbean and rice–wheat–maize sequences in northern Bangladesh. Thai J. Agric. Sci. 35:51–62.

Sánchez, P.A. 1976. Properties and management of soil in the tropics. John Wiley and Sons, New York.

Sharma, P.K., and S.K. de Datta. 1986. Physical properties and processes of puddled rice soils. Adv. Soil Sci. 5:139–178.

Singh, K.B. 1998. Quantification of water balance components in rice–wheat system as influenced by soil and water management. Ph.D. diss. Dep. of Soil Science, Punjab Agricultural University, Ludhiana, India.

Singh, R.B. 2000. Environmental consequences of agricultural development: A case study from the Green Revolution state of Haryana, India. Agric. Ecosyst. Environ. 82:97–103.

Sur, H.S., S.S. Prihar, and S.K. Jalota. 1981. Effect of rice–wheat and maize–wheat rotations on water transmission and wheat root development in a sandy loam of the Punjab, India. Soil Tillage Res. 1:361–371.

Timsina, J., and D.J. Connor. 2001. Productivity and management of rice–wheat systems: Issues and challenges. Field Crops Res. 69:93–132.

Tracy, P., B.D. Sims, S.G. Hefner, and J.P. Cairns. 1993. Guidelines for producing rice using furrow irrigation. Dep. of Agron. Univ. of Missouri, Columbia, MO.

Tranggono, R., and S.T. Willatt. 1988. The growth of maize after wetland rice in East Java. p. 903–908. *In* Proc. 11th ISTRO Conference. 10–15 July 1988. Edinburgh, UK.

Utomo, W.H., B. Prastowo, T. Adisarwanto, E.V. Dacanay, G. Kirchhof, and H.B. So. 1996. Soil puddling and rice growth. p. 90–94. *In* G.S. Kirchhof and H.B. So (ed.) Management of clay soils for rainfed lowland rice-based cropping systems. ACIAR Int. Workshop. Bureau of Soil and Water Management. Quezon City, Manila. 20–24 Nov. 1995. ACIAR Proc. No. 70. Aust. Cent. for Int. Agric. Res., Canberra, Australia.

Willatt, S.T., and R. Tranggono. 1987. Puddling, its effect on soil physical properties. *In* K.J. Coughan and P.N. Truong (ed.) Effect of management practices on soil physical properties. Conference and Workshop Ser. QC87006. Queensland Department of Primary Industries, Brisbane, Australia.

Wright, G.C., M.A. Foale, and D.A. Charles-Edwards. 1985. Nitrogen nutrition of grain sorghum under sprinkler and furrow irrigation in the tropical dry season. I. Plant establishment, nitrogen uptake, and grain yield. Field Crops Res. 12:203–222.

Yadav, R.L., B.S. Dwivedi, and P.S. Pandey. 2000. Rice–wheat cropping systems: Assessment of sustainability under green manuring and chemical fertilizer inputs. Field Crops Res. 65:15–30.

Yadvinder-Singh, and Bijay-Singh. 2001. Efficient management of primary nutrients in the rice–wheat system. J. Crop Prod. 4:23–86.

Yadvinder-Singh, Bijay-Singh, O.P. Meelu, and C.S. Khind. 2000. Long-term effects of organic manuring and crop residues on the productivity and sustainability of rice–wheat cropping system in northwest India. p. 149–162. *In* I.P. Abrol et al. (ed.) Proc. Long-term Soil Fertility Experiments in Rice–Wheat Cropping Systems. RWC Paper Series 6. Rice–Wheat Consortium for the Indo-Gangetic Plains, New Delhi, India.

10 The Researcher–Farmer Interface in Rice–Wheat Systems: Moving from Agricultural Productivity to Livelihoods

John L. Gaunt and Stephanie K. White

*Rothamsted Research,
Harpenden, Herts, UK*

John R. Best

*The University of Reading
Reading, UK*

Alistair J. Sutherland

*Natural Resources Institute, Chatham,
Kent, UK*

Patricia Norrish

*Independent Consultant, Development Communication
Bucks, UK*

Elizabeth J.Z. Robinson

*St. Hugh's College
Oxford, UK*

Zahid Hossain

*Proshika, A Centre for Human Development
Dhaka, Bangladesh*

Richard J. Palmer-Jones

*University of East Anglia
Norwich, UK*

ABSTRACT

This chapter explores some of the implications of adopting a research approach that focuses on people and their livelihoods in the rice–wheat system of the Indo-Gangetic Plains. We draw on information from a study undertaken by the authors in Bangladesh and

---

then consider the transferability of our findings to other situations. We conclude that if our research is to bridge the researcher–farmer interface, ongoing technical research must be supported by research that explores how institutional, policy, and communication strategies determine livelihood outcomes. The challenge that now faces researchers is to move beyond their involvement in participatory research to understand how to facilitate a process in which they provide information and products for others to test. Building capacity at various levels for openness in sharing information and products—seeing research as a public good for all—seems to be a prerequisite for more effective dissemination of the available information and technologies.

South Asia accounts for 22% of the global population, or 42% if China is considered. A significant proportion of these people (59% of the population of India, Pakistan, Bangladesh, and Nepal combined) depend on agriculture for their livelihoods. The rice–wheat system is an important cropping system covering 32% of total rice area and 42% of the total wheat area, with a total area of 13.5 million ha (an additional 10.5 million ha in China) (Ladha et al., 2000). Although rice–wheat cultivation has been practiced for thousands of years, it emerged as a dominant cropping system in the mid 1960s, largely as a consequence of the Green Revolution.

Green Revolution technologies were central to national strategies intended to achieve food security, equality of health provision, rural development, conservation of natural resources, and alleviation of poverty. The Green Revolution focused on securing increased production through the use of improved high yielding-varieties (HYVs) of crops and subsidies on power and fertilizer inputs. Simultaneously, the expansion of surface irrigation systems and private investments in tubewells led to an increase in the area under irrigation on which rice and wheat could be grown. Typically, the market for the agricultural produce was assured through price support with an emphasis on achieving national food security through commodity-based production targets.

The benefits of the Green Revolution technologies are clear. Lipton and Longhurst (1989) highlighted that beyond the obvious achievements in area grown with HYVs and increased levels of production which allowed the survival of many millions of people, the characteristics of modern varieties seemed favorable to the poor. They argued that compared with varieties previously grown, modern varieties were less labor-intensive, reduced risk, and resulted in more affordable food staples.

However, there is a complex relationship among investment in agriculture, increases in production and productivity, and levels of rural poverty. Thirtle et al. (2001) show that an increase in agricultural output leads to a reduction in poverty. The analysis by Thirtle et al. (2001) of data from Asia suggests that a 1% increase in total factor productivity (TFP) lessens the proportion of the population living on less than $1 per day by 1.3%. It has further been argued by Thirtle et al. (2001) that, in the long term, agricultural supply response is not possible at the sector level without growth in productive efficiency as well as output. Productivity can be measured in several ways, including output per unit of land or labor. Total factor productivity provides an aggregate measure of output, with appropriate weighting, per unit inputs (again aggregated and weighted).

Datt and Ravallion (1998) pointed out that the rate of rural poverty reduction at a given rate of agricultural growth is influenced by the initial conditions of human resource development and rural infrastructure. It is clear that investment in infrastructure, particularly roads and electricity, was central to the development of agriculture in the Indian Punjab. Fan et al. (1999) estimated the impact on agricultural TFP and poverty reduction of different investments in rural India. They found that improvements in TFP and reduction in poverty, driven by investment in roads, agricultural research and development, and extension, were typically manifested through wage increases and lower food prices.

Thirtle et al. (2001) further disaggregated the factors that contribute to TFP. They distinguished technical progress from changing efficiency. This analysis suggested that only efficiency change led to a significant effect on poverty. While it is counterintuitive that research and development provide new technologies without a significant effect on poverty, the authors pointed out that in the past the technological frontier was defined by more advanced countries such as Malaysia and Thailand. The significant elasticity observed in efficiency may suggest that technology transfer rather than development alone results in an efficiency change. Their limited data suggest that a 1% gain in efficiency could lead to a 2% reduction in poverty.

It can be argued that national strategies to implement Green Revolution technologies reflected a general urban bias in government policies, focusing on providing a reliable food supply to the urban population at reasonable cost. These observations apply even in countries where the majority of the population is engaged in agriculture (such as Bangladesh), and are reflected in the lack of relationship between the price of rice and wheat and the productivity of the rice–wheat system (Ladha et al., 2000). Investment in agriculture and associated public-sector research and development has also tended to favor those areas considered to be of "high potential" productivity.

Some authors have expressed concern that increases in production and a reduction in poverty may have involved a trade-off against long-term sustainability of production (Asaduzzaman, 1995; Brandon, 1995; Pagiola, 1995). The existing evidence on productive sustainability in Asia is somewhat conflicting. Long-term rice–wheat experiments examined by Duxbury et al. (2000) showed that rice yields were declining in eight out of 11 experiments and wheat yields in three. However, Dawe et al. (2000) analyzed long-term yield trends in 47 long-term experiments in rice–rice and rice–wheat systems and argued that yield decline was not very common, particularly at yield levels achieved by farmers.

These apparent differences are worrying, but to assess the performance of the system as a whole, it is necessary to look beyond yield alone. As described above, TFP provides a measure of output relative to inputs. When such measures are used, some studies have suggested that TFP is declining, but analysis of district-level data from the Indian Punjab suggests that early fears may have been exaggerated (Murgai, 1999, as cited in Ladha et al., 2000). These findings appear to confirm the findings of Dawe et al. (2000).

However, such analyses represent a narrow view of sustainability. Work in the Middle East and Africa has documented the relationship between natural capital and economic development. Earle (2001) cited examples to suggest that countries pass through a series of phases in the management of their natural resources.

As an economy develops, it starts to deplete its base of natural resources until they become degraded; however, by this stage, the economy should be strong enough to diversify to limit damaging activities and pursue income from more sustainable sources before moving to a phase of natural resource reconstruction. It seems reasonable to hypothesize that those states that received higher levels of early investment are likely to be demonstrating signs of environmental degradation. Ali and Byerlee (2000) found substantial deterioration of soil and water quality in the Pakistan Punjab, and similar concerns arise in the Indian Punjab. Such deterioration will only be captured by a measure such as TFP when it results in an increase in the direct costs associated with agricultural production.

Such survey-based analyses concentrating on production, income, or consumption provide a limited view of poverty. For instance, a survey may provide data that enable description of the access of individuals to the resources necessary to meet their material needs. It does not capture the entitlements and capabilities that determine the nature of an individual's life or livelihood (Sen, 1981, 1985; Drèze and Sen, 1989). For example, in a study of two villages in Rajasthan, assessing the change in the situation of 36 households from 1964–1966 to 1982–1984, the households were on average better off according to their own criteria, despite being 5% worse off in terms of per capita income. The criteria used by the households included indicators of social and economic independence such as not having to migrate for work and not having to skip a third meal (Jodha, 1998).

In the following sections, we explore the implications of adopting a research approach that focuses on people and their livelihoods, drawing on a study undertaken by the authors in Bangladesh, and then consider the transferability of our findings to other locations within the rice–wheat system of the Indo-Gangetic Plains.

THE IMPLICATIONS OF ADOPTING A LIVELIHOODS APPROACH

What Is a Livelihood?

In the introduction, we noted that analyses based on income or consumption provide a limited picture of poverty. In recognizing this limitation, the UK Government Department for International Development (DFID) has proposed the following definition of livelihood:

> A livelihood comprises the capabilities, assets, and activities required for a means of living. A livelihood is sustainable when it can cope with and recover from stresses and shocks and maintain or enhance its capabilities and assets both now and in the future, while not undermining the natural resources base.
>
> (Ashley & Carney, 1999)

The livelihoods approach draws on entitlements and capabilities assessment approaches of poverty and well-being (referred to above), the extended entitlements and vulnerability analysis (Swift, 1989, 1993; Davis, 1993), gender analysis of livelihoods (Grown & Sebstadt, 1989), and poverty–environment interactions (Reardon & Vosti, 1995). Many publications focus on livelihoods approaches coupled to agricultural and rural development problems (Carney, 1998). These publications generally focus on the endowment and asset status of the poor, the productivity of their assets, the damage inflicted (directly or indirectly) on these assets by technical or institutional innovation, and the access of the poor to them. This approach is intended to draw attention to the key opportunities and vulnerabilities of the poor, and contrasts with approaches that cast them as destitute, assetless victims.

Five types of assets (*capital*) can be identified—natural, physical, financial, human, and social (Fig. 10–1)—that people can draw on selectively to pursue desired livelihood outcomes. Elaboration of these capitals other than financial capital is in order, since these are not widely known terms. Natural capital comprises land, water, and the atmospheric environment and the biophysical resources that depend on them (e.g., forests, fisheries) or are embodied within them (e.g., minerals, soil fertility). Natural capital can be directly or indirectly augmented as well as used or degraded by human activity and nonanthropogenic trends. Physical capital represents assets such as livestock, equipment, and so on. Human capital reflects the health, skills, knowledge, and education of the population. Social capital broadly corresponds to the cultural resources in society that facilitate collective action. The relationship between activities that draw on these capitals and desired outcomes will be moderated, or indeed determined, by policies, institutions, and processes. The structures and processes involved may be formal or otherwise.

These definitions are subject to debate. Baumann (2000), for example, argued that a sixth capital—political capital—exists. This would capture the influence of politics and power relations that are not encompassed in the structures and processes. Information or knowledge is generally considered as a facet of human capital. However, Garforth et al. (2001) argued that information, knowledge, and communication issues permeate most of the elements of the framework. Communication can be considered as the set of processes by which information is sought, accessed, exchanged, and strengthened; knowledge is developed and modified; social capital is maintained and strengthened; and claims on assets are transacted.

Livelihoods analysis takes a holistic approach that encompasses nonmarket as well as market activities, especially those conventionally viewed as household or domestic work. These activities are often neglected or marginalized in conventional approaches. This is achieved in two ways: (i) by viewing the household as an integrated system framework or (ii) by focusing on the individual within the household context. Both approaches draw attention to the contributions and needs of women, children, and the old rather than assuming that their needs and contributions are taken care of by fair intrahousehold processes. Livelihoods analysis emphasizes nonmarket activities, involving the use of common and open-access resources and the use of domestic labor. The list of work done by women, children, and the old is substantial in total but tends to get neglected in a focus on conventional market-oriented economic activity.

Livelihoods analysis has been applied more frequently to semiarid and nonirrigated environments where, especially for the poor, livelihoods tend to be diverse, drawing on many resources and activities. These areas have a relatively short agricultural season, followed by a longer season during which agriculture cannot provide much employment or activity. Of necessity, people in these areas move into other sectors, at least for the nonagricultural season, to gain their livelihood. Some

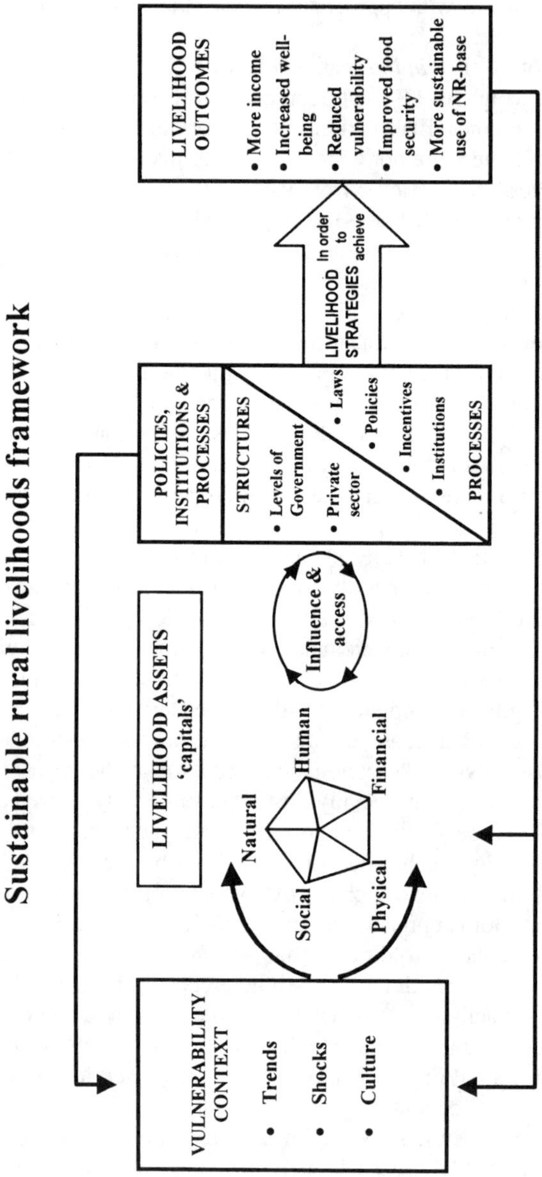

Fig. 10–1. Sustainable rural livelihoods framework, modified from Ashley and Carney (1999).

researchers perceive that in intensively irrigated areas, agriculture provides year-round employment and that the diversity in livelihoods may be less. Our experience and even casual observation in the field seem to suggest that this presumption is not valid, at least for Bangladesh and areas of eastern India.

This consideration signals the type of factors that need to be analyzed in agricultural research if we are to be clear on how our research and the technologies we seek to deliver affect the livelihoods of our intended beneficiaries.

Agriculture from the Farmers' Perspective

In our work in Bangladesh, we explored farmer perceptions of their agricultural enterprises. This was not a detailed livelihoods analysis, rather an elaboration by farmers of issues related to a commonly raised problem, the low income from rice production. Consultations were held separately with groups of men and women and with groups commanding different levels of resources (based on participants' own definition of whether they are small or medium farmers, sharecroppers, or landless workers). We used two techniques: (i) scored causal diagrams and (ii) participatory farm budgets (Galpin et al., 2000) with small groups of farmers at three locations (Rangpur, Comilla, and Kustia). The scored causal diagram is based on the flow chart, with the important modification of a scoring procedure. It helps to examine in detail the causes and effects of problems and to identify the root causes that need to be examined. The scoring enables the *relative* importance of particular problems and their causes to be analyzed. The participatory budget is used to identify resource-use patterns (e.g., labor and draft power peaks), cash flows, and critical seasonal points (e.g., water availability, pests, diseases).

As mentioned above, the exercise focused on rice cultivation. This was because rice is the predominant commodity crop grown in these areas. Sites were chosen to be representative of the major systems within Bangladesh. Construction of scored causal diagrams began with the identification of an end problem. In all our cases, this was agreed (by the participants) to be "low income from rice production". The facilitator played a role in encouraging acceptance of this as the starting point so as to achieve some comparability among the five separate groups. An example of the scored causal diagrams is presented (Fig. 10–2), and the responses of the five groups of farmers in our study are summarized in Table 10–1. We found a keen awareness among farmers of the nonsustainability of parts of their farming systems and of the need for better management of soil fertility and pests. The interactions and integrated natures of their livelihoods were clear. Each group identified the same three "proximate" causes of low income: (i) the low ("unfair") price that they received for rice sold soon after harvest, (ii) the high cost of inputs, and (iii) poor yield. In two cases (of poor and medium income), the demand for rice made by the household (leaving relatively little available for sale to raise cash income) was cited as an additional cause.

The farmers traced the cause of low price and high input costs to problems with marketing and supply systems, and to inefficiencies or corruption. Lack of storage capacity, rather than need for money, was mentioned as the reason for rice being sold soon after harvest at prices determined by the buyers. A strength of the scored causal diagram is that it provides a visual tool for problem analysis at the farm level.

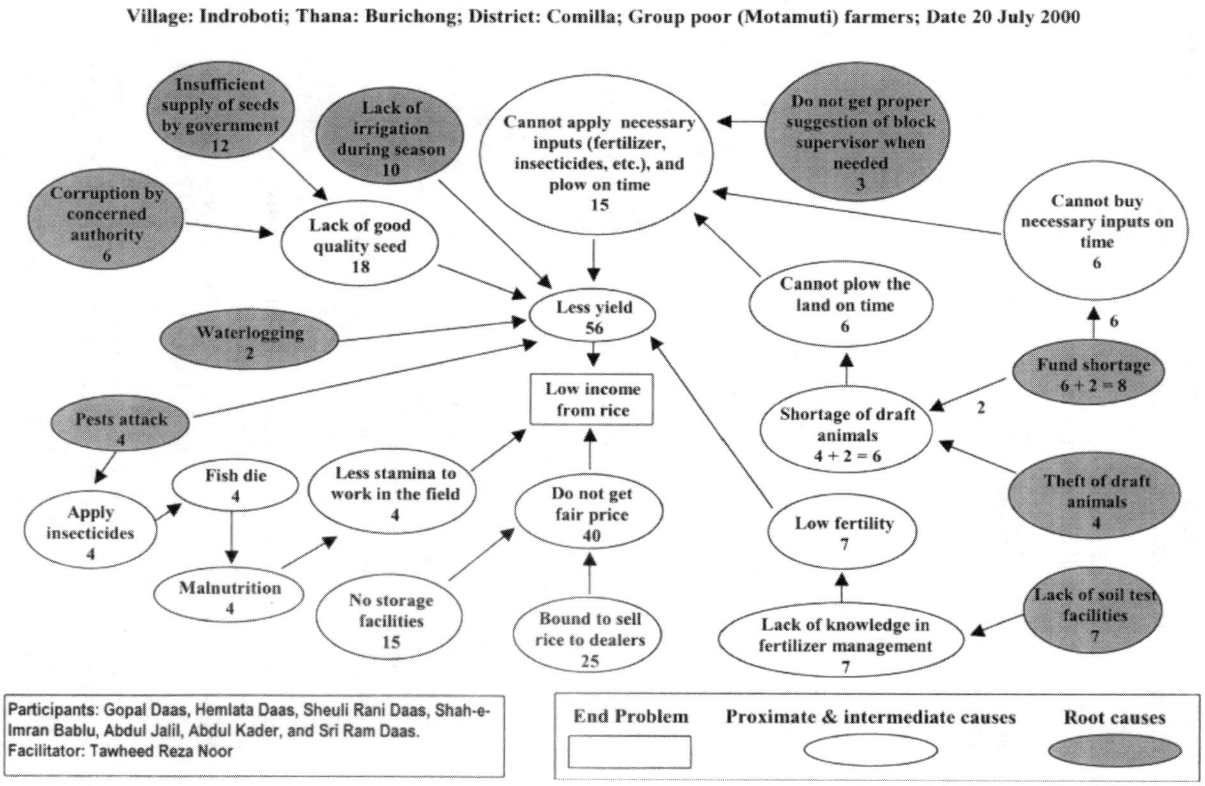

Fig. 10–2. Example of a scored causal diagram.

Table 10–1. Summary of proximate, intermediate, and root causes of low income in rice production identified by five groups of medium- and small-scale farmers in Khustia, Rangpur, and Comilla districts, Bangladesh.

Proximate causes	Intermediate causes				Root causes	
Low price of rice •••••†	Bound to sell to agents •••••				Lack of storage •••	Corruption ••
High cost of inputs (incl. water) ••••	Poor supply •				Corruption •••	Low water table •
Less rice available for sale ••					Household consumption needs ••	Crop damage •
Poor yield •••••	Low fertility •••••	Poor knowledge of fertilizer use •••••	Excessive use of chemical fertilizer ••••		Lack of extension guidance •••	
		Lack of organic fertilizer •••	Few cattle—also linked to nutrition •••	Lack of pasture ••	Demand for crop land ••	High crop production needed for livelihood •••
	Poor-quality seed •••				Lack of effective supply by govt. •••	Corruption •
	Cannot buy inputs on time •••				Scarce funds •••	Lack of timely and accessible credit ••
	Waterlogging •					
	Lack of new high-yielding varieties •					
	Shortage of irrigation water ••••		Scarcity of diesel and power •••		Corruption ••	High price of equipment •
	Poor health/work efficiency ••••	Shortage of protein ••••	Fewer fish ••••	High fertilizer and pesticide use ••••	Pest attack •••	
	Plastic garbage pollution •				Increased use of plastic packaging •	
	Cannot plow land timely ••	Lack of draft power ••	Few cattle (see above)		Theft of cattle ••	

† Number of dots in each cell record number of observations (max. = 5).

As shown in this example, its use enabled problems relating to such things as marketing infrastructure to be separated from agronomic and extension- or information-related problems.

A plethora of intermediate and root causes was identified for low yield at the farm level. One of the causes, low soil fertility, was traced to both the excessive use of inorganic fertilizer and the lack of organic fertilizer. The latter was identified as resulting from a relative decline in cattle numbers, with an increasing proportion of land used for cropping. Interestingly, all groups identified their lack of knowledge as a reason for their failure to use inorganic fertilizer efficiently.

All groups identified a "loop" that begins with poor health affecting work efficiency, which in turn leads to a poorer quality diet as fish become less abundant in rice fields and water bodies because of high levels of pesticide and fertilizer use. Other causes of poor rice yield lead back to the root causes in marketing and supply logistics: poor quality of seed, lack of fuel to power irrigation pumps, and the lack of credit that would enable the timely purchase of inputs. From this, it can be seen that a discussion that started with a question directly related to rice production led to consideration of related livelihood problems as root causes.

DELIVERY OF INFORMATION AND AGRICULTURAL SUPPORT

In light of the observation above that all farmers identified their lack of knowledge as a reason for their failure to use inorganic fertilizer efficiently, we looked further at the delivery of information and access to agricultural support. Throughout the Indo-Gangetic Plains since the mid 1970s, the underlying model for extension was based on the "training and visit" (T&V) system. This system was highly structured and restricted public extension services to basic information transfer, advisory, and training functions (Christoplos et al., 2001). In Bangladesh, the T&V system was readily incorporated into the government bureaucracy. Visits by field extension staff were made to deliver centrally determined messages to farmers. While amenable to management and supervision, the T&V system militated against local planning and responsiveness to clients' needs. As a consequence, T&V failed to engage with the existing patterns of communication and information exchange that are integral to interactions among members of the rural community (Ramirez, 2000).

In considering the prospects for public-sector agricultural extension services in Bangladesh, Garforth (1995) argued that several themes should guide the design of support services for farmers. These included a more clearly delineated focus on farmers' information needs, stronger extension–research linkages, the application of systems thinking to agricultural extension and technology development, the search for increased efficiency and cost-effectiveness, cost recovery, decentralization of extension planning and management, and collaboration among government, nongovernment, and commercial-sector service providers.

Considering the role of extension, Christoplos et al. (2001) drew attention to the considerable transaction costs faced by the poor in accessing the means of production. Thus, they went on to argue that the core objective of extension is to help people reduce transaction costs in productive activities through the supply of information and enhancement of people's capacity to process information and act on it.

As part of our work in Bangladesh, we brought together 22 representatives from various organizations delivering information and agricultural support in Bangladesh, with the objective of sharing ideas, among other things, on the nature of the interactions among farmers, researchers, and intermediaries. From their discussion on the role and specific tasks of intermediaries in relation to agriculture, it was very clear that most follow a farming systems approach. This suggests that this community recognizes the need for systems thinking in agricultural extension and technology development, an element of an extension strategy for Bangladesh proposed by Garforth (1995). However, the group agreed that its strategies were, in general, to target groups of farmers with sufficient homogeneity defined either on the basis of socioeconomic or environmental criteria to constitute a recommendation domain. The means by which recommendation domains are defined, and extension activities prioritized, varied.

Despite such efforts, there is a widely articulated belief among intermediaries that technologies developed by the research community are not reaching farmers. This seems to confirm the findings of Thirtle et al. (2001) that technology transfer rather than the technology frontier limits the impact of research on poverty. It also seems to suggest a dismissal or lack of awareness of the capacity for innovation within the communities within which these organizations work.

Recognizing that technologies being produced by researchers were not reaching farmers, the United Nationas Development Program supported the Agriculture Technology Transfer Program (ATTP) coordinated by PROSHIKA (a nongovernmental organization [NGO] in Bangladesh). This program provides resources through projects that enable intermediary organizations to validate and promote technologies developed within the research arena. The mechanism used to promote technologies varies from project to project. However, the ATTP project was underpinned by a compilation of technologies from the national research institutes that were judged by scientists to be appropriate for dissemination. Thus, although an important effort, this program did not seek to capture technologies developed elsewhere, which Thirtle et al. (2001) argued may be important to achieve advances.

It is also important at this point to consider that seeking to establish recommendation domains as discussed above compromises a basic principle of the livelihoods approach—that one should start from the place in which the individual is located. We noted earlier that existing patterns of communication and information exchange are integral to interactions among members of the rural community. It is through these interactions that people seek to negotiate and create opportunities to fulfill their needs and pursue their own interests (Ramirez, 2000). In terms of agricultural development, the various stakeholders can be collectively perceived as a social organization (Engel, 1994, as cited in Ramirez, 2000). It has also been argued that extension must recognize that knowledge and information provide a common denominator among farmers, extension workers, and researchers (Christoplos & Nitsch, 1993).

AN INSTITUTIONAL PERSPECTIVE

We introduced the importance of policies, institutions, and processes as one of the elements of sustainable livelihoods analysis. Sen (1999) considered freedom of choice to be a basic criterion for development. If we accept such a position, then the ability to participate in the market through ready access to information, freedom of movement, and access to the sociopolitical and institutional infrastructure that underpins the market become important, rather than solely considering improved production or productivity alone.

It is clear that the rural poor do not enjoy equitable access to physical inputs and financial services. Christoplos et al. (2001) suggested that extension must be firmly nested within a wider service context that includes credit input, supply, processing, and marketing. Matin et al. (1999) (as cited in Christoplos et al., 2001) described the evolution of rural finance:

> Previously we saw the poor as small or marginal (male) farmers needing subsidised agricultural credit. As the 'micro-finance' revolution spread, our image changed and we came to see the poor as (largely female) micro-entrepreneurs with no collateral to pledge but with a business world to conquer with 'micro-credit'. Now a new understanding is emerging: the poor are a diverse group of vulnerable households with complex livelihoods requiring a full set of micro-financial services.

Large and influential NGOs in Bangladesh have emerged in the areas of microfinance, the Grameen Bank being one of the more widely known. Indeed, the local NGOs in Bangladesh are particularly strong. They constitute large national organizations, with infrastructures and resources that match those of government ministries. This situation has arisen, at least in part, owing to the substantial support to national NGOs that has been made available by bilateral and multilateral donors who see these organizations as a more cost-effective mechanism for delivery of services to the rural poor (Garforth, 1995).

However, within Bangladesh, as elsewhere, some problems appear to mitigate against providing support to vulnerable households with complex livelihoods. Institutional stakeholders in agriculture in Bangladesh are diverse government ministries, national government research institutes, local universities, local and international NGOs, and international research organizations. Moreover, partnerships are often supported by bilateral or multilateral development aid. There appears to be a lack of integration, particularly among the priorities of the agricultural, fisheries, and water management sectors. Within each sector, the agenda of those donors and international research organizations supporting agricultural research and extension appears to contribute to the fragmentation of efforts.

The strength of these NGOs has led to tension between the national government research institutes and NGOs. During 1997, this tension was particularly apparent, both in the open discussions held at various conferences and workshops and in discussions with individual representatives of these organizations. In the subsequent four years, the situation has changed: it appears that a mutual respect has developed. We have not documented these changes in detail, but they are reflected in the inclusion of NGOs in government research planning. Increasingly, research projects funded by donors have supported partnerships between NGOs and scientists in national agricultural research and extension systems (NARES). Through this process, scientists have been encouraged to understand and analyze the needs of farmers.

We worked with PROSHIKA, a large national NGO in Bangladesh, to examine the implications, for its organization, of developing a focus that included diversification of the cropping base (including niche cash crops and linked processing and marketing opportunities) in response to client demand, along with continued attention to the sustainable production of staples. Sustainable production of staple crops demanded attention to expanding options in terms of variety choice for rice and wheat, soil and organic matter management and fertilizer or nutrient use efficiency, integrated pest and disease management, and a wider range of crop combinations, relays, and rotations. We called this *integrated crop management* (ICM). Concerns were raised within the organization (Fig. 10–3) that ICM, with its client focus, implies a more complex system for delivery of technologies to farmers. Indeed, it was felt that ICM would require reorganization to develop a structure with appropriate disciplinary and institutional integration.

We can see that the challenge we face is to identify research opportunities that relate to improving the provision of rural services (such as developing methods to reduce the costs associated with delivery of micro-credit), distinguishing them from issues that are best analyzed by others. The example of the implications for ICM is an interesting case—do the findings indicate a research opportunity? Or merely organizational issues that would need to be analyzed if the organization were to get into ICM? It is not possible to answer these questions at this stage, but what is clear is that in moving forward to develop ICM, analyzing organizational issues would be as important as answering technical questions.

ARE THESE FINDINGS TRANSFERABLE?

This topic is often discussed under the heading of "scaling." However, we are uncomfortable with this term. It seems rooted in geography and—since it is perceived and defined in many different ways—subject to misinterpretation. We also need to emphasize again that analysis of the local situation should be the starting point in implementing a livelihoods approach. Reductionism, where we assume similarities on the basis of biophysical or socioeconomic parameters and seek to apply models, compromises the basic principle of the livelihoods approach that one should start from the point where an individual is located.

Ashley and Carney (1999) recognized that while community-level institutions and processes are emphasized in the sustainable livelihood framework, a core aim of the livelihoods approach is to understand and facilitate the link from micro to macro scales. As we have discussed above, livelihood outcomes are determined by the ability of an individual to influence and access the transforming structures and processes.

TRANSFER OF RESEARCH RESULTS FROM ONE LOCATION TO ANOTHER

To overcome the problem in undertaking farming systems research, which was location-specific, the concept of a recommendation domain was developed to iden-

PROSHIKA perceives the Agriculture Technology Transfer Project (ATTP) as an intermediary step toward integrated crop management (ICM) in which integration has occurred within PROSHIKA at an organizational level but not in terms of the messages that are given to farmers. ICM was seen by PROSHIKA staff as requiring integration of the delivery mechanism. Strength, weaknesses, opportunities, and threats (SWOT) analyses were undertaken by PROSHIKA staff to examine each step toward their vision of ICM.

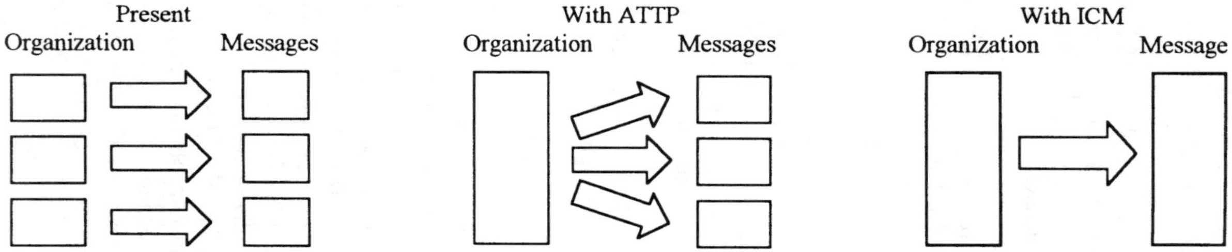

Present

STRENGTHS Integrated program; Skilled, multidisciplinary workers; Easy communication; Less time consuming; Ecological agriculture; Strong skill development, training cell; Other related program exists; Area Development Center (ADC) infrastructure	**WEAKNESSES** Feedback—lack of beneficiaries, need-based know-how; Program overlap; Program staff occupied with other activities; Management burden—hard to implement any program properly
OPPORTUNITIES Resource person developed from group member; Present where infrastructure is developed; Grass-roots organization with group federation is strong; Popular theatre	**THREATS** Activities (some) not oriented; Single program emphasis; Existing program threatened

ATTP

STRENGTHS Skilled and committed person power from different related disciplines; Structural set-up; Organized farmers' groups; Network with partner NGO's for technology dissemination; Professional linkage with different agricultural research agencies (domestic and international); Available material support; Good integration with different programs within PROSHIKA; Most components are being practiced by ATTP	**WEAKNESSES** Shortage of person power; Shortage of skilled person power at the implementation level of partner NGOs; Lack of sufficient information regarding ICM; Need of logistic support such as vehicles; Lack of full-time trusted staff at field level
OPPORTUNITIES Eagerness of farmers throughout Bangladesh; Available NGOs to implement this project; Available sources of necessary information; Appropriate technologies; Sufficient national and international support; Scope of required funds; Favorable government policies	**THREATS** Farmers' acceptance yet to be measured; Natural hazards and calamities; Political unrest; Change in government policies; Dependency on donors for funds

ICM

STRENGTHS Area coverage of PROSHIKA; Technical and trained staff; Existing physical infrastructure; Existing different programs	**WEAKNESSES** Lack policy; Grass-roots management systems; Lack support material
OPPORTUNITIES Holistic development; Coordinating farmers and groups through activities; Existence of some policy	**THREATS** Government and donor policies; Market policy; Natural calamities

Fig. 10–3. An analysis by PROSHIKA, a large national NGO in Bangladesh, of the organizational changes required in adopting an integrate (crop management) approach to the delivery of services.

tify others for whom findings may be relevant. The concept was developed during the 1970s (Jolly, 1988). The recommendation domain concept was also amenable to a centralized dissemination system (whether disseminating knowledge only or knowledge plus inputs, credit, and marketing support). Recommendation domains have been used as a way of targeting research and or dissemination strategies based on the extrapolation of findings from elsewhere.

Agroecological zone (AEZ) models can be applied to analyze potentials of land for various kinds of use. The AEZ classification attempts to capture the biophysical limitations on agricultural possibilities. Thus, AEZs are relatively homogeneous geographical areas that have characteristics that suit them for the design and use of similar agricultural technologies. The AEZs include models for the calculation of length of growing period, irrigation requirements, crop biomass, land suitability, and land productivity (FAO, 2002). These AEZs may not be completely "natural" in that human modification of the environment (clearance, cultivation, terracing) before the development of agricultural technologies and investments in infrastructure, especially irrigation and drainage, have important effects on potential cropping patterns and their productivity. While not following the AEZs defined above, environmental characteristics have informed prioritization of research objectives within the rice–wheat system (Gupta, 2002; Ladha et al., 2000).

Harrington et al. (2000) point out that, while rainfall, water control, and soil texture follow an east–west gradient, as described by the AEZs, at a specific location this general trend is overlain by variation that is determined by the position (i.e., lower, middle, and upper terraces) within a toposequence. Harrington et al. (1993, as cited in Harrington et al, 2000) further argue that while local terminology may differ, the influence of land type on management is often similar. These relationships are generally well understood, for example, the Nirdeshika program in Bangladesh uses environmental characteristics (land type, flood depth, and soil type) to determine the crops that can be grown.

It is also important to consider that while we seek to provide an optimum solution or technology for a particular situation, it may be that what affects the transfer of a technology is the range of critical values within which a technology will perform satisfactorily (Reece & Sumberg, 2003). Reece and Sumberg (2003) used the term *solution space* to denote this range.

We highlighted earlier the apparent conflict between the concept of recommendation domains and the principle that analysis of the local situation should be the starting point in implementing a livelihoods approach. We also identified that it is important to understand how our research findings can feed into and support existing patterns of communication and information flow.

A key question that remains in the context of livelihoods research is, to what extent should scientists continue to seek to actively filter information and knowledge, or specify a technology, based on biophysical or farming system parameters (e.g., the household resource base, economic situation, gender, access to markets, food preferences)? Should we rather focus our efforts on ways to facilitate access to the information, providing tools that allow those accessing it to judge its relevance to their situation?

MICRO–MACRO TRANSFER

In this section, we are not seeking to discuss upscaling of results in situations where findings have not been derived at the appropriate scale. For example, findings on improved water use efficiency derived from a research plot may not lead to improved water use at the level of the whole irrigation system; likewise, experiences obtained working with small numbers of groups or individual farmers may not be relevant when wishing to work with larger numbers of groups or individuals. We are dealing here with linking from the situation of an individual to macrolevel policies, institutions, and processes as discussed above in the context of livelihoods.

It must be recognized that there simply are not enough researchers, or research funds, to achieve a development impact across large areas. Further, it is important to recognize that participatory methods that work well in a research context may not be transferable. For example, Quizon et al. (2001) highlighted the fiscal unsustainability of agricultural extension programs based on the very strongly established farmer field school approaches. It is also not appropriate for research to bypass those agencies involved in the delivery of rural services (NGOs, government extension, and the private sector) to demonstrate "development impact". It is these agencies that will facilitate the link from micro to macro scales, which Ashley and Carney (1999) argue is needed, and deliver a sustainable impact. However, an important prerequisite is that these agencies be committed to providing support and rural services to meet the needs of the rural poor. Where this commitment exists, the challenge is to provide support that enables these agencies to undertake problem-oriented adaptive research, backed up by the formal research sector.

This moves us beyond various models of "farmer participatory research," which have been in vogue since the 1980s. We are arguing that first we must ensure that methodologies developed for working with the poor must be appropriate to the level of operation of such agencies. The example of ICM discussed above clearly showed the perceived implications of a research concept from the perspective of an organization involved in service delivery. Such implications need to be recognized at the onset of research. Indeed, working out ways to overcome constraints that emerge may become a research topic in its own right.

We suggest rather that a form of capacity building is required that will enable others to undertake their own research, drawing on scientists only when needed. Such approaches will challenge existing functional boundaries between organizations and will require an open sharing of skills and knowledge. The centers of the Consultative Group on International Agricultural Research (CGIAR) may have a key role to play, perhaps as facilitators who freely give out information and products for others to test. Building a capacity for openness in sharing information and products at various levels, seeing research as a public good for all, seems to be a prerequisite for more effective dissemination and access to available information and technologies. This is different from a commercial model, whereby new products are sold through market channels. There is a need to differentiate clearly between the types of information and products that farmers are prepared to pay for and the private sector is able to supply on the one hand and those that require sustained public-sector input on the other.

DEALING WITH EXTERNALITIES

In considering the problems of sustainability, we recognize that while immediate measures such as TFP may be stable or increasing, they may mask a situation that is unsustainable. It is possible to capture this lack of sustainability in economic terms by considering the change in total economic value to capture externalities, such as climate change, degradation of water resources, and so on. Total economic value takes into account not only changes in direct use values (crop yields and revenues) but also indirect use values (environmental services, including C sequestration and improved water quality) and nonuse values (which might include greater access, increased biodiversity, or a premium for organic farming, for example). If account were taken of indirect use values and nonuse values, the resulting implications could lead to significant shifts in the form of agriculture practiced in areas considered as the rice–wheat systems.

We used the example of water quality earlier and refer readers to a review by Buresh et al. (2001) for a consideration of other relevant environmental problems. Here we consider the interesting example of climate change. The Kyoto protocol represents an international agreement on targets for reduced CO_2 emissions, which contribute to climate change. The protocol and alternatives proposed by the United States allow for both reductions in emissions and C removed from the atmosphere in soils or biomass to be offset against emissions. A consequence of these agreements is that markets for trading C emissions are being developed. The market innovations and specific policy requirements of this market, particularly the view taken on C sequestration in agricultural soils are not yet clear. Exploring how individuals might gain access to this market is a clear example of linking the micro situation of individuals to macro policy issues.

We have outlined that for an individual's livelihood to be sustainable, it is necessary to take account of externalities, which have previously been ignored, and to explore mechanisms that provide appropriate incentives for individuals to deliver services that will ensure sustainability.

CONCLUSIONS

Political moves toward decentralization and market diversification are likely to be the immediate drivers for agriculture across the Indo-Gangetic Plains. Recognizing that the current economic situation differs widely across the Indo-Gangetic Plains, we have sought to identify the underlying problems that are of importance at the interface between research and the farmer.

We have argued the need to move beyond a focus on productivity gains as an outcome of research, to look for and address beneficial livelihood outcomes for the rural and urban poor. These outcomes must be sustainable.

We have elaborated that livelihoods comprise the capabilities, assets, and activities required for a means of living. Further, we have considered the need to address access to both information and political capital, assets that capture the considerable opportunity that exists to inform livelihood strategies and improve livelihood outcomes. This encompasses improvements to the communication of information, and the influence of policies and power relations in a way that benefits poor farmers, particularly those who have had few benefits to date from the technologies promoted by the Green Revolution. However, we argue strongly that, if our research is to bridge the researcher–farmer interface, the ongoing technical research must now be supported by research to explore institutional, policy, and communication dimensions that are so crucial to achieving developmental impact. These include

1. Better use of information by the various agents communicating with farmers to aid implementation of integrated agricultural strategies.
2. Opportunities for improving the communication of information, concerning farmers needs, between farmers' and those responsible for responding to their needs.

We recognize that policy support through targeted subsidies and favorable credit packages is likely to remain very important to maintain low food prices and lower producer risks and sustain investment in farm productivity. However, the best model cannot be presumed to be the current one. Indeed, existing policies may maintain differences in development within the areas covered by the rice–wheat system.

Researchers will not be everywhere. Participatory methodologies that tend to assume the researcher as a required participant in development at all locations are not realistic. The challenge that now faces researchers is to move beyond their involvement in participatory research to understand how to facilitate a process in which they provide information and products for others to test. Building capacity at various levels for openness in sharing information and products, seeing research as a public good for all, seems to be a prerequisite for more effective dissemination of the available information and technologies. This will require appropriate information communication technology (ICT) to support its access.

There is a need to resolve the direction implied above with existing commercial models for dissemination, and particularly to differentiate clearly between the types of information and products that farmers are prepared to pay for, and the private sector is able to supply, and those that require a sustained public-sector input.

ACKNOWLEDGMENTS

This chapter is an output of projects R7600, R7583, and R7839 funded by the UK Department for International Development (DFID) for the benefit of developing countries. The views expressed are not necessarily those of DFID. Many people contributed to the projects. It is not possible to name all those people here. We thank you for your enthusiasm and willingness to contribute to our work. In the case of R7600, Dr. Noel Magor and Mr. Ahmad Salahuddin (IRRI-PETRRA) provided assistance and logistical support and Tawheed Rezah Noor (PRA Promoters Society) undertook facilitation of farmer validation activities. Projects R7583 and R7839 are undertaken in collaboration with the Indian Council for Agricultural Research Research Complex for the Eastern Region and Cirrus Management Services, under the leadership of Dr. A.K. Sikka and Mr. M.S. Ashok, respectively. In preparing this chapter, we have made inferences based on information gathered during our studies. Inevitably, there will be things we have missed and differences of interpretation. Thus, we present this work to stimulate discussion and we invite comments and feedback.

REFERENCES

Ali, M., and D. Byerlee. 2000. Productivity growth and resource degradation in Pakistan's Punjab. Policy Research Working Paper 2480. World Bank, Washington, DC.

Asaduzzaman, M. 1995. Resource degradation and sustainable development in Bangladesh. Bangladesh Inst. of Development Studies, Dhaka, Bangladesh.

Ashley, C., and D. Carney. 1999. Sustainable livelihoods: Lessons from early experience. DFID, London.

Baumann, P. 2000. Sustainable livelihoods and political capital: Arguments and evidence from decentralization and natural resource management in India. ODI Working Paper 136. Overseas Development Institute, London.

Brandon, C. 1995. Environmental degradation and agricultural growth in Bangladesh: Instilling a sense of urgency. Bangladesh agriculture in the 21st century workshop, Dhaka. 5–6 Nov. 1995. Bangladesh Ministry of Agric., Dhaka, Bangladesh and World Bank, Washington, DC.

Buresh, R.J., D. Dawe, T.P. Tuong, J.K. Ladha, B. Bouman, R. Lantin, S. Peng, M. Mortimer, and J.E. Hill. 2001. Sustainable soil and water management of irrigated rice ecosystems. Workshop on Integrated Management for Sustainable Agriculture, Forestry and Fisheries, 28–31Aug 2001. Available at http://www.ciat.cgiar.org/inrm/workshop2001/docs/titles/2-2BPapertRBuresh.pdf [verified 20 Dec. 2002.] CIAT, Cali, Colombia.

Carney, D. (ed.) 1998. Sustainable rural livelihoods: What contribution can we make? Natural Resources Advisers' Conference. July 1998. Dep. for International Development, London, UK.

Christoplos, I., J. Farrington, and A.D. Kidd. 2001. Extension, poverty and vulnerability: Inception report of a study on the Neuchâtel Initiative. ODI Working Paper 144. Overseas Development Institute, London.

Christoplos, I., and U. Nitsch. 1993. Changing extension paradigm. IRDC Currents 6:22–26.

Datt, G., and M. Ravallion. 1998. Why have some Indian States done better than others at reducing rural poverty? Economica 65:17–38.

Davis, S. 1993. Knowledge is power? The use and abuse of information. IDS Bull. 24:60–72.

Dawe, D., A. Dobermann, P. Moya, S. Abdulrachman, P. Lal, S.Y. Li, B. Lin, G. Panaullah, O. Sariam, Y. Singh, A. Swarup, P.S. Tan, Q.X. Zhen. 2000. How widespread are yield declines in long-term rice experiments in Asia? Field Crops Res. 66:175–193.

Drèze, J., and A. Sen. 1989. Hunger and public action. Clarenden Press, Oxford, UK.

Duxbury, J.M., I.P. Abrol, R.K. Gupta, and K.F. Bronson. 2000. Analysis of long-term soil fertility experiments with rice–wheat rotations in South Asia. p. vii–xxii. *In* I.P. Abrol et al. (ed.) Long-term soil fertility experiments with rice–wheat rotations in South Asia. Rice–Wheat Consortium Paper Ser. 6. Rice–Wheat Consortium for the Indo-Gangetic Plains, New Delhi, India.

Earle, A. 2001. The role of virtual water in food security in Southern Africa. Occasional Paper 33. Water Issues Study Group. School of Oriental and African Studies, University of London, London.

Engel, P.H. 1994. Facilitation innovation: An action-oriented approach and participatory methodology to improve innovative social practice in agriculture. Ph.D. diss. Dep. of Communication and Innovation Studies. Wageningen Agric. Univ., Wageningen, The Netherlands.

Fan, S., P. Hazell, and S. Thorat. 1999. Linkages between government spending, growth and poverty in rural India. IFPRI Research Report 110. International Food Policy Research Institute Washington, DC.

FAO. 2002. AEZ—Agro-ecological zoning. Available at http://www.fao.org/waicent/faoinfo/agricult/agl/agll/aez.stm [last update Dec. 2002; verified 23 Dec. 2002] Land and Water Dev. Div., FAO, Rome, Italy.

Galpin, M., P. Dorward, and D. Shepherd. 2000. Participatory farm management methods for agricultural research and extension: A training manual. Department of Agric./Agric. Extension and Rural Development Dep. The University of Reading, Reading, UK.

Garforth, C.J. 1995. Building on diversity. Prospects for public-sector agricultural extension in Bangladesh. Working Paper 95/1. Agric. Extension and Rural Development Dep. The University of Reading, Reading, UK.

Garforth, C., Y. Khatiwada, and D. Campbell. 2001. Communication research for knowledge interventions: Case studies from Eritrea and Uganda. IRD Seminar Oct. 2001. Int. and Rural Dev. Dep., The University of Reading, Reading, UK.

Grown, C.A., and J. Sebstadt. 1989. Introduction: Toward a wider perspective on women's employment. World Dev. 17:937–952.

Gupta, R.K., P.R. Hobbs, J.K. Ladha, and S.V.R.K. Prabhakar. 2002. Resource-conserving technologies: Transforming the rice–wheat systems of the Indo-Gangetic Plains. Asia-Pacific Association of Agricultural Research Institutions. FAO Regional Office for Asia and the Pacific, Bangkok, Thailand.

Harrington, L., S. Fujisaka, P. Hobbs, C. Adhikary, G.S Giri, and K. Cassaday. 1993. Rice–wheat cropping systems in Rupandehi district of the Nepal Terai: Diagnostic surveys of farmers' practices and problems, and needs for further research. Sustainablility of the rice–wheat systems in South Asia. CIMMYT/NARC/IRRI, Mexico, D.F.

Harrington, L., J. White, P. Grace, D. Hodson, D. Hartkamp, K. Vaughan, and C. Meisner. 2000. Taking it higher: Thoughts on scaling up and scaling out within problem-solving approaches to research on INRM. Available at http://www.inrm.cgiar.org/documents/workshop_2000.htm [verified 20 Dec. 2002]. CIFOR, Jakarta, Indonesia.

Jodha, N.S. 1998. Poverty debate in India: A minority view. Econ. Polit. Weekly 23(45–47):2421–2428.

Jolly, C.M. 1988. The use of action variables in determining recommendation domains: Grouping Senegalese farmers for research and extension. Agric. Admin. Extension 30:253–267.

Ladha, J.K., K.S. Fischer, M. Hossain, P.R. Hobbs, and B. Hardy. 2000. Improving the productivity of and sustainability of rice–wheat systems of the Indo-Gangetic Plains: A synthesis of NARS-IRRI partnership research. IRRI Discussion Paper Ser. 40. IRRI, Makati City, Philippines.

Lipton, M., and R. Longhurst. 1989. New Seeds and poor people. Unwin Hyman, London.

Matin, I., D. Hulme, and S. Rutherford. 1999. Financial services for the poor and poorest: Deepening understanding to improve provision. Finance and Development Programme Working Paper Series 9. Institute for Development Policy and Management, Manchester, UK.

Murgai, R. 1999. The green revolution and the productivity paradox: Evidence from the Indian Punjab. Policy Research Working Paper No. 2234. World Bank, Washington, DC.

Pagiola, S. 1995. Environmental and natural resource degradation in intensive agriculture in Bangladesh. Environmental Economics Ser. Paper 15. World Bank, Washington, DC.

Quizon, J., G. Feder, and R. Murgai. 2001. Fiscal sustainability of agricultural extension: The case of the farmer field school approach. J. Int. Agric. Ext. Educ. 8:13–24.

Ramirez, R. 2000. Understanding farmers' communication networks: Combining PRA with agricultural knowledge systems analysis. Gatekeeper 66. International Institute for Environment and Development, London.

Reardon, T., and S.A. Vosti. 1995. Links between poverty and the environment in developing countries: Asset categories and investment poverty. World Dev. 23:1495–1506.

Reece, D., and J. Sumberg. 2003. More clients, less resources: Towards a new conceptual framework for agricultural research in marginal areas. Technovation (in press).

Sen, A.K. 1981. Poverty and families: An essay on entitlement and deprivation. Oxford University Press, Oxford, UK.

Sen, A.K. 1985. Commodities and capabilities. North Holland, Amsterdam, The Netherlands.

Sen, A.K. 1999. Development as freedom. Oxford University Press, New York.

Swift, J. 1989. Why are poor people vulnerable to famine? IDS Bull. 20:8–15.

Swift, J. 1993. Understanding and preventing famine and famine mortality. IDS Bull. 24:1–16.

Thirtle, C., I. Xavier, L. Lin, V. McKenzie-Hill, and S. Wiggins. 2001. Relationship between changes in agricultural productivity and the incidence of poverty in developing countries. DFID Rep. 7946. Department for International Development, London.